Field Confirmation Testing
for Suspicious Substances

Field Confirmation Testing
for Suspicious Substances

Rick Houghton

CRC Press
Taylor & Francis Group
Boca Raton London New York

CRC Press is an imprint of the
Taylor & Francis Group, an **informa** business

CRC Press
Taylor & Francis Group
6000 Broken Sound Parkway NW, Suite 300
Boca Raton, FL 33487-2742

First issued in paperback 2017

ISBN-13: 978-1-4200-8615-7 (hbk)
ISBN-13: 978-1-138-11203-2 (pbk)

This book contains information obtained from authentic and highly regarded sources. Reasonable efforts have been made to publish reliable data and information, but the author and publisher cannot assume responsibility for the validity of all materials or the consequences of their use. The authors and publishers have attempted to trace the copyright holders of all material reproduced in this publication and apologize to copyright holders if permission to publish in this form has not been obtained. If any copyright material has not been acknowledged please write and let us know so we may rectify in any future reprint.

Library of Congress Cataloging-in-Publication Data

Houghton, Rick.
    Field confirmation testing for suspicious substances / Rick Houghton.
        p. cm.
    "A CRC title."
    Includes bibliographical references and index.
    ISBN 978-1-4200-8615-7 (hard back : alk. paper)
    1. Chemical detectors. 2. Assaying apparatus. 3. Chemistry, Analytic. 4. Hazardous substances. 5. Chemical terrorism--Prevention. I. Title.

TP159.C46H68 2009
604.7--dc22   2009010932

Visit the Taylor & Francis Web site at
http://www.taylorandfrancis.com

and the CRC Press Web site at
http://www.crcpress.com

Table of Contents

2   Chemical Confirmation Tests   75

Author

Rick Houghton is a retired Michigan firefighter, paramedic, and a hazardous materials technician from the Lansing Fire Department. He has been an instructor at Michigan State University since 1992. Rick holds an associates degree in Fire Science and a bachelors degree in Public Safety Studies.

Early in his career, Rick specialized in qualitative analysis as it applies to field operations in hazardous materials emergency response. His experience in hazmat chemistry and his instructional skills have been useful in various projects for several clients. He has provided instructional or consulting services to clients such as the U.S. Customs and Border Protection Advanced Training Center, the U.S. Environmental Protection Agency National Decontamination Team, Michigan State Police, Michigan Department of Agriculture, Michigan State University, Ford Motor Company, several fire departments, hazmat teams, and others. He has designed and facilitated county- and statewide full-scale exercises and delivered several training programs on hazard identification, characterization, and identification of unknown materials.

Rick is currently a member of the Livingston County Hazardous Material Response Team and continues to provide training, consultation, and response.

He is the author of *Emergency Characterization of Unknown Materials* (CRC Press, 2007) and maintains a Web site at www.HazardID.com.

Acknowledgments

My thanks go to my son, Shane Houghton, who helped produce many of the images that appear in this book. His ability to accurately represent key test results, especially flame colors, has made this a stronger book. I would also like to thank my father, Larry Houghton, a reliable proofreader and author himself.

The Lansing Police Department provided me access to Officers Matt Ramsey and Beth Larabee for the section on canines. I appreciate their candor and accessibility while they worked with me to explore new ways of testing suspicious substances. Their commitment as canine handlers is exemplary.

My editor, Becky McEldowney Masterman, remains the best (if only!) editor I have ever had. She has been a supportive partner throughout this endeavor, as well as my previous book, and her positive attitude certainly adds quality to the books she oversees.

Introduction

This book was written to help fulfill the need by those who work with hazardous materials to test suspicious substances and to either confirm or deny the labeled identity. Materials that are unlabeled often have circumstantial clues that suggest an identity or class of substance that can be confirmed by unrelated tests. Two, preferably three, field tests confirming the same result provide a high degree of confidence and allow those charged with managing an incident to plan accordingly. Materials that are accidentally mislabeled or mixed in a single container can be tested to deny the original label information. The same applies to materials that are intentionally mislabeled by someone who is less than forthright.

Tests are divided into three chapters. Chapter 1, "Physical Confirmation Tests," involves methods that are mostly physical, such as measurement of temperature, vapor density, radioactivity, and so on. Some tests may involve a small amount of chemistry, but they are essentially physical tests and it seems logical to present them as such. Chapter 2, "Chemical Confirmation Tests," includes over 400 chemical tests, most of which provide a colorimetric result. The tests were included based on selectivity and adaptability to field use. Some tests identify individual compounds while others identify groups or families of compounds. Chapter 3, "Instrumentation," provides an overview of the technologies used to analyze materials. The overview presents strengths and weaknesses of the technology so the corresponding weak or strong result can be used in the overall analysis. The appendix provides two detailed sections on drug and explosives tests.

To use the book, consider first any situational clues and label information, if present. Formulate an identity of the suspicious substance and then choose tests to confirm or deny the identity.

These field tests are not meant to replace laboratory resources. However, these tests can immediately generate valuable information in the field, which in turn can be used to provide life safety, property conservation, and environmental protection as well as to support law enforcement activities.

Physical Confirmation Tests

1

Introduction

This first chapter describes physical confirmation tests that are practical for field use, even though some chemistry might be involved. The physical tests involve manipulation of the sample followed by observation of the reaction. Physical confirmation tests in this chapter are presented in this order:

- Air Reactivity
  - Confirming Air Reactive Alkali Metals
  - Confirming Air Reactive Hydrides
  - Confirming Air Reactive Phosphorous
- Appearance and Odor
  - Human Senses
  - Canine Senses
    - Considerations for the Use of Canines
- Boiling Point
  - Confirming Boiling Point
  - Variables for Boiling Point Measurement
    - Pressure Variations Due to Weather
    - Pressure Variations Due to Elevation
  - Compensating for Pressure Variations
- Density
  - Specific Gravity
    - Confirming Specific Gravity
  - Vapor Density
    - Confirming Vapor Density
- Flash Point
  - Confirming Flash Point
- Heating by Torch
- Melting Point
  - Confirming Melting Point of a Solid Sample
  - Confirming Freezing Point of a Liquid Sample
- Physical State
- Radioactivity
  - Confirming Radiological Material
    - Confirming Radioisotopes
    - Confirming Radioactive Emission
    - Confirming Transportation Activity
    - Confirming Specific Activity

- Solubility
  - Confirming Water Solubility
  - Confirming Vapor Solubility
- Vapor Pressure, Volatility, and Evaporation Rate
  - Confirming Vapor Pressure or Evaporation Rate
- Viscosity
  - Estimating Viscosity
  - Confirming Viscosity
- Water Reactivity
  - Confirming Water Reactivity

Air Reactivity

Air reactive compounds are determined by simply exposing them to open air.

Pyrophoric materials will burn when exposed to air. Self-heating materials or spontaneously combustible materials will react when exposed to air without an external energy source. The U.S. Department of Energy (DOE) defines pyrophoric materials as those that will immediately burst into flame when exposed to air. DOE defines spontaneously combustible materials as those that will ignite after a slow buildup of self-generated heat. Material that fumes when exposed to air may be reacting with moisture in the air and is actually water reactive and not air reactive.

The U.S. Department of Transportation (DOT) defines a pyrophoric material as a solid or liquid material that will burn without an ignition source within 5 minutes of exposure to air according to United Nations specifications. DOT defines self-heating materials as those that exceed 200°C or self-ignite within 24 hours under certain test criteria. These defined air reactive materials are described broadly as air reactive alkali metals, hydrides, and phosphorous. Table 1.1, "Reaction Products of Air Reactive Materials," summarizes several air reactive materials and the products formed when exposed to air.

Because there are relatively few air reactive materials as defined by DOE or DOT, their characteristics and specific confirmation tests are included below from a previous work, "Emergency Characterization of Unknown Materials" (CRC Press, 2007), in which air reactivity is first determined and then confirmation tests are applied. Applicable confirmation tests are described in Chapter 2.

Air reactive materials are sometimes packaged in gas cylinders, although they may not be gases themselves, or they may be packaged under nitrogen or some other inert atmosphere. In general, air reactive materials are identified by exposing a small amount of substance to the air and observing. Rapid discoloration is suggestive of reaction with air. Some of these materials may not ignite immediately, but an increase in temperature can be detected with a thermometer, preferably an infrared model. Expect an air reactive compound to emit toxic fumes or smoke.

Confirming Air Reactive Alkali Metals

Alkali metals, such as cesium, rubidium, potassium, sodium, and lithium are air reactive and are described in order of decreasing reactivity. Cesium and rubidium react vigorously with air at room temperature to form the metal oxide, which results in a self-sustaining metal fire that rapidly heats up to almost 2000°C. Potassium is less reactive at room

Table 1.1   Reaction Products of Air Reactive Materials

Air Reactive Material	Reaction Product(s)[a]
Air Reactive Alkali Metals	**Metal Hydroxide (Strongly Basic pH)**
Cesium	Spontaneous ignition and basic cesium hydroxide.
Rubidium	Spontaneous ignition, basic rubidium hydroxide.
Potassium	Basic potassium hydroxide. Larger volumes produce detectable natural gamma radiation.
Sodium	Basic sodium hydroxide.
Lithium	Basic lithium hydroxide.
Zirconium	Pyrophoric at < 3 microns. Mild pH.
Hafnium powder or sponge	Pyrophoric if very finely divided.
Calcium	Pyrophoric if very finely divided. Basic calcium hydroxide.
Aluminum	Pyrophoric if very finely divided. Mild pH.
Uranium	Radioactive. Pyrophoric if very finely divided.
Plutonium	Radioactive. Pyrophoric if very finely divided.
Nickel	Pyrophoric if very finely divided. Mild pH.
Zinc	Pyrophoric if very finely divided. Mild pH.
Titanium	Pyrophoric if very finely divided. Mild pH.
Air Reactive Hydrides	**Possibly Pyrophoric**
Arsine (arsenic hydride)	Not pyrophoric. Other hazardous gas may be present in container.
Diborane	Pyrophoric in moist air. Easily ignited and forms a green flame.
Phosphine	Not pyrophoric.
Silane	Likely to be pyrophoric and produce white soot.
Hydrazine	Basic pH.
Plutonium hydride	Radioactive. Basic plutonium hydroxide.
Barium hydride	Basic barium hydride. Pyrophoric if finely divided.
Diisobutylaluminum hydride	Basic fume.
Phosphorous	**Phosphine**
Phosphorous, yellow or white	Pyrophoric above 30°C (86°F) and forms phosphine.
Phosphorous, red	Not pyrophoric. Ignites easily and forms phosphine.

[a] See text for more detail.

temperature and generally will not ignite spontaneously. Bulk sodium and lithium will not result in a fire because they are even less reactive and oxidation occurs slowly. However, if these less reactive metals are finely divided or heated (the rate of reaction doubles for every 8° to 11°C [15° to 20°F] increase in temperature according to the U.S. Department of Energy) and then exposed to air, spontaneous ignition and self-sustaining fire may occur. When these metals burn, the resulting metal oxide immediately condenses to form a dense, white fume that is highly corrosive to the lungs, eyes, and skin. These metal oxides form metal hydroxides that can be detected by placing a small amount in water and testing with a pH test strip. A pH test strip wetted with deionized or distilled water will indicate strong base if placed in the fume emitting from the burning sample.

Under various circumstances, alkali metals if cut or scraped may react to form unstable, higher oxides (e.g., peroxides or superoxides) that may react if disturbed. These higher oxides can react with the base metal or organic materials in an explosive manner or can begin burning. In some cases, they may be shock sensitive. The higher oxide crust may be white, yellow, or orange. If a peroxide or superperoxide is suspected, it could possibly be detected with a wet potassium iodide starch test strip or a wet peroxide test strip. Refer to the section on reactive materials related to peroxide detection.

Sodium oxidizes rapidly in moist air, but spontaneous ignitions have not occurred unless the sodium is finely divided. Sodium ignites at about 880°C (1616°F), burns vigorously, and forms a dense white cloud of caustic sodium oxide fume. A pH strip wetted with deionized or distilled water will indicate strong base if placed in the fume emitting from the burning sample.

Potassium is more reactive than sodium and will react with atmospheric oxygen to form three different oxides: potassium oxide (K_2O), potassium peroxide (K_2O_2), and potassium superoxide (KO_2). These higher oxides can react with the base metal or organic materials and burn intensely or they could be shock sensitive and explode. Use very small amounts of sample when testing. A pH strip wetted with deionized or distilled water will indicate strong base if placed in the fume emitting from the burning sample.

NaK (commonly pronounced "nak") is any number of sodium-potassium alloys, which are liquids or low-melting-point solids. The potassium oxides named previously can form a crust over the NaK. If the crust is pierced and the potassium oxides mix with the NaK, a high-temperature reaction similar to a thermite reaction can occur. This observation is more likely to be experienced if the crust is accidentally pierced and not necessarily through careful manipulation. Test small amounts of crust using a method described previously.

Lithium is not pyrophoric, but is listed here as an alkali metal. Lithium ignites and burns at 180°C (356°F) as it melts and flows. A pH strip wetted with deionized or distilled water will indicate strong base if placed in the fume emitting from the burning sample.

Zirconium in bulk form can dissipate high temperature without igniting, but a cloud of 3 micron dust particles can ignite at room temperature. A layer of 6 micron dust particles will ignite at 190°C (374°F). Metal dust fires have ignited when the dust is disturbed through maintenance and cleaning processes. Finely divided metal dusts such as nickel, zinc, and titanium may be air reactive. These metals are more likely to react if disturbed or mixed into the air.

Hafnium in dust form reacts comparably to zirconium. In sponge form hafnium may spontaneously combust. Calcium, if finely divided, may ignite spontaneously in air as will barium and strontium. Aluminum in powdered form can ignite spontaneously in air.

Uranium may be pyrophoric if finely divided. The source and products of combustion will be radioactive. Plutonium has pyrophoric characteristics similar to uranium, but plutonium dust ignites more readily. According to lessons learned from the 1969 Rocky Flats Plant fire, spontaneous pyrophoricity of plutonium is unpredictable, even in bulk form. Use a gamma-detecting instrument to screen metals for radiation. Any of these metals will leave a basic fume or residue that can be detected with a pH test strip. Some of these metals will burn with a characteristic flame color; some emit intense UV light and eye protection may be necessary.

Confirming Air Reactive Hydrides

Hydrides are air reactive and often begin burning if exposed to the atmosphere. Examples include arsine, barium hydride, diborane, and diisobutylaluminum hydride.

Arsine (AsH_3 or arsenic hydride) is a gas and will generally not ignite in air unless at elevated temperatures, but it can be detonated by a suitably powerful initiation (heat source, shock wave, electrostatic discharge). Arsine may be carried in other substances. The ignition temperature of many of these arsine-containing substances is lower than that of arsine, causing ignition in air even at low temperatures (below 0°C, 32°F). All arsine compounds should be considered pyrophoric until they are properly characterized. Arsine may be detected with a colorimetric air monitoring tube. Some colorimetric tubes are sensitive to both phosphine and arsine.

Diborane (B_2H_6) is a highly reactive gas with a flammable range of 0.9% to 98% and an ignition temperature of 38° to 52°C (100° to 125°F). Diborane will ignite spontaneously in moist air at room temperature. Diborane is normally stored at less than 20°C (68°F) in a well-ventilated area segregated from other chemicals. A sample of diborane can be ignited easily in air with a hot wire or other low-temperature source. The flame should be green, but the presence of other materials may obscure the color.

Phosphine (PH_3) is a highly toxic colorless gas. Phosphine may ignite spontaneously at 212°F. Diphosphine (P_2H_4) is also pyrophoric. Both of these materials are detectable with an AP2C air monitor that is highly sensitive to phosphorous and sulfur compounds. The response of the monitor should favor the G-agent alarm, although both light columns may illuminate. If using an air monitor, be sure to begin with a dilute fume sample so as to not overload the monitor. Phosphine can also be detected by the use of a colorimetric air-monitoring tube. Some colorimetric tubes are sensitive to both phosphine and arsine.

Silane (SiH_4 or silicon tetrahydride) and disilane (Si_2H_8) are gases that might ignite in air. The presence of other hydrides as impurities causes ignition always to occur in air. Nearly pure silane ignites in air and mixtures of up to 10% silane may not ignite. Hydrogen liberated from its reaction with air often ignites explosively. All silanes should be considered pyrophoric until they are properly characterized. Silanes will leave a residue similar to the soot of a diesel fuel fire with one striking exception: the soot is white, not black (silicon oxides are essentially glass).

Hydrazine (N_2H_4) is a clear, oily liquid resembling water in appearance and possesses a weak, ammonia-like odor. It may be in aqueous or anhydrous solution. Higher concentrations fume in air. The flash point of hydrazine is 38°C (100°F), and the flammable range is 4% to 100%. Hydrazine is detected by its high pH if even a minute amount of water is present. Colorimetric air-monitoring tubes are available for hydrazine, although some do not discern hydrazine from ammonia and amines (all forming basic gas or vapor) because the tube uses basic pH as an indicator.

Plutonium hydride (PuH_2 or PuH_3) in finely divided form is pyrophoric. Both the hydride and products of combustion will be radioactive. This material should be screened with a gamma-detecting instrument. Like the other hydrides, its fume will present a basic pH.

Barium hydride (BaH_2) is pyrophoric in powdered form. Diisobutylaluminum hydride is air sensitive and water reactive. Both of these materials produce a basic fume, detectible with water-moistened pH test strips. A colorimetric spot test for barium and aluminum would identify them if identification is necessary. All of these hydrides will react similarly and pose similar hazards with the exception of radioactive plutonium hydride.

Confirming Air Reactive Phosphorous

Phosphorous may be pyrophoric. White (also called yellow) phosphorous is a colorless to yellow, translucent, nonmetallic solid and ignites spontaneously on contact with air at or above 30°C (86°F). Red phosphorous is not considered pyrophoric but ignites easily to form phosphine, a pyrophoric gas described previously. A solid that is able to ignite from contact with a warm wire should have the smoke analyzed for phosphine with a colorimetric air-monitoring tube or other air monitor sensitive to phosphorous and sulfur compounds. Use a dilute fume sample so as to not overload the monitor.

Other reactions may be observed when the sample is exposed to air. Some of these metals may only discolor in air at room temperature, but if finely divided or heated, they may spontaneously ignite.

Appearance and Odor

Human and canine senses can be used as clues to verify the identity of a material. Obviously, exposure to a hazardous material should never be intentional, but descriptions from people inadvertently exposed may be useful in confirming a substance.

Human Confirmation

Odor cannot be relied on to prevent overexposure. Human sensitivity to odors varies widely. Some chemicals cannot be smelled at toxic concentrations or sometimes cannot be smelled at all. Some odors can be masked by other odors and some compounds can rapidly deaden the sense of smell. Likewise, appearance can be affected by many factors. Many observers may not readily know the differences used to describe materials. The materials may change form from that listed in shipping resources due to heat, cold, exposure to water or soil, lighting, and other variables that might exist outside the storage container. Color blindness in some individuals and memories formed under stress can affect the perception of a witness and make information less reliable.

Because odor and appearance are difficult to define and describe in a manner that would be helpful, the characteristics are purposefully vague. The NIOSH Pocket Guide to Chemical Hazards contains a heading called "Physical Description" for each listing, which very briefly describes the appearance and odor of each substance, whether it can be shipped as a liquefied compressed gas, and whether it is used as a pesticide. For example, 2,4-D (2,4-dichlorophenoxyacetic acid) is described as "white to yellow, crystalline, odorless powder [herbicide]." However, most people would recognize it in its more common form as a brown water solution (<1%) with other ingredients and a recognizable odor that is used to kill dandelions and other broadleaf weeds on a lawn, such as Ortho® Weed-B-Gon MAX® Weed Killer.

CHRIS and other available databases work by searching a database of descriptive terms (Table 1.2). The search is only as accurate as the terms used to describe individual hazardous materials as legally shipped on waterways. This is not to say the information is not important; to the contrary, accurate observations can provide significant information when characterizing or identifying a material. You are responsible for formulating a response by using the database as a tool; do not allow the tool to tell you what to do.

**Table 1.2   Appearance and Odor Search Options for Chemical Hazards
Response Information System (CHRIS)**

Appearance		Odor	
Color	Physical State	Strength	Likeness
Any color	Amorphous solid	Aromatic	Alcohol
Amber	Aqueous solution	Characteristic	Almond
Black	Compressed gas	Choking	Ammonia
Blue	Crystalline solid	Faint	Apple
Brown	Flakes	Irritating	Banana
Buff	Gas	Mild	Benzene
Clear	Heated liquid	Nauseating	Bleach
Colorless	Liquefied compressed gas	Odorless	Burning
Dark	Liquefied flammable gas	Penetrating	Butter
Gold	Liquefied gas	Pungent	Camphor
Gray	Liquid	Sharp	Chemical
Green	Liquid resin	Slight	Chlorine
Orange	Liquid solution with water	Strong	Chloroform
Peach	Molten solid	Weak	Coconut
Pink	Oil		Creosote
Red	Oily liquid		Detergent
Rose	Paste		Disagreeable
Silver	Pasty liquid		Disinfectant
Straw	Pellets		Ether
Tan	Soft solid		Ethereal
White	Soft solid under kerosene		Ethyl
Yellow	Solid		Fatty
	Solid beads		Fishy
	Solid in solution		Flowery
	Solid crystals		Fragrant
	Solid granules		Fruity
	Solid powder		Fuel
	Thick liquid		Garlic
	Viscous liquid		Gasoline
	Water solution		Geranium
	Waxy solid		Goat-like
	Wet crystals		Hydrochloric acid
	Wet solid		Kerosene
			Lemon
			Lube
			Medicinal
			Mercaptan
			Mothballs
			Mushrooms
			Musky
			Mustard
			Oil
			Onion
			Paint

(continued)

Table 1.2   Appearance and Odor Search Options for Chemical Hazards Response Information System (CHRIS) (*Continued*)

Appearance		Odor	
Color	Physical State	Strength	Likeness
			Peanut
			Pear
			Peculiar
			Peppermint
			Phenolic
			Pineapple
			Pine tree
			Pleasant
			Pyridine
			Rancid
			Rose
			Rotten egg
			Rubber
			Skunk
			Soapy
			Sour milk
			Stale
			Sulfur
			Sweet
			Tobacco
			Turpentine
			Unpleasant
			Vinegar
			Wintergreen

Source: Adapted from Chemical Hazards Response Information System (CHRIS), Coast Guard Headquarters, Washington, DC, March 1999.

For example, searching the "physical state shipped" field in CHRIS for "aqueous solution" produces two records: chromic acetate and sodium 2-mercaptobenzothiazol solution. Searching for "water solution" produces three records: ammonium thiosulfate, calcium chloride, and methylamine solution. Understanding the limits and methods of a database search will improve use of the database to process clues and will help prevent tunnel vision. To be sure, CHRIS is a very useful database and should be used in the practical manner in which it was designed.

Many odors are recognizable by humans, but often in controlled settings. The perception of odors is highly subjective and should not be relied upon as a defining test; rather, odor characteristics of unknown material should be used as suggestive evidence as to the identification of one or more components in a compound. Likewise, appearance characteristics of pure material can be helpful in identifying particular substances. Be skeptical of typically characteristic appearance if it is possible the unknown material has been altered in any way. Never intentionally smell a sample. Use odor clues as they may be reported to you by people who may have inadvertently been exposed. Use appearance clues to determine if a material may be relatively pure or to corroborate the possible alteration of a pure compound.

Canine Confirmation

Canines have been used as biological sensors by humans to provide warning and protection for thousands of years and have been used as detectors for more than 50 years. Dogs have been proven to be very effective at detecting trace to gross amounts of specific material, mixtures, and hidden substances.

While dogs can be trained to detect and indicate almost any substance, they are most commonly used to detect the following materials.

- Explosives
- Illegal drugs
- Accelerants (flammable liquids used for arson)
- Live humans (search and rescue, tracking, guarding)
- Cadavers

Explosive-sensing dogs are typically trained to detect nitrate, perchlorate, chlorate, and peroxide compounds. The recognition of these scents allows dogs to recognize a wide range of explosive material as well as many precursors. Additional scents are added based on mission requirements. Since these dogs are trained to seek out the scent from a hiding place, they will not typically respond to the same scents in a normal environment. For example, an explosives dog could discriminate the scent of ammonium-nitrate-based explosive material hidden in a car parked on a lawn recently treated with ammonium nitrate fertilizer.

Drug dogs are commonly trained to detect highly purified heroin, cocaine, methamphetamine, ecstasy, and marijuana. By learning these scents they can also detect other illegal drugs that contain smaller amounts of these five drugs as common ingredients or their derivatives. Of course, dogs can be trained to learn other scents and would search and alert in the same manner.

Arson dogs are trained to detect trace amounts of flammable liquid residue in postfire and unburned environments and could be adapted to detect the combustible characteristic of a sample, but simply igniting a small sample would be much easier and eliminates exposing the dog to unknown material. Arson dogs are trained to detect light, medium, and heavy petroleum distillate as shown in Table 1.3.

Table 1.3 Categories and Examples of Accelerants Detected by Arson Dogs

Light Petroleum Distillate (LPD)	Medium Petroleum Distillate (MPD)	Heavy Petroleum Distillate (HPD)
• Lighter fluid	• Charcoal lighter fluid	• Kerosene
• Paint thinner	• Citronella fluid	• Diesel fuel
• Lacquer thinner	• Paint thinner	• Jet fuel
• Gasoline octane booster	• Lacquer thinner	• Lamp oil
• Aviation gasoline	• Fresh gasoline	

Dogs used to detect live or dead human scent, although truly valuable and worthy in their missions, are not likely to be used to verify the identity of a material. Other dogs are trained to detect illegally imported fruit and pirated compact discs as well as other specialized materials. Highly specialized teams can detect chemical and biological warfare agents. The difference between detectable and lethal concentrations may be small and the dog may be lost as a cost of early and sensitive detection. Dogs are less sensitive to some agents designed to harm humans. For example, dogs are 500 to 1000 times more resistant to *Bacillus anthracis* than humans.

The canine sense of smell is much more acute than its sense of sight. One estimate claims a dog's sense of smell is up to 10 million times more sensitive than that of a human. A canine snout contains about 440 times more scent cells than a human nose and the mucosa supporting the canine's scent cells has a surface area greater than the dog's skin. Canine-detection limits vary by substance, but some materials can be detected by a dog at 500 parts per trillion (0.0005 ppm).

Dogs are always trained to work as a team with a human handler, from which the dog receives direction and reward. Various training methods are used, such as reward by toy, food, and so on. Whatever the training method, the handler is trained to be aware of changes in the dog's behavior, which changes when the scent is detected. Often the handler will notice one or more behavioral changes when the dog is "onto something" and a second behavioral change when the dog believes it has located the source. Figures 1.1 to 1.4 depict a typical search and find of illegal drugs.

Dogs can be trained to recognize an individual scent out of a range of odors, such as when a dog is used for tracking an individual person. Dogs can also be trained to detect certain substances, such as a specific drug, or a wider range of substances, such as flammable liquid accelerants, live human scent, and a wide range of explosive material that includes gelatin (dynamite or nitroglycerine gel), nitroglycerin, ammonium

Figure 1.1 A handler and a dog trained to detect illegal drugs approach a suspect vehicle. A small plastic bag of a drug is hidden next to the fuel tank cap (foreground).

Figure 1.2 The dog has already indicated a scent and is turning back toward the fuel cap while localizing the scent. Changes in key behaviors, such as a sudden change in direction or the way the dog's head and ears are pointed, indicate the dog is detecting a target scent but has not yet located the source. The handler must learn to read the dog's body language in order to be effective.

Figure 1.3 The dog narrows the search pattern as it localizes the source.

Figure 1.4 The dog sits as a passive indicator that it believes it has located the source. Reward in the form of a tennis ball will soon follow.

nitrate, TNT, smokeless powder, C-4 and other plastic explosives, primer cord, and others. Explosive-detecting dogs have a detection reliability of 95%. Overall, detection canines have an average successful detection rate for discriminating specific odors in excess of 85%.

Dogs learn to depend upon the most abundant vapor constituents of a substance for identification of that substance, thus they learn to identify a class of substances by using only a few compounds common to a group. The training is specific, and a single dog cannot be trained for all the specialties. Specifically, dogs are not cross trained in both explosives and drug detection.

Notably, dogs can detect a target odor in a "noisy" environment when interfering substances might produce false positives or negatives with an electronic air monitor; however, canine interferences do exist. One U.S. provider of working dogs, K9 SOS, claims a false positive rate of only 0.003% in over 550,000 canine searches. Other studies suggest confidence of 85% to 95%. Significantly, a dog cannot "alert" to a material that does not contain a compound it has been trained to detect.

The handler regularly trains the dog in a manner that discourages the dog from a false positive indication. Through uniform and consistent training methods, the dog learns reward never comes from a false positive indication, which is a modification of the dog's behavior.

Dogs are living, thinking detectors that work in tandem with the handler. As they train, the dog learns to locate the target scents that are typically hidden and may initially overlook an obvious sample placed in the open. Before testing a sample with a dog, make a plan by consulting the handler. The handler holds final approval authority over the plan since the handler alone is responsible for the safe and efficient use of the dog. Figures 1.5 to 1.9 show a handler and explosives dog screening an unknown sample.

Figure 1.5 Officer Beth Larabee and K9 Sarge of the Lansing Police Department approach the sample (arrow). The sample is a pesticide gas device that contains a nonexplosive mixture of potassium nitrate (45%), sulfur (45%), carbon (8%) and inert ingredient (2%). Sarge has not previously been exposed to this specific mixture. The dog is in work mode and begins his normal mission of searching for hidden explosive material. Typically the search begins from downwind and the dog moves forward to localize the source.

Figure 1.6 Sarge has been trained to detect hidden explosive material. Possibly because the sample is too obvious, Sarge passes over the sample and continues forward while searching into the wind. This is an important safety consideration if you are using a dog to characterize any unknown sample. It may be helpful to place the sample in an obstructed area to slow the dog's search. Always make a detailed plan in consultation with the handler.

Figure 1.7 Officer Larabee directs Sarge to the area of interest. Sarge is now displaying behavior that signals his interest in the sample.

Figure 1.8 Sarge localizes the scent from this small, obvious object while Officer Larabee backs away to give him room to work without distraction.

Figure 1.9 By sitting and looking to his handler, Sarge signals a positive result for the presence of explosive material. Passive signals by the canine such as sitting or lying are a protective measure in the presence of dangerous material.

Considerations for the Use of Canines

The performance of most dog/handler teams will deteriorate if the handler becomes emotionally involved in the dog's success and this becomes an important consideration under stressful conditions. Several other factors can influence detection. Environmental conditions, especially temperature, humidity, and wind direction, affect the accuracy of canine detection. A good handler can compensate for some of these influences in the way the dog is managed. Additionally, a dog is a living, biological detector and as such, lack of food, lack of sleep, too much or too little exercise, and stress can adversely affect detection. The presence of an interfering scent is difficult to detect, but is an ever present concern in field response. Finally, sampling protocol can influence detection and the handler should be consulted before subjecting the dog/handler team to a sample. For example, a glass jar containing a sample that contains no flammable liquid could be examined by an arson detection dog. The dog would be likely to alert if the jar was handled by a firefighter's glove that had previously absorbed even a small amount of flammable liquid that contaminated the outside of the jar.

When using a dog in a characterization strategy you must be aware of the targets the dog is trained to detect and the capability of the dog and handler. You must also be aware of any intentional or coincidental odors that could be present. The handler can provide much of this information on a case-by-case basis. If you are a handler of an explosives or drug dog, you should be familiar with the field detection technologies for explosives and drugs in case your dog eats the evidence or is otherwise exposed. Field characterization information will be very helpful to veterinary staff as your partner will require immediate veterinary intervention.

Table 1.4 lists detection requirements and options for explosives and drugs as described by a few of the many organizations with interest in working dogs. Use these basic requirements in conjunction with the handler's knowledge to determine if characterization of an unknown material through a canine is a confident test. There are several response organizations not listed that are not willing to detail their canine capabilities, probably due to concerns about operational security. If you have occasion to work with a team from Alcohol, Tobacco and Firearms (ATF), Transportation Security Administration (TSA), or others, make a detailed sampling plan and interview the handler completely in order to understand their detection capability.

Boiling Point

The boiling point of a liquid is the temperature at which the vapor pressure of the liquid equals the atmospheric pressure. Higher atmospheric pressure requires higher temperature to induce the liquid to boil.

References will state boiling point for liquid materials at various pressure, but often 760 mmHg (101.325 kPa, the previous standard atmosphere) or ~750.01 mmHg (100 kPa, the IUPAC standard since 1982). However, field measurement of boiling point of a substance is affected by nonstandard atmospheric pressure.

Boiling point information is readily available for pure substances and many uniform mixtures and commercial products. In the case of mixtures, the boiling point is usually given as the lowest temperature at which the mixture will begin to boil; boiling may stop if lower fractions are boiled away, leaving only higher boiling point compounds. A boiling point of a sample measured in the field that deviates more than a few degrees from the referenced boiling point should be assumed to be improperly labeled or mixed with a substance other than the labeled material.

Confirming Boiling Point

Boiling point is determined by evenly heating a liquid sample until a steady boil is maintained and then measuring the temperature with a probe, nonmercury thermometer, or infrared thermometer. A hot plate is likely to give the most controllable heat, while a torch or open flame may be necessary for higher boiling point liquids. Remember the necessity of adequate ventilation and control of ignition sources when choosing equipment for heating combustible materials.

Covering the heating container with a lid will speed the heating process and help maintain an even heat inside the container. Be sure not to seal the lid; the container must be able to vent pressure or the container may burst (Figure 1.10). Smaller samples will heat more quickly and produce fewer vapors than samples in larger containers, but use enough sample to ensure the liquid can boil long enough to provide a confident result. If using a thermometer, insert it before heating since insertion of a cold thermometer will temporarily halt the boiling. Apply heat evenly by constant movement of the sample or heat source so that the sample is not thermally degraded by hot spots on the container wall. When a constant, steady boil is maintained, wait for the thermometer to register a steady temperature and then record it as the boiling point. Compare the boiling point to that listed in reference material.

Table 1.4   Canine Drug and Explosive Detection Required by Certifying Organizations

Certifying Organization	Required Explosive Detection	Required Drug Detection
National Narcotic Detector Dog Association (NNDDA)	Six basic odors: • Powders (black and smokeless) • Commercial dynamite or ammonium nitrate • C-4 or Flex-X • TNT or military dynamite • Primer cord • Slurries or water gel 0.25 lb minimum sample	Marijuana and cocaine detection required. Other narcotics are optional. 10–28 g cocaine sample. 0.25–2 oz marijuana sample. Optional narcotics may be, but are not limited to, heroin, methamphetamine, or opium. 10–28 g sample.
National Police Canine Association (NPCA)	Six commercial-grade explosive materials containing: • RDX • PETN • TNT • Powder • ANFO • Water gel 0.25–1 lb sample	Samples must contain controlled dangerous substances as classified by law. Two of the finds shall be soft drugs (marijuana) and two finds shall be hard drugs (cocaine or crack cocaine). 8–28 g sample.
North American Police Work Dog Association (NAPWDA)	Six required odors: • Powders • Commercial dynamite • Plastic explosive • TNT or military dynamite • Primer cord or Det cord • Water gel or slurries Greater than 0.5 lb sample. Additional odors are optional. Optional peroxide-based explosives: • TATP (triacetone triperoxide) • HMTD (hexamethylene triperoxide diamine) • MEKP (methyl ethyl ketone peroxide, a mixture of methyl ethyl ketone, hydrogen peroxide, and strong acid) 1–2 g sample size for safety.	Basic narcotic odors (and their derivatives): • Marijuana • Cocaine • Heroin • Methamphetamine 1 g minimum sample.
The Scientific Working Group on Dog and Orthogonal Detector Guidelines (SWGDOG). This organization makes recommendations and does not certify.	Mandatory detection of: • RDX (RDX-based Det cord) • PETN (PETN-based Det cord) • TNT (military TNT) • Dynamite (containing EGDN and NG) • Black powder (free flowing, time fuse, or safety fuse) • Double-base smokeless powder Optional substances as required by mission or specific threat: • Ammonium nitrate • Black powder substitutes (Pyrodex, Triple Seven) • Blasting agents	Teams must detect • Marijuana • Cocaine • Heroin • Methamphetamine • Other substances as required 5 g minimum sample size. Additional sample sizes must vary by factors of 10.

(continued)

Table 1.4　Canine Drug and Explosive Detection Required by Certifying Organizations (*Continued*)

Certifying Organization	Required Explosive Detection	Required Drug Detection
	• Cast boosters • Composition B • Emulsions • Nitromethane • Photoflash/fireworks/pyrotechnic powders • Plastic explosives (unmarked and marked with detection agent) • Semtex • Single-based smokeless powder • Slurries • Tetryl • Water gels • Homemade Explosives (HME) 　• Chlorate-based mixtures 　• Nitrate-based mixtures 　• Perchlorate-based mixtures 　• Urea nitrate 　• Peroxide-based explosives 　• Hexamethylene triperoxidediamine (HMTD) 　• Triacetone triperoxide (TATP) 　• Emerging threats such as liquid explosives • Taggants (nitro compounds used to mark explosive legally manufactured in the U.S. after April 24, 1997) 0.25 lb minimum sample size. Maximum size to be determined based on mission requirements.	
United States Police Canine Association (USPCA)	Must detect any four from: • Chlorates • RDX • TNT • Nitrates • PETN 8 oz minimum sample size. No maximum size specified.	Samples must contain controlled dangerous substances as classified by law. Two of the finds shall be soft drugs (marijuana, hashish, or certified derivative), and two finds shall be hard drugs (cocaine, heroin, methamphetamine or any certified derivative). 5 g sample minimum.

A water bath may be used for material with boiling points less than that of water. Glycerol (CAS 56-81-5) can be used instead of water to increase the bath temperature to 290°C (554°F). Figure 1.11 shows a sample within an Erlenmeyer flask that is immersed in glycerol being heated by the hot plate. Heat is well controlled and applied evenly over a large area of the container. The temperature probe may be submerged in the liquid or, if a reflux column is used, the temperature probe is suspended just above the liquid.

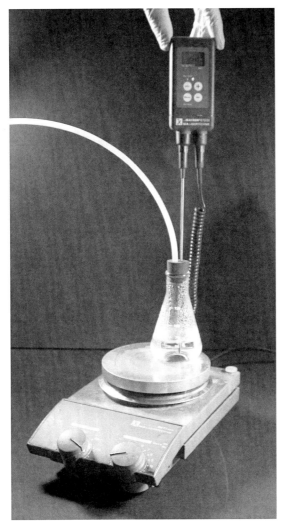

Figure 1.10 An Erlenmeyer flask contains a liquid sample that is being heated by a hot plate and agitated by a magnetic stir bar. The stopper contains two holes. One hole is used to hold a temperature probe, but could hold a glass thermometer in the liquid. The other hole is used to safely vent gas away from the operator through tubing that leads to controlled ventilation or a neutralizing solution. If using an infrared thermometer, point it so that the "spot" contains only the sidewall of the flask that is contacting boiling liquid. Erroneous results will occur if the "spot" includes the hot plate surface or an area of the flask that is not in contact with the boiling liquid.

Maintain a steady boil and wait for the column to clear of condensation. Once the temperature gauge stabilizes, record the boiling point and compare it to the reference value.

Variables for Boiling Point Measurement

Atmospheric pressure changes from weather seldom cause more than a 2°C variation in boiling point. Significant changes in elevation can markedly affect boiling point since

Figure 1.11 The hot plate is heating a glycerol bath, which in turn heats the sample in the Erlenmeyer flask. The temperature probe is immersed in the liquid sample. Sample vapors are removed through tubing and are condensed in an ice bath. Glycerol (CAS 56-81-5) is used instead of water to increase the bath temperature to 290°C (554°F).

pressure decreases exponentially as altitude increases. For example, the boiling point of water will vary about 5°C from sea level to 5000 ft (1524 m); many organic compounds will vary more than water.

Atmospheric pressure readings in the United States are provided by the National Weather Service in units of inches of mercury (inHg) and in other units throughout the world. The National Weather Service uses a defined standard atmosphere as 29.92 inHg (760 mmHg) and is not equivalent to the International Union of Pure and Applied Chemistry (IUPAC) standard atmosphere of 100 kPa (751.9 mmHg). Additionally, references for boiling point data may use various units of pressure when specifying a temperature. In any case, be sure to convert values accordingly. Table 1.5 is useful in converting values.

Pressure Variations Due to Weather
High and low pressure systems move throughout the atmosphere and influence weather. Normal atmospheric pressure at sea level is defined in the United States as 29.92 inHg

Table 1.5   Approximate Pressure Conversions for Field Use

	Pound per square inch (psi)	Atmosphere (atm)	Torr (mmHg)	Kilopascal (kPa)	Inches of mercury (inHg)
1 psi =	1	0.068	51.715	6.895	2.04
1 atm =	14.696	1	760	101.325	29.92
1 mmHg =	0.01934	0.00132	1	0.1333	25.4
1 kPa =	0.14504	0.00987	7.5	1	0.295
1 inHg =	0.019	0.033	25.40	3.34	1

Note: In 1982 the International Union of Pure and Applied Chemistry (IUPAC) defined standard pressure as precisely 100 kPa (~750.01 mmHg or 29.53 inHg) rather than the 101.325 kPa value of one standard atmosphere.

(760 mmHg). A typical strong low pressure system may drop the pressure to 28.90 inHg (734 mmHg), and a typical strong high may raise pressure to 30.10 inHg (765 mmHg). While more extreme pressure variations have been recorded, it is doubtful anyone will be measuring boiling point in the eye of a hurricane. Accordingly, Table 1.6 shows the result of moderate atmospheric pressure fluctuation of +/– 25 mmHg on boiling point. As can be seen, variations tend to be 1° to 2°C.

The calculations used to develop Table 1.6 are based on Trouton's rule, which states nearly all liquids have nearly the same value for entropy of vaporization. This means boiling points can be estimated for changes in pressure if a single reference point is known. An applet is available from the University of Cambridge, Department of Chemistry, that is helpful in estimating boiling points at various atmospheric pressures. The applet link is accessible through www.HazardID.com.

Some substances deviate from Trouton's rule due to intermolecular forces. Hydrogen bonding observed in strongly polar compounds, such as water and alcohols, is most likely to cause slight deviation from Trouton's rule and will produce higher boiling points. London dispersion forces and dipole–dipole interactions also cause deviation. Conversely, compounds with low intermolecular forces, including most organic compounds, are more likely to vaporize. The applet noted previously can modify calculations to be more water like or methane like. Field tests for boiling point will always be done in atmospheric pressures close to standard atmospheric pressure, which helps to reduce uncertainty when estimating boiling point with the applet. In any case, deviation of boiling point from a stated reference should not be more than +/– 4°C for any substance likely to be encountered in a field test, and most should be accurate to +/– 2°C.

Pressure Variations Due to Elevation

Change in elevation is more likely to affect boiling point than is change in weather. For example, the boiling point of water will decrease about 5°C from sea level to 5000 ft (1524 m); the boiling point of many organic compounds will decrease more than water.

Pressure decreases exponentially as altitude increases. However, in the lower portion of the atmosphere where field testing is likely to occur, the decrease can be considered linear rather than exponential. Atmospheric pressure is calculated by

Table 1.6   Boiling Point Variation Due to Weather-Induced Atmospheric Pressure

Compound	Referenced Boiling Point	Boiling Point (converted to °C at 760 mmHg)	Boiling Point at 735 mmHg (°C)	Boiling Point at 785 mmHg (°C)	Boiling Point Variation Due to Weather (°C)
Acetaldehyde	20.8°C at 760 mmHg	20.8	19.9	21.7	+/− 0.9
Acetic acid	117.9°C at 760 mmHg	117.9	116.7	119.1	+/− 1.2
Benzaldehyde	178.1°C at 760 mmHg	178.1	176.7	179.4	+/− 1.4
Benzene	80.1°C at 760 mmHg	80.1	79.0	81.1	+/− 1.1
Butanone (methyl ethyl ketone)	79.6°C at 760 mmHg	79.6	78.5	80.6	+/− 1.1
Chlorine	−34.6°C at 760 mmHg	−34.6	−35.4	−33.9	+/− 1.3
Cyclohexane	80.74°C at 760 mmHg	80.7	79.6	81.7	+/− 1.1
Diethyl ether	34.51°C at 760 mmHg	34.5	33.5	35.4	+/− 1.1
Formic acid	100.7°C at 760 mmHg	100.7	99.5	101.8	+/− 1.2
Hexane	68.95°C at 760 mmHg	69.0	67.9	70.0	+/− 1.0
Methanol	64.96°C at 760 mmHg	65.0	63.9	66.0	+/− 1.1
Phenol	181.75°C at 760 mmHg	181.8	180.4	183.1	+/− 1.4
Sulfur	444.674°C near 1 atm	444.7	442.6	446.7	+/− 2.1
Toluene	110.6°C at 760 mmHg	110.6	109.4	111.7	+/− 1.2
Triethylamine	89.3°C at 760 mmHg	89.3	88.2	90.4	+/− 1.1
Water	100.0°C at 760 mmHg	100.0	98.8	101.1	+/− 1.2

$$P_{elv} = 760 \text{ mmHg} \times [(288 - 0.0065 \times h_m)/288]^{5.2561}$$

P_{elv} is the atmospheric pressure at elevation in mmHg

h_m is the elevation in meters

Table 1.7 shows normal atmospheric pressure for several elevations based on the previous calculation.

Compensating for Pressure Variations

After determining a boiling point with a field test method, accuracy will be improved if the result is adjusted for pressure variation.

Table 1.7 **Normal Atmospheric Pressure at Various Elevations**

Elevation Above Sea Level (m)	Elevation Above Sea Level (ft)	Normal Atmospheric Pressure (mmHg)
0	0	760
100	328	751
200	656	742
300	984	733
400	1312	724
500	1640	716
600	1969	707
700	2297	699
800	2625	691
900	2953	682
1000	3281	674
1100	3609	666
1200	3937	658
1300	4265	650
1400	4593	642
1500	4921	634
1600	5249	626
1700	5577	619
1800	5906	611
1900	6234	604
2000	6562	596

As shown previously, atmospheric pressure variations are not likely to produce more than +/– 2°C variation for most organic compounds and not more than +/– 4°C for highly polar compounds. Simply apply this range to the result.

Adjust for elevation differences by using Table 1.7. (Local, current barometric pressure can be used from a weather station or a web-based service; however, the value will be corrected for sea level and must be converted to pressure at the local elevation.) After the local elevation is determined, the converted pressure reading can be used regularly.

Use the current local atmospheric pressure and the University of Cambridge applet (www.ch.cam.ac.uk/magnus/boil.html) to determine the expected boiling point. Allow +/– 2°C for accuracy; +/– 4°C for strongly polar compounds. Compare the result to the reference boiling point.

Example

A container of bright green liquid is labeled as a "low tox" antifreeze containing propylene glycol (CAS 57-55-6 20). The material safety data sheet describes the boiling point as 96.1°C. A small amount of sample is heated and the boiling point is determined to be 89°C. The local elevation is 600 m. Is the boiling point consistent with the label?

A reference gives the boiling point of "low tox" propylene glycol as 188.2°C. The boiling point of conventional antifreeze, ethylene glycol, is 197.3°C. The sample is a commercial product, not a pure compound, and probably contains water (which can be determined with a water test from Chapter 2), so the mixture must be assumed to have a boiling point different than the pure material.

The labeled boiling point of 96.1°C is assumed to be determined at 760 mmHg. Using Table 1.7, the ambient atmospheric pressure is adjusted to 707 mmHg. These data are inputted to the University of Cambridge applet (www.ch.cam.ac.uk/magnus/boil.html), and the boiling point is estimated to be 87.5°C. This is 1.5°C less than the observed boiling point. Glycols are strongly polar, so the variation from weather changes and the accuracy of the applet is considered to be +/− 4°C. Therefore, the observed boiling point confirms the labeled boiling point.

Density

Density is measured by weighing a volume of the sample using a simple scale and a standardized container. Density is mass divided by volume and is often expressed as grams per milliliter (g/ml). Density values of pure materials are found in reference materials. Density values of most manufactured products may be found in the material safety data sheet or by contacting the manufacturer.

The density values of many materials are compared to water and termed specific gravity. Since a 1 milliliter volume of water weighs 1 gram, its density or specific gravity is 1 (1 g/1 ml). Materials occupying the same volume with less mass have a specific gravity less than 1; those with greater mass have a specific gravity greater than 1.

Common equipment is used to determine density of a sample. Select a container for the liquid or solid sample that is easily filled exactly to a known volume. This might be a graduated cylinder or some other precisely marked vessel with the size dependent on the sample and situation. A scale is needed, preferably a digital scale that will auto-zero and that is sensitive to at least 0.01 g.

With the scale on a level surface, place the empty container on the scale and push the auto-zero button (sometimes labeled "tare"). When the scale reads zero add liquid sample to the container exactly to the mark and observe the weight (Figure 1.12). Solid samples may need to be packed tightly to fill the volume of the container; do this off the scale and then place it back on the scale to determine the weight. The density of the material will be the weight displayed on the scale divided by the volume of the filled container.

Density of irregularly shaped, water-insoluble solids may be determined by a slightly different method. Zero the scale, weigh the irregularly shaped solid, and then record the weight. Use a graduated cylinder and add water to a mark high enough that the solid can be submerged completely in the water. Place the solid in the water and then determine the volume of the solid by subtracting the original water volume from the second "high water" mark (Figure 1.13). The volume of water-soluble solids can be measured with a suitable liquid, such as light oil.

In either case, the volume of the sample (V) is total volume (V_t) minus the water volume (V_w). The density (D) of the irregular solid is the mass of the solid (m) divided by the total volume of water displaced by the solid:

$$D = m/V$$

or

$$D = m /(V_t − V_w)$$

This method is used commercially to determine volume of one or several small objects, such as gems. If any two values of density, mass, or volume are known, the third is easily

Figure 1.12 A container of known volume is filled with liquid sample and weighed. A simple ratio is used to determine the density of the sample expressed as g/ml. This method also works (with lower accuracy) for solids, which should be evenly packed in the container. This 10 ml sample was determined to weigh 6.62 g, which is consistent with the referenced specific gravity of hexane, 0.657 at 25°C.

calculated. Once the density is calculated, it can be compared to the listed value for confirmation. Remember that this method is not likely to be as precise as the manufacturer's laboratory method, but it should be satisfactory for field use.

Specific Gravity

Specific gravity is a ratio of the weight of a solid or liquid to the weight of an equal volume of water and is used to describe buoyancy of a material in a liquid. Water is the standard and has a defined value of 1. Something with a specific gravity greater than 1 will sink in water; less than 1 will float.

Solid organic materials may be heavier or lighter than water. Most solid inorganic material is expected to sink in water.

Most organic liquids are lighter than water, with the exceptions being those that are halogenated as shown in Figure 1.14. There are only a few inorganic liquids and they possess unusual characteristics, for example, bromine and mercury.

Confirming Specific Gravity

Specific gravity may be quickly estimated by placing the sample in water. Water-insoluble solids and liquids will either float or sink, and the specific gravity can be assigned as either greater than or less than 1 (Figure 1.15). Water-soluble solids and liquids might be estimated by carefully watching the addition to see if the solid or liquid briefly floats or sinks before dissolving into the water (Figure 1.16).

Figure 1.13 Density of a previously weighed irregular solid is determined by submersion in water. The density is the mass of the sample divided by the volume of water displaced. This irregularly shaped sample was found to displace 13 ml of water when fully submerged.

If a more exact value is necessary, use the weight method described previously in the section on density.

Vapor Density

Relative vapor density and water solubility of gases and vapors can be estimated with a simple field test. The test uses a visible fume mixed with the gas or vapor that is then released from a beaker. Lighter than air gases and vapors will rise from the beaker and appear to lift the "smoke" with it; heavier than air vapors will appear to drag the "smoke" downward. The most practical way to produce a visible "smoke" is by using an air flow indicator from a glass tube similar in appearance to a colorimetric air monitor tube.

Sensidyne, Inc. manufactures the Sensidyne Tube 5100 Smoke Tube (Air Flow Indicator Tube). The Sensidyne smoke tube contains stannic chloride that reacts with humidity in the air to form a visible white "smoke" containing hydrochloric acid mist and tin oxide fume. The "smoke" is designed to be approximately neutral buoyancy for use in tracing air currents, but in practicality it appears to be just slightly heavier than air. The stannic chloride tube can also be used in the Irritant Smoke Respirator Qualitative Fit Test Protocol (Occupational Safety and Health Administration, 29 CFR 1910.134).

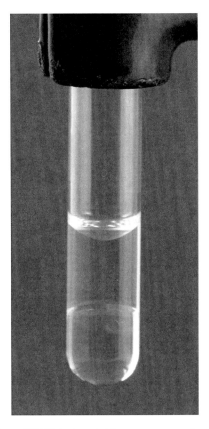

Figure 1.14 Trichloroethylene (TCE) is insoluble in water and even after agitation will separate and fall below the water to form a distinct layer.

Drager also manufactures the Air Current Detector Tube (CH 00216). The Drager tube contains iodine pentoxide, potassium permanganate, and fuming sulfuric acid that produces a visible white fume when it contacts humidity in the air. The Drager fume appears to be slightly denser than the Sensidyne fume; the smoke fume will tend to slightly increase the relative density of the sample gas or vapor.

Additionally, Drager makes the Flow Check Air Flow Indicator (6400761). This hand-held device uses a hot wire and a solution of long-chain alcohols to produce a visible smoke similar to that produced by a theater smoke machine or smoke generators used in fire academy training. It is difficult to control the amount of smoke injected into the sample while keeping the observation area free of smoke. The vapor density of the partially combusted alcohol could not be confirmed.

It is helpful to practice the procedure first with room air and fume from the indicator tube so you will notice any influence of the fume in the test. This will also allow you to determine if subtle room air currents are present. This test must be done in still air in a manner that protects the breathing zone of the operator. Ideally, you could do this near an exhaust fan that can be turned on to clear the air of gas, vapors, and fume when they are eventually released from the beaker. Do not breathe the smoke fume.

Relative vapor density is affected by temperature, atmospheric pressure, and humidity. Since this test is used to estimate relative vapor density in the field for immediate

Figure 1.15 Motor oil floats on water. The specific gravity of this particular motor oil is 0.87, but this method of simply adding the oil to water can only determine that the specific gravity is less than 1. Weighing 1 ml of motor oil would yield a more precise specific gravity.

application, atmospheric pressure and humidity can be considered stable for the time of the test. Humid air is lighter than dry air because the molecular weight of water (about 18) is less than the average molecular weight of air (about 29). Expect heavier than air vapors to settle into low areas and be more persistent in humid conditions. Day-to-day variations in conditions at a site can affect relative vapor density.

Temperature is easily variable, and it is important that the sample, container, and air are the same temperature. Vapor that is warmed or cooled in the container relative to ambient air will markedly affect vapor buoyancy. Before testing the sample, determine the temperature of the bottom of the container with an infrared thermometer. Next, wave a sheet of paper through the air several times to assure the temperatures of the paper and air are identical and then measure the temperature of the paper with the infrared thermometer. Proceed only when the temperatures of the container and paper are equal.

Confirming Vapor Density
To perform the test, break the ends off an air flow indicator tube and insert one end of the tube into a rubber bulb. Test the fume production by squeezing the bulb once in the area of the test site. Observe the fume in air to assure the room air is still. The images below were made using Sensidyne Tube 5100 Smoke Tubes (Air Flow Indicator Tubes).

Figure 1.16 Sugar initially sinks in water and has a specific gravity of about 1.3, but this method of simply adding sample to water can only determine that the specific gravity is greater than 1. Weighing 1 cc of sugar would yield a more precise specific gravity, but is not totally accurate because of the air spaces around the sugar crystals.

Add a few drops of liquid sample to the bottom of a 250 ml beaker, cover with a lid, and roll the beaker to spread the liquid across the glass in a manner that maximizes the surface area of the liquid (Figure 1.17). If the sample is a gas, inject a stream of gas into a 250 ml beaker. Keep the top covered as much as possible with a lid, such as a watch glass or some flat, inert material, in order to trap the vapor or gas. Insert the air flow indicator tube into the beaker through the spout while keeping the beaker covered as much as possible. Fully squeeze the bulb once and hold it; do not aspirate the vapor sample from the beaker into the tube (Figure 1.18). Gently center the cover on the beaker and allow the contents to stabilize.

Use an infrared thermometer to compare the temperature of the beaker and lid to the surrounding temperature (Figure 1.19), including the surface on which the beaker is resting. If the surface supporting the beaker is not the same temperature as the beaker, place an insulator such as a towel or cardboard between them. Do not use your hand without an insulating glove to hold the beaker as it will cause an increase in the temperature of the beaker and sample. Evaporating liquids will cool the bottom of the container slightly. Injecting fume from the air current tube will warm the contents slightly due to the reaction that creates the fume. Allow the temperature of the beaker and contents to stabilize.

In still air, gently lift the beaker and remove the lid by gently sliding the lid horizontally away from the beaker (Figure 1.20). Lighter than air gases and vapors will begin to rise

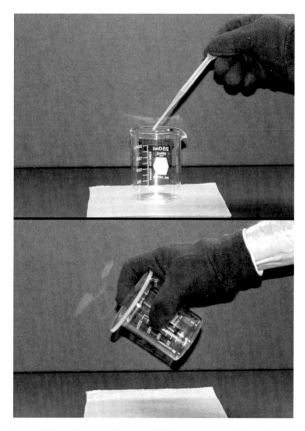

Figure 1.17 A few drops of liquid sample are added to a 250 ml beaker. Acetone (vapor density 2.0) is used in this sequence. The liquid is rolled within the covered beaker to increase surface area and vapor production.

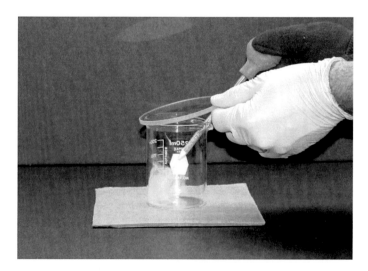

Figure 1.18 The vaporized sample is injected with fume from an air current tube. Do not allow the bulb to expand while the tube is in the beaker. This will prevent sample vapors from being diluted through both the movement of vapor into the tube and the subsequent movement of air into the beaker.

Figure 1.19 Use an infrared thermometer to determine the temperature of the container and surroundings. Allow the container to rest until temperatures are equal.

Figure 1.20 Gently lift the beaker and slide the lid horizontally away from the beaker.

from the beaker along with some of the white "smoke." Heavier than air gases and vapors will rest within the beaker, although any slight air movement can draw some fume from within the top of the beaker. Gently turn the beaker almost 90° (tilting less than 90° will prevent spilling any liquid sample remaining in the beaker) so the opening is perpendicular to the floor and observe the movement of the white fume (Figure 1.21). Slightly denser than air vapors and gases will move downward in a slow, tumbling movement. Very dense vapors and gases will appear to plummet (Figure 1.22); some may continue down to the bench top and spread laterally. Relative density of the sample can be estimated based on experience with known samples (Figure 1.23). Compare the vapor density of six samples to the test result images of Figures 1.22 to 1.27.

Flash Point

Flash point for a material can only be estimated in the field but can still be used to help verify a material.

Flash point is the lowest temperature of a flammable liquid that can form an ignitable mixture with air, and it has formal definitions from National Fire Protection Association (NFPA) and National Institute of Occupational Safety and Health (NIOSH). Flash point is "the minimum temperature at which a liquid or a solid emits vapor sufficient to form

Figure 1.21 Tilt the beaker almost 90° and observe the flow of the fume and acetone vapor.

an ignitable mixture with air near the surface of the liquid or the solid" (NFPA). NIOSH defines flash point as "the temperature at which the liquid phase gives off enough vapor to flash when exposed to an external ignition source." The DOT defines flammable and combustible liquids by class (Table 1.8), which includes parameters of temperature, phase, and so on. The vapor pressure of the flammable liquid determines, in part, the production of flammable vapor. The vapor pressure is influenced mainly by the liquid temperature.

Flash point is determined in a laboratory using a device with the method denoted as Cleveland, Pensky-Martens, Setaflash, or others. The device may use a closed cup (c.c.) or an open cup (o.c.) to hold the liquid as it is heated. A field test for flash point would be more closely related to the open cup method, although the precision of open cup and closed cup methods is obliterated by nonlaboratory conditions.

Confirming Flash Point

To verify the flash point of a liquid, soak a cotton swab with a sample of the liquid. Hold the swab horizontally and bring a match or lighter flame toward it from a line slightly lower than the swab. Relative flammability can be estimated by the distance fire flashes between the swab and flame (Figure 1.28). Combustible liquids will require heating by the lighter

Figure 1.22 Trichloroethylene vapor (vapor density 4.5) remains in the beaker when the lid is removed and falls rapidly when the beaker is tipped.

Figure 1.23 1,1,1,2-tetrafluoroethane has a vapor density of 3.6 and falls rapidly but does not plummet as quickly as the trichloroethylene in Figure 1.22.

flame before it ignites. Both flammable and combustible liquids will burn on the cotton swab and sustain combustion when the lighter is removed. The liquid will burn away and the cotton should remain unburned for some time. Water-based liquids will obviously not ignite and the water will quench the cotton even when the lighter flame is held under it. Mixtures of water and organics will ignite relative to concentration. Some of these mixtures may flash as the organic portion is distilled away from the water.

Heating by Torch

Feigl reports a set of characteristics resulting from smooth, quiet heating "to redness" of a sample on a spoon or micro crucible. A 13 mm borosilicate test tube heated by a torch may also be used with about a gram of solid material. Mixtures may undergo reactions between components that are only possible at higher temperatures. However, use of heating as a confirmation test can often support the presence of certain materials. Apply heat slowly; watch for changes as temperature steadily increases. Observations and indications include:

The sample remains unchanged in color or form: Volatile compounds such as mercury, ammonium salts (produce a basic vapor), and carbonates (produce

Figure 1.24 Hexane has a vapor density of 3.0 and drifts downward with less energy than heavier vapors. This sample contains only atmospheric moisture. Compare to the image of isopropyl alcohol.

carbon dioxide gas) are absent, as are organic compounds (charring or burning). Compounds containing water of crystallization (e.g., copper sulfate pentahydrate [$CuSO_4 \cdot 5H_2O$]), which briefly form a condensation on the upper wall of the tube, are absent. Finally, combinations of oxidizing and reducing agents are absent.

The sample melts: Possibilities include nitrates (NO_3^-), nitrites (NO_2^-), carbonates (CO_3^{-2}), formates (HCO^-), chlorates (ClO_3^-), or others of alkali metals (Li^+, Na^+, K^+, Rb^+, Cs^+, Fr^+); nitrates of alkaline earth metals (Be^{+2}, Mg^{+2}, Ca^{+2}, Sr^{+2}, Ba^{+2}, Ra^{+2}); silver chloride, and silver bromide. Many organic compounds will melt (but often ignite with continued heating). Many hydrated, crystallized compounds melt at lower temperatures.

Charring occurs: Organic material including metallo-organic compounds, metal formats, and oxalates ($C_2O_4^{-2}$).

Charring begins and disappears, sample may deflagrate: Organic material mixed with an oxidizer such as nitrite, nitrate, or chlorate.

A color change occurs without charring: Certain heavy metal compounds. Darkening color is caused by formation of oxides from metal salts of volatile, non-heat-resistant acids such as Cu^{+2} to copper oxide. Lightening of color is caused by decomposition of higher oxides, such as PbO_2 to PbO. A dark color change when hot that returns to the original, lighter color is characteristic of many metal oxides.

Figure 1.25 Isopropyl alcohol, 91%, contains water and produces a denser fume. The vapor density of isopropyl alcohol is 2.1, and the vapor density moist air is less than dry air. This test measures the net buoyancy of the sample vapor and fume in air. Compare to hexane image with a lesser water content.

A dark metal is finely dispersed after heating: Noble metal (gold, silver, tantalum, platinum, palladium, and rhodium) salts of carbonates, oxides and organic compounds. The metal may be visible initially and then form its oxide if in contact with air.

A blue color forms and remains or changes: Ammonium salts of molybdic acid, phosphomolybdic acid, tungstic acid, molybdite, or tungstite form lower oxides of molybdenum or tungsten that may oxidize upon additional heating.

Yellow or orange changes to green: Possibly indicates a chromate or bichromate reacting to chrome oxide.

Water condenses on the walls of the tube: Water may be evident from hydrated salts, excess moisture in the sample, metal hydroxides, and oxyhydrates and ammonium salts. Water may also be distilled from soluble organic compounds.

Acid condensate: Acid salts; sulfur trioxide from heavy metal sulfates; sulfur dioxide from sulfites, sulfur, or polysulfides in the presence of oxidizers or air.

Basic condensate: Ammonium salts of heat-stable, nonoxidizing acids; complex metal-amine salts; pyridine salts.

Clear sublimate forms on the tube wall: Some organic compounds may condense on the wall of the tube, for example, phenol.

Figure 1.26 Starter fluid is a mixture of carbon dioxide (vapor density 1.53), ethyl ether (vapor density 2.6) and heptane (vapor density 3.5). The material safety data sheet for this product describes the vapor density only as "heavier than air." This test can be used to determine relative vapor density of mixtures after temperatures stabilize to that of ambient air.

White sublimate: Ammonium salts of some arsenic, antimony, and mercury compounds; oxalic acid.

Yellow sublimate: Sulfur, arsenic sulfide, mercuric iodide, polysulfides, and heavy metal thiosulfates.

Gray sublimate: Mercury from higher compounds, arsenic from tervalent and quinquevalent arsenic compounds reacting with reducing compounds, iodine from iodides reacting with oxidizing compounds, or acid oxides.

Yellow or brown vapors: Bromine in oxidizing environment or nitrogen oxide from nitrite or nitrate. Bromine will test positive on a potassium iodide-starch test strip while nitrogen oxide will not. Some iodine compounds in the presence of moisture will produce yellow to brown vapor.

Violet vapors: Iodine from iodides and iodates with acid, mercury, or silver iodide. Some iodine compounds in the presence of moisture will produce yellow to brown vapor.

Atoms and molecular fragments of thermally degraded substances emit unique light signatures. Flame photometry is a method of burning material and measuring light emitted by hot atoms, usually using an optical reader. Some color signatures are bold enough to be seen by the human eye in the torch flame downstream from the sample. The following

Figure 1.27 Ammonium hydroxide vapor and air current tube fume rise immediately upon removal of the lid. Ammonia has a vapor density 0.6 but dissolves into water vapor that also has a lighter-than-air vapor density.

colors suggest the associated element **(See color images 1.1 through 1.10 following page 234.):**

Blue: Arsenic
Gray-Blue: Lead
Green: Boron or copper
Lavender: Potassium
Orange-Red: Calcium
Red: Lithium
Yellow-Green: Barium
Yellow-Orange: Sodium

Melting Point

Melting point, also known as freezing point, may be determined for a solid or liquid sample. Solids may be heated until the sample is roughly half liquid and half solid while

Table 1.8 Classification of Flammable and Combustible Liquids

Class	Flash Point Range and Boiling Point Range
IA flammable liquid	FP < 73°F and BP < 100°F
IB flammable liquid	FP < 73°F and BP at or > 100°F
IC flammable liquid	FP ≥ 73°F and < 100°F
II combustible liquid	FP ≥ 100°F and < 140°F
IIIA combustible liquid	FP ≥ 140°F and < 200°F
IIIB combustible liquid	FP ≥ 200°F

Source: Adapted from OSHA criteria.

gently stirring. Liquids may be cooled until the sample is roughly half liquid and half solid. Temperature in both cases can be determined with an infrared thermometer.

Material that consists of a single, nearly pure compound or a consistently formulated mixture can be verified by referencing the melting point or freezing point. The NIOSH pocket guide, material safety data sheets, and often the label of a chemical container will contain melting point or boiling point information. Contaminating material will alter the temperature at which a phase change will occur, so the sample material must be fairly pure, such as reagent-grade materials or consistently formulated commercial products.

This method can be especially useful to a waste site worker, laboratory technician, or others who suspect a reagent may be mislabeled. For example, a white powder in a brown glass container that is labeled acetylsalicylic acid should have a melting point of 275°F (135°C). If the melting point is not within a few degrees of the referenced value, the material is either grossly contaminated or indeed mislabeled. This principle also applies to law enforcement in the case of smuggled material.

Below are two methods for field determination of melting point and freezing point of unknown material. The techniques are ways to determine the freezing point of a material that is liquid at room temperature and the melting point of a material that is solid at room temperature. Determination of the condensation point of a material that is a gas at room temperature is not practical in the form of a field test.

Heat energy will be dispersed more evenly throughout the sample if heating or cooling is applied gently while also gently stirring the sample. Phase change requires energy, which will halt the change in temperature until the phase change is complete. When the phase change is complete and heat energy is added to (or removed from) the sample, the temperature will again increase (or decrease).

Confirming Melting Point of a Solid Sample

Melting point is determined by placing a solid sample into a container and heating until roughly half the sample is liquid and half remains as solid. Heat should be applied gently and evenly while the sample is continuously stirred. Before heating a suspicious material, apply a flame to a small amount to determine if the sample is heat sensitive. Explosive material should not be tested in a container that may fragment or cause explosive injury to the operator.

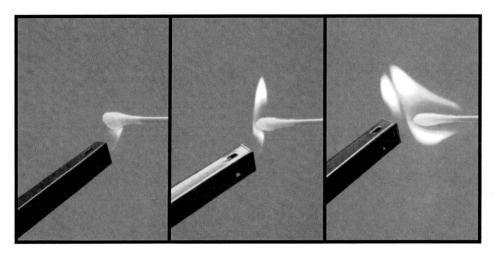

Figure 1.28 Characterization of a fire hazard from a liquid sample on a cotton swab. Left: a water-based liquid that does not contain a high concentration of polar organic solvent (e.g., acetone, methanol) will not burn. Note that the cotton swab is not burning because the heat of the flame is quenched by the water. Middle: a combustible liquid with a high flash point (charcoal lighter in the image) must be heated directly by the flame before the sample sustains combustion. Right: a combustible liquid with a low flash point (petroleum ether in the image) will flash as the flame appears to jump through the air.

A simple method for determining melting point involves the use of a hot plate as a heat source and a beaker to contain the sample. The beaker should be of a size that will allow the "spot" of an infrared thermometer to be placed on the surface of the sample and not on the sidewall of the beaker. The solid material is heated and continuously stirred, occasionally lifting liquid onto the surface of any floating solid. The heat source can be adjusted based on the response of the sample. If a glass thermometer is used in place of an IR thermometer, do not stir with the thermometer because it can easily break. Rather, stir with a tool designed to stir while the glass thermometer is placed in a liquid area. Allow the temperature reading to stabilize and determine the melting point before the phase change is complete.

Use an Erlenmeyer flask with a two-hole stopper if hazardous vapors must be removed with tubing to a scrubber or ventilation system. Use one hole for a thermometer or probe and the other for tubing.

If a torch is used as the heat source, keep the sample and container moving relative to the flame. Use a low flame and avoid overheating any one spot. Heat until about half the sample is liquid while half remains as solid. Ice directly from a freezer is a good material to practice the method, as is paraffin wax (food-grade wax that is used for canning will have a well-determined melting point listed in the material safety data sheet).

Confirming Freezing Point of a Liquid Sample

Firefighters know that even on the hottest summer day a carbon dioxide fire extinguisher can be used to cool a six-pack of their favorite beverage within a minute. Discharge of

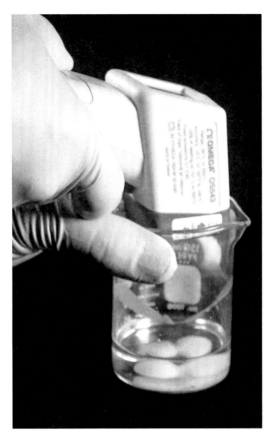

Figure 1.29 This substance, labeled as paraffin canning wax, was heated and swirled until most of the sample melted. Two lumps of solid are visible within the liquid. An infrared thermometer was used to determine a melting point of 132°F, consistent with the MSDS referenced melting point of 128°F.

pressurized liquid carbon dioxide from the extinguisher produces "dry ice" as a visible white solid and invisible carbon dioxide gas. As the pressurized liquid in the extinguisher is released to the atmosphere, relatively warmer molecules change to the gas phase, leaving the relatively cooler molecules as a solid. The dry ice will sublime to gas as it warms. The temperature of dry ice at its freezing point is –109°F (–78°C). The visible white vapor is condensed water vapor from the air.

The melting point of unknown liquids can be roughly determined with a simple field test using carbon dioxide from a fire extinguisher as a refrigeration source. Recommended additional items include an infrared thermometer, a cloth towel, a large watch glass, or similar thin-walled container for the liquid sample and an enclosure such as a cardboard box.

Remember to take adequate precautions to protect yourself and others from the unknown material you are testing. Additionally, dry ice is cold enough to produce frostbite injury by freezing any tissue it contacts, so use insulated gloves or tools to handle any cold objects, including the fire extinguisher. Carbon dioxide is an asphyxiating gas, so work in a well-ventilated area.

Figure 1.30 Carbon dioxide from a fire extinguisher deposited on a cotton towel. An alternate method is to obtain dry ice from a nearby vendor.

Place a cloth towel in the bottom of the box or enclosure. The towel will help catch the solid carbon dioxide as it blows out of the extinguisher and will aid insulation. Next, hold the horn of the carbon dioxide fire extinguisher about an inch (2.5 cm) above the towel and slowly release carbon dioxide. Since the hose and horn are warm initially, only gas will escape. Soon more solid carbon dioxide will collect on the towel (Figure 1.30). Do not spray the gas too quickly or it will blow the dry ice off the towel. With some practice you should be able to deposit about ½ inch (1 cm) of dry ice on the towel.

Place a large watch glass or other dish-like container on the mound of dry ice. The watch glass in the image is 5 inches (12.5 cm) wide. A thin-walled vessel is preferred since its lower mass will consume less dry ice as it cools. The first time you try this use an infrared thermometer to monitor the temperature of the watch glass. Glass does not conduct heat as quickly as might be suspected. Additionally, the center of the watch glass will be cooler than the edges. Next, place a pool of liquid sample on the watch glass (Figure 1.31). Make the pool surface about twice as large as the spot that will be "seen" by the infrared thermometer; too large volume of liquid will require a greater amount of dry ice and time in order to cool to the melting point. Cover the box to help cool the sample more quickly.

The sample should cool until it is partially frozen. When the sample is mostly solid but still has a little liquid flowing, remove the watch glass and place it on an insulated surface. The edges of the watch glass may now be colder than the center sample area since the lower mass of the rim area may have cooled faster than the center (Figure 1.32). If the entire sample freezes completely solid, the method will still work but it will take longer for the sample to warm to its melting point (Figure 1.33).

As a material moves through a solid-to-liquid phase change, the temperature of the material will be stable at its melting point. The sample and watch glass will assume a more uniform temperature, thus reducing relatively hot or cold spots if the solid and liquid

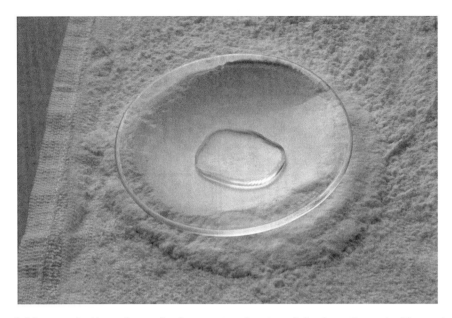

Figure 1.31 A pool of liquid sample about twice the size of the "spot" required by an infrared thermometer is placed on the watch glass.

sample is swirled gently over the glass and a little time is allowed for the watch glass and sample temperatures to stabilize. Use the infrared thermometer to determine several temperature readings by slowly moving the sensor over several areas of the watch glass and sample. The most accurate reading will be observed near the center as a large area that contains both solid and liquid. The result will be more accurate if you average several readings a

Figure 1.32 The sample may freeze unevenly, usually from the edges toward the center due to uneven distribution of the sample mass in the curved watch glass.

Figure 1.33 The sample slowly warms on an insulated surface, corrugated cardboard in this case. This sample is roughly half solid, half liquid and is at its freezing point.

few seconds apart. Variables that can affect the temperature registered by the IR thermometer are size and shape of the sample container, operating range of the thermometer, minimum spot size, operating ambient temperature range, and spectral response. A few results from this method are shown in Table 1.9, "Melting Point Determinations Using Carbon Dioxide." Practice this method several times using water as the sample. This will help you become familiar with your particular equipment and method. After you are familiar with the technique, move on to other known substances, such as those in Figure 1.34.

This method is not precise for laboratory use and is intended for field use. Accuracy can be affected by several factors, such as sample vapors that may interfere with infrared emissions, atmospheric water vapor absorbed into the sample, and uneven heating and cooling based on individual equipment. Commercial equipment for the determination of melting point is available but is more suitable for lab use.

Melting point is particularly useful in confirming the identity of a pure substance. For example, if a container is labeled as containing a laboratory-grade substance the melting

Table 1.9 Melting Point Determinations Using Carbon Dioxide

Material	Melting Point (°F)	Melting Point (°C)	Determined (°C)
Water	32	0	0
Acetic acid, glacial	63	17	12
Dimethyl sulfoxide	64	18	15
Conc. sulfuric acid	37	3	7
2,2'-thiodiethanol	5	−15	−13

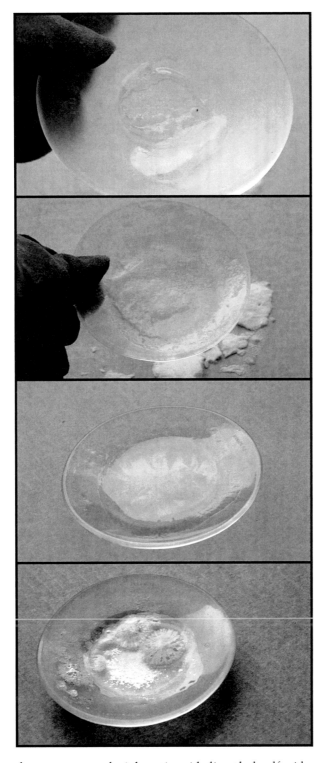

Figure 1.34 Top to bottom: water; glacial acetic acid; dimethyl sulfoxide; concentrated sulfuric acid at melting point.

point can be referenced from the material safety data sheet, the manufacturer's web site, or sometimes the container label. Field determination of melting point should be within a few degrees of the expected value, but if not, suspicion as to the legitimacy of the label is increased.

Although this test result may vary a few degrees from referenced melting and freezing points, a database search allows +/– 5°C for 2483 organic compounds. The Organic Compounds Database (www.colby.edu/chemistry/cmp/cmp.html) has been compiled by Harold M. Bell at Virginia Tech (Virginia Polytechnic Institute and State University).

Physical State

Confirming the physical state of a suspicious material as solid, liquid, or gas is a simple test that should be intuitive, but might also easily be overlooked. Physical state will be influenced by pressure and temperature of the container. Most often these changes are influenced by container pressure and ambient temperature.

Container types, even if insulated, should be measured for temperature with a suitable instrument, such as an infrared thermometer. Determine the phase of the material for the measured temperature and compare it to the design of the container.

For example, butane would be a gas at normal temperature and pressure, but should be a liquefied gas in a container similar to one designed to hold liquefied petroleum (LP) gas. As a liquid, it should be felt to move as a liquid within the container. Another example is a container that appears to hold a liquid and is found to contain a solid with a melting point slightly above ambient temperature. If the container is designed with heating coils, the owner may intend to ship the material at ambient temperature and then heat it and pump it as a liquid at a later time.

Some dangerous solids or liquids may be transported or stored in containers that appear to hold a gas, such as air reactive materials stored under an inert gas. Others may contain a solid stored under inert oil.

Solids and gases may appear as liquids if dissolved in a water or organic solvent. Liquids and gases may appear as solids if absorbed in silica gel, carbon powder, or other solids.

In any case, a moment of observation often can reveal obvious contradiction to the purported identity of a material.

Radioactivity

Lastly, colorimetric test results are listed in Table A.3, "Final Colors Produced by Reagents A.1 to A.12 by Various Drugs and Other Substances." Radioactive material cannot be sensed by humans, and instrumentation is required to verify the presence of radiation. Ionizing radiations vary in mass and energy characteristics and are summarized in the following.

- Gamma rays are weightless packets of energy with a neutral electrical charge. With no mass and no charge gamma rays can penetrate 6 inches of lead or 3 ft of concrete. Most gamma rays can pass completely through a human, causing ionizing destruction along the way. Gamma rays are emitted from the nucleus of a source atom and alpha and beta particles may be present. Gamma rays are emitted from substances such as cobalt-60, cesium-137, and uranium.

- X-rays have essentially the same properties of gamma rays but have lower energy. A few millimeters of lead can stop most x-rays. While gamma rays are emitted from the nucleus and are often accompanied by alpha and beta particles, x-rays are produced outside the nucleus. X-rays are most often produced by machines and are unlikely to be encountered from a suspicious substance.

- Beta particles are high-energy, lightweight particles with a negative or positive charge. A negatively charged beta particle is an electron, and a positively charged beta particle is a positron. Due to some amount of mass as opposed to no mass, beta particles can be stopped by a few layers of aluminum foil or clothing, although clothing is not necessarily approved personal protective equipment. The range of a beta particle is dependent on its energy and the material it is traveling in. For example, a 32P beta has a range in air of about 20 ft (7 m). Beta particles are emitted from substances such as 14C, 90Sr, and tritium.

- Alpha particles consist of two protons and two neutrons and are about 7000 times more massive than a beta particle. The increased mass means alpha particle radiation can be stopped by a sheet of paper or the dead layer of skin cells on a human. Comparatively, alpha particle radiation is not much of a hazard to the skin, but there are serious inhalation, ingestion, and injection hazards. Epithelial cells, which line the respiratory and gastrointestinal tract, have no dead layer of cells for protection. A radioisotope dust that is swallowed or inhaled will settle directly on live cells and cause local irradiation injury. The same type of injury can occur with eye contact. Because alpha particles are large and prone to collision while delivering energy over a short distance, they have potential to cause severe biological damage. Alpha radiation can only travel a few centimeters through air. Alpha particles are emitted from substances such as plutonium, radon, and radium.

- Neutron radiation is present in background cosmic radiation. It is also produced spontaneously or during a fission reaction. The most significant neutron hazard to be encountered is that which occurs as a result of a nuclear bomb blast or a fission reactor accident. In both cases other types of radiation would be present. 239Pu in a shielded container is capable of emitting only neutron radiation. Unshielded 239Pu would also emit alpha, beta, and gamma radiation. If 239Pu were suspected inside a heavily shielded container, a neutron detector would be helpful in identifying the source through the sidewall of the container. Other industrial sources can emit neutron and other radiation.

Radioisotopes are known to often emit more than one type of radiation. Detection of certain types of radiation by field instrumentation is suggestive of certain isotopes. For example, a field radiation test might indicate emission of alpha and gamma radiation from a sample. This suggests commonly encountered [241]Am, Th, or U are possible, although other radioactive material, including a mixture of pure alpha and pure gamma emitters, are possible.

Some isotopes occur naturally from the creation of the universe and others are decay products of original isotopes. Others are manmade. Primordial isotopes may be detected in larger samples of common materials. For example, a bag of potassium chloride water softener salt will produce a higher-than-background result due to concentration of naturally occurring ($< 0.02\%$) ^{40}K within the package. Human-produced isotopes may be found as radiopharmaceuticals (inside and outside humans) and in other commercial products.

Confirming Radiological Material

A radionuclide identifier can identify hundreds of individual radionuclides. It can measure the spectrum produced by an ionizing radiation source and compare it to a library within the instrument. Spectra are defined by decay characteristics of spin and parity, gamma peak energy, alpha and beta transitions, and so on. The ability of the identifier to determine the identity of the source depends on the type and range of sensors in the device as well as the presence of a matching spectrum. Consult the manual if this device is available. A radionuclide identifier is the best tool to verify the presence of labeled radioactive material. If an identifier is not available, other methods follow that are helpful in confirming radioactive material.

Confirming Radiation Isotopes

A radiation isotope identification device (RIID) can be used to identify hundreds of individual radionuclides. The device can measure the spectrum produced by an ionizing radiation source and compare it to a library of spectra stored in the device. Most RIID are able to sort and identify mixtures of isotopes.

Using an RIID is not always as simple as "point and shoot." As isotopes decay, daughter products are produced that may or may not interfere with the analysis. An RIID may compensate for decay in the spectrum of a sample. Some analysis is presumptive; an example is the analysis of cesium-137 (^{137}Cs). As ^{137}Cs decays, ^{37m}Ba is produced. ^{37m}Ba has a strong spectrum and is easily identified. ^{137}Cs does not have a strong spectrum. An RIID may presumptively identify ^{137}Cs based on the presence ^{37m}Ba. Pure ^{37m}Ba could then be misidentified as ^{137}Cs. An RIID can be a very powerful tool in the hands of a skilled operator, such as a health physicist who is well versed in decay paths and other details of a particular isotope. Those with less skill who have access to an RIID should view the initial result with some skepticism while awaiting consultation with reach back support or an expert.

Approximately 5000 natural and artificial radioisotopes have been identified according to the U.S. Nuclear Regulatory Commission. Many of these have extremely short half-life and exist only in research laboratories. Interpretation of spectra can be complex and may not be as simple as overlaying a sample's spectrum over a spectrum already resident in the library. Sometimes a skilled interpreter is necessary. However, isotopes likely to be encountered with field testing can be divided into groups: special nuclear material (SNM), medical isotopes, naturally occurring radioactive material (NORM), industrial isotopes, neutron sources, and material of special interest for use in radiological dispersion devices (RDD).

SNM is material defined by the U.S. Atomic Energy Act of 1954 that could be used in a nuclear weapon. These materials include:

- Americium
- Neptunium-237
- Plutonium

- Uranium enriched in Uranium-235
- Uranium-233

Medical isotopes may be injected into the bloodstream, ingested or inhaled as a medical treatment. Some medical isotopes are implanted as a pellet or needle in order to target a specific area. Radiopharmaceuticals often have a half-life measured in hours to months; older packages may have decayed below the activity noted on the label. Generally, medical isotopes used for diagnostic purposes emit gamma radiation; medical isotopes used for therapeutic purposes may emit gamma and/or beta radiation.

Medical isotopes include:

- Gallium-67
- Germanium-68
- Indium-111
- Iodine-123
- Iodine-125
- Iodine-131, the most common iodine medical isotope
- Palladium-123
- Samarium-153
- Strontium-82
- Strontium-85
- Technicium-99m, the most common medical isotope
- Thallium-201
- Xenon-131, a radioactive inert gas
- Yttrium-90
- Zinc-65

NORM may be detected via emission of alpha particles, beta particles, or gamma rays. NORM includes:

- Potassium-40 (beta and gamma emissions)
- Radium-226 (alpha, beta, and gamma emissions)
- Thorium-232 (alpha, beta, and gamma emissions)
- Uranium-238 (alpha, beta, and gamma emissions)

Common isotopes used in industry, such as radiographic sources, are dangerous if misused. Without detection by screening instrumentation, these materials could be scrapped and mixed with other metals. Common industrial isotopes include:

- Americium-241
- Barium-133
- Cesium-137
- Cobalt-57
- Cobalt-60
- Iridium-192
- Radium-226
- Thorium-232

Neutron sources are used in some geological applications, such as soil density measurements and well drilling. Other neutron sources exist, but common industrial isotopes and sources for neutrons include:

- Americium-241/Beryllium (^{241}AmBe)
- Californium-252
- Plutonium-238/Beryllium (^{238}PuBe)
- Polonium-210/Beryllium (^{210}PoBe)

Isotopes of any type could theoretically be used in a radiological dispersion device; however, isotopes of particular concern for use in an RRD include:

- Americium-241
- Californium-252
- Cesium-137
- Cobalt-60
- Iridium-192
- Plutonium
- Radium-226
- Strontium-90

Confirming Radioactive Emission

A shielding material between the sample and detector can be useful in determining emissions as alpha, beta, or gamma radiation. Shielding is easily accomplished with a sheet of paper and a 1/8 inch (32 mm) thick piece of aluminum.

Survey meters and other detectors are available with various detection abilities. Be sure to use an instrument that will detect the type of radiation suspected: alpha, beta, gamma, or neutron. A survey meter with a pancake probe can detect many, but not all, forms of alpha, beta, and gamma emissions. Most personal dosimeters and radiation pagers detect gamma and sometimes neutron radiation; alpha and beta are undetected. If the source is a low-energy beta emitter, the radiation hazard may not be detected by a Geiger-Mueller (GM) tube. Examples of low-energy beta emitters are hydrogen-3 (tritium), carbon-14, and sulfur-35, all of which are difficult or impossible to detect in a field test using only a GM tube. These low-energy beta emitters are detectable only by more specialized and sensitive detectors, such as liquid scintillation counting instruments or open window gas proportional detectors. Some RIID detect more than one form of radiation while others only utilize gamma radiation spectroscopy. Be certain that the instrument specifications match the task.

Alpha radiation is determined by holding the detector very close to the sample and sliding a sheet of paper between the sample and detector. An observed drop in activity indicates alpha radiation. Alpha particles will only radiate a few inches in air, so slight increases in distance may show a marked decrease in activity.

Beta radiation is determined by holding the detector very close to the sample and sliding a 1/8 inch (32 mm) sheet of aluminum between the sample and detector. An observed drop in activity indicates beta radiation. Beta radiation is capable of traveling up to about 30 ft in air.

Gamma radiation is assumed for any activity that remains after the source is shielded with the aluminum plate.

Bremsstrahlung is x-rays produced when fast electrons (beta) pass through matter and are abruptly halted. Bremsstrahlung (German for "slowing-down" or "braking" radiation) energy varies from zero to the energy of the electron and is caused by beta particles colliding with dense shielding and emitting a gamma ray. This is the operating principle for a simple medical x-ray machine; a beam of high-speed electrons collides with a high-density tungsten target to generate x-rays. Bremsstrahlung is lessened by low-density shielding because the "slowing down" is not as abrupt as with high-density shielding. Bremsstrahlung will be detected as gamma radiation from some beta sources when shielded by the aluminum plate, although at a lesser rate than the original beta activity. This shielding test will not differentiate a high-energy beta source with Bremsstrahlung from a source producing both beta and gamma radiation.

Neutrons are most effectively shielded with hydrogenous material: gamma by lead. However, neutrons are detected by an entirely different method from gamma radiation and this shielding test does not apply to separation of neutrons from other forms of ionizing radiation. A neutron detector is a tube filled with boron trifluoride (BF_3) or helium-3 (^3He) and has a high voltage applied to it. When neutron radiation interacts with the gas in the tube a detectable particle is emitted; BF_3 emits a helium-4 nucleus and ^3He emits a proton. The emission pulse is modified through electrical circuitry and is outputted as a visual meter reading or an audible sound. Through the use of filters and electronic circuitry, high-energy gamma, beta, and alpha radiation is rejected and the reading represents neutron radiation as measured in rem. A multichannel analyzer is available to determine the identity of the neutron source. Neutron radiation can also be detected with a lithium iodide scintillator (LiI). Some instruments combine the LiI scintillator with another scintillating material in a single instrument capable of detecting gamma and neutron radiation. This has application in "pagers," among others.

Table 1.10 describes for illustrative purposes only some common radioisotopes and the type of radiation emitted. It is important to note that an isotope may emit more than one type of radiation, but some types may be of such low proportion or low energy as to be very difficult to detect with field test equipment. Identification of radiation as alpha, beta, or gamma should be used for tactical purposes and not as a method of identifying the isotope(s). As important as identifying that alpha, beta, gamma, or neutron radiation is present is the confirmation that certain types of radiation are *not* present.

More detail for the emissions of a particular isotope is available through reference material describing the emission energy and the corresponding percentage of each emission. Table 1.11 shows emission energies for ^{137}Cs. For example, a survey meter is selected that uses an internal, high-range energy compensated GM detector with an energy response of +/– 15% true value of 60 keV – 3 MeV. When the detector specifications are compared to the emission energies listed in Table 1.11, only the E3 gamma radiation (shaded) will be detected. The E1 and E2 gamma emissions are below the lower limit of the detector range (60 KeV). The E1 and E2 gamma emissions are also a low percentage of the total gamma radiation from ^{137}Cs.

In a second example, the survey meter described previously is coupled to a sidewall GM detector with a rotary beta shield for general purpose survey. The manual lists the beta cutoff at approximately 200 keV when the window is open and the detector is energy dependent. When these detector specifications are compared to the energies in Table 1.11,

Table 1.10   Commonly Encountered Radioisotopes

| | Atomic | Radiation Emission | | |
Radioisotope	Number	Alpha	Beta	Gamma
Americium-241	95	•		•
Cesium-137	55		•	•
Cobalt-60	27		•	•
Iodine-129 and Iodine-131	53		•	•
Plutonium	94	•	•	•
Radium	88	•		•
Radon	86	•		
Strontium-90	38		•	
Technetium-99	43		•	•
Tritium	1		•	
Thorium	90	•		•
Uranium	92	•		•

Source: Adapted from EPA.
Note: Tritium is a specific isotope, ^{3}H.

it is apparent that the open sidewall detector will respond to the E1 and E2 beta emissions (bold) in addition to the E3 gamma emission. Some auger electrons (beta particles emitted from the shell rather than the nucleus) are also detected (bold), but they are a relatively low proportion of the energy detected by the probe.

Confirming Transportation Activity

Certain packages are labeled for radioactivity during transport. The DOT requires all shipments of radioactive material have two identifying warning labels, with the exception of low specific activity (LSA) material and packages of limited quantities. The labels, in order of increasing hazard, are White I, Yellow II, and Yellow III. Each label displays the radioactive content (isotope) and the activity in curies (Ci). The yellow labels also display the Transportation Index (TI), which is the maximum dose in mrem/h at 1 m from the package. The requirements for transportation are summarized in Table 1.12.

A White I label (Figure 1.35) is commonly used for radiopharmaceuticals and other low-level material. A maximum reading of 0.5 mrem/h at the surface of the package is allowable for transport.

Table 1.11   Emission Energies for Cesium-137

| | Gamma/X-ray | | Beta (E max) | | Electrons | | Alpha | |
Emission	E(keV)	%	E(keV)	%	E(keV)	%	E(keV)	%
E1	32	6	**512**	**95**	624	8		
E2	36	1	**1173**	**5**	656	1		
E3	662	85			660	<1		
% omitted		<1		0		<1		

Source: Adapted from *A Basic Guide to RTR Radioactive Materials, Revision 3,* Sandia National Laboratories.

Table 1.12   U.S. DOT Label Categories for Class 7 (Radioactive) Materials Packages

Transport Index (mrem/hr at 1 meter)	Max. Level at Any Point (surface level)	External Label
If the measured TI is <0.05, the value may be considered to be zero.	Less than or equal to 0.005 mSv/h (0.5 mrem/h)	WHITE-I
More than 0 but not more than 1	Greater than 0.005 mSv/h (0.5 mrem/h) but less than or equal to 0.5 mSv/h (50 mrem/h)	YELLOW-II
More than 1 but not more than 10	Greater than 0.5 mSv/h (50 mrem/h) but less than or equal to 2 mSv/h (200 mrem/h)	YELLOW-III
More than 10	Greater than 2 mSv/ h (200 mrem/h) but less than or equal to 10 mSv/h (1000 mrem/h)	YELLOW-III (Must be shipped under exclusive use provisions; see173.441(b) of this subchapter)

Source: Adapted from Code of Federal Regulations 49CFR §172.403 Class 7 (radioactive) material.

A Yellow II label (Figure 1.36) is also commonly used for low-level materials; however, the material is more active or is of greater quantity. A maximum reading of 50 mrem/hr at the surface of the package and 1 mrem/hr maximum at 1 meter from the package.

A Yellow III label (Figure 1.37) is used for higher-level materials. A maximum reading of 200 mrem/hr at the surface of the package and 10 mrem/hr maximum at 1 m from the package is allowable for transport. The Yellow III label is also required for Fissile Class III or large quantity shipments, regardless of radiation levels.

The measurement taken 1 meter from a Yellow II or Yellow III package is called the TI. The total of all the transport indices on one carrier may not exceed 50 mrem/hr and is used to limit exposure from a single carrier. TI values apply to gamma sources only.

Figure 1.35 Example of a White I label. Label background image courtesy of Labelmaster.

Figure 1.36 Example of a Yellow II label. Label background image courtesy of Labelmaster.

To perform the activity test, place the probe of an appropriate gamma detector 1 meter from the side of the package and determine the result after allowing time for the meter to stabilize. The test result should not exceed the labeled TI value, if present. A measurement that exceeds the labeled activity (Ci) at the package surface or exceeds the TI (mrem/hr) 1 meter from the package can mean:

- The package is mislabeled and contains a more active source than declared.
- The package has been breached.
- The survey probe is being affected by another nearby radioactive package.

Figure 1.37 Example of a Yellow III label. Label background image courtesy of Labelmaster.

Table 1.13 describes gamma dose rates at 1 m for 1 curie of material; however, the material is not packaged. The table can be used to presumptively identify the isotope based on label information. The dose rate in the table can be adapted to individual label information applying value proportionately. The expected dose rate at 1 m (D_1) from the unshielded source can be calculated by converting the table dose rate to mrem/hr and multiplying by the labeled activity.

Table 1.13   Gamma Dose Rates of Selected Isotopes at 1 Meter for 1 Curie

Isotope	(rem/hr)/ Ci @ 1 m	Isotope	(rem/hr)/ Ci @ 1 m	Isotope	(rem/hr)/ Ci @ 1 m
Actinium-227	0.22	Gold-198	0.23	Potassium-43	0.56
Antimony-122	0.24	Gold-199	0.9	Radium-226	0.825
Antimony-124	0.98	Hafnium-175	0.21	Radium-228	0.51
Antimony-125	0.27	Hafnium-181	3.1	Rhenium-186	0.2
Arsenic-72	1.01	Indium-114m	0.02	Rubidium-86	0.05
Arsenic-74	0.44	Iodine-124	0.72	Ruthenium-106	0.17
Arsenic-76	0.24	Iodine-125	0.07	Scandium-46	1.09
Barium-131	0.30	Iodine-126	0.25	Scandium-47	0.056
Barium-133	0.024	Iodine-130	1.22	Selenium-75	0.20
Barium-140	1.24	Iodine-131	0.22	Silver-110m	1.43
Beryllium-7	0.03	Iodine-132	1.18	Silver-111	0.02
Bromine-82	1.46	Iridium-192	0.48	Sodium-22	1.20
Cadmium-115m	0.02	Iridium-194	0.15	Sodium-24	1.84
Calcium-47	0.57	Iron-59	0.64	Strontium-85	0.30
Carbon-11	0.59	Krypton-85	0.004	Tantalum-182	0.68
Cerium-141	0.035	Lanthanum-140	1.13	Tellurium-121	0.33
Cerium-144	0.04	Lutecium-177	0.009	Tellurium-132	0.22
Cesium-134	0.87	Magnesium-28	1.57	Thulium-170	0.0025
Cesium-137	0.33	Manganese-52	1.86	Tin-113	0.17
Chlorine-38	0.88	Manganese-54	0.47	Tungsten-185	0.05
Chromium-51	0.016	Manganese-56	0.83	Tungsten-187	0.30
Cobalt-56	1.76	Mercury-197	0.04	Uranium-234	0.01
Cobalt-57	0.09	Mercury-203	0.13	Uranium-235	0.339
Cobalt-58	0.55	Molybdenum-99	0.18	Uranium-238	0.076
Cobalt-60	1.32	Neodymium-147	0.08	Vanadium-48	1.56
Copper-64	0.12	Nickel-65	0.31	Xenon-133	0.01
Europium-152	0.58	Niobium-95	0.42	Ytterbium-175	0.04
Europium-154	0.62	Osmium-191	0.06	Yttrium-88	1.41
Europium-155	0.03	Palladium-109	0.003	Yttrium-91	0.001
Gallium-67	0.11	Platinum-197	0.05	Zinc-65	0.27
Gallium-72	1.16	Potassium-42	0.14	Zirconium-95	0.41

Source: Adapted from *Emergency Response to Terrorism: Weapons of Mass Destruction (WMD) Radiation/ Nuclear Course for Hazardous Materials Technicians.*

$$D_1 = (0.22 \text{ rem/hr/Ci}) \times 1000 \times 0.2 \text{ Ci}$$

For example, a label states the package contains actinium-227 with an activity of 0.2 Ci. In this example the measured dose rate at 1 m from the package should not exceed 44 mrem/hr. A lower dose rate is expected for packages with heavier shielding. If the dose exceeds the expected dose rate and no other sources are present, the package has leaked or is mislabeled. The same applies if the measured dose exceeds the TI, if so labeled. Additionally, if a TI is present on the label, the label should comply with the requirements in Table 1.12.

Confirming Specific Activity

The specific activity of an isotope is defined by the number of curies (Ci) or becquerels (Bq) of activity per mass (g) of pure isotope. This allows simple conversion from mass to activity. Specific activity equals the activity of the isotope divided by the mass of the isotope. Activity is usually expressed in becquerels, and mass is usually expressed in grams and therefore specific activity is measured as Bq/g. However, DOT labeling requirements utilize the Ci for activity. Tables 1.14 and 1.15 display activity in both TBq and Ci. Specific activity is listed for pure isotope. One terabecquerel (TBq or 10^{12} Bq) equals 27.03 Ci.

$$\text{TBq} = \text{Ci} \times 27.03$$

A molecule that incorporates a radioactive atom can be proportioned by mass. The same is true of mixtures. For example, what mass of pure cobalt-60 has an activity of 0.5 Ci? Table 1.14 provides the activity of ^{60}Co as 1.13E + 03 or 1.13×10^3 or 1130 Ci/g. The 0.5 Ci source in the package therefore is 0.000442 g or 0.442 mg.

$$0.5 \text{ Ci} \div 1130 \text{ Ci/g} = 0.000442 \text{ g}$$

As a second example, what is the expected activity of uranium-238 in a 500 g container labeled uranyl nitrate hexahydrate [$UO_2(NO_3)_2 \cdot 6H_2O$]? The proportion of uranium in the compound is the molecular weight of uranium divided by the molecular weight of the compound.

$$238 \div 502 = 0.475$$

The uranium present is the mass of the material in the container multiplied by the fraction of uranium:

$$500 \text{ g} \times 0.475 = 238 \text{ g}$$

Table 1.14 lists the specific activity of ^{238}U as 1.24E – 08 TBq/g (3.35E – 07 Ci/g). The estimated activity of 500 g of uranium nitrate hexahydrate is:

$$238 \text{ g} \times 1.24\text{E} - 08 \text{ TBq/g} = 2.95\text{E} - 06 \text{ TBq}$$

Table 1.14   Specific Activity and Half-Life of Select Isotopes by Atomic Weight

Isotope	Half-Life[a, b]	Specific Activity (TBq/g)[a]	Specific Activity (Ci/g)[a]
^{3}H	12.35 y	3.57E + 02	9.65E + 03
^{7}Be	53.44 d	1.29E + 04	3.49E + 05
^{11}C	20.38 m	3.10E + 07	8.38E + 08
^{13}N	9.97 m	5.37E + 07	1.45E + 09
^{14}C	5730 y	1.65E − 01	4.46E + 00
^{16}N	7.20 s	3.62E + 09	9.78E + 10
^{18}F	109.74 m	3.52E + 06	9.51E + 07
^{22}Na	2.602 y	2.31E + 02	6.24E + 03
^{24}Na	15.00 h	3.22E + 05	8.70E + 06
^{32}P	14.29 d	1.06E + 04	2.87E + 05
^{35}S	87.44 d	1.58E + 03	4.27E + 04
^{36}Cl	301,000 y	1.22E − 03	3.30E − 02
^{41}Ar	1.827 h	1.55E + 06	4.19E + 07
^{42}K	12.36 h	2.23E + 03	6.03E + 04
^{45}Ca	163 d	6.58E + 02	1.78E + 04
^{46}Sc	83.83 d	1.25E + 03	3.38E + 04
^{51}Cr	27.704 d	3.42E + 03	9.24E + 04
^{54}Mn	312.5 d	2.86E + 02	7.73E + 03
^{55}Fe	2.7 y	8.91E + 01	2.41E + 03
^{56}Mn	2.579 h	8.03E + 05	2.17E + 07
^{57}Co	270.9 d	3.13E + 02	8.46E + 03
^{59}Ni	75,000 y	2.99E − 01	8.08E + 00
^{59}Fe	44.529 d	1.84E + 03	4.97E + 04
^{60}Co	5.271 y	4.18E + 01	1.13E + 03
^{63}Ni	96 y	2.19E + 00	5.92E + 01
^{64}Cu	12.70 h	1.43E + 05	3.87E + 06
^{65}Zn	243.9 d	3.05E + 02	8.24E + 03
^{65}Ni	2.52 h	7.08E + 05	1.91E + 07
^{72}Ga	14.10 h	1.14E + 05	3.08E + 06
^{76}As	26.32 h	5.79E + 04	1.57E + 06
^{82}Br	35.30 h	4.00E + 04	1.08E + 06
^{86}Rb	18.66 d	3.01E + 03	8.14E + 04
^{89}Sr	50.50 d	1.07E + 03	2.89E + 04
^{90}Sr	29.12 y	5.05E + 00	1.37E + 02
^{90}Y	64.00 h	2.01E + 06	5.43E + 07
^{92}Y	3.54 h	3.56E + 05	9.62E + 06
^{92}Sr	2.71 h	4.65E + 05	1.26E + 07
^{95}Zr	63.98 h	7.95E + 02	2.15E + 04
^{95}Nb	35.15 d	1.45E + 03	3.92E + 04
^{99}Mo	66.00 h	1.77E + 04	4.78E + 05
^{99m}Tc	6.02 h	1.94E + 05	5.24E + 06
^{103}Ru	39.28 d	1.19E + 03	3.22E + 04
^{106}Ru	368.2 d	1.24E + 02	3.35E + 03
^{110m}Ag	249.9 d	1.76E + 02	4.76E + 03

(continued)

Table 1.14   Specific Activity and Half-Life of Select Isotopes by Atomic Weight   (*Continued*)

Isotope	Half-Life[a, b]	Specific Activity (TBq/g)[a]	Specific Activity (Ci/g)[a]
^{110}Ag	24.60 s	1.54E + 08	4.16E + 09
^{123}I	13.20 h	7.14E + 04	1.93E + 06
^{125}I	60.14 d	6.42E + 02	1.74E + 04
^{129}I	1.57E+07 y	6.53E − 06	1.77E − 04
^{131}I	8.040 d	4.59E + 03	1.24E + 05
^{132}Te	78.20 h	1.12E + 04	3.03E + 05
^{132}I	2.30 h	3.82E + 05	1.03E + 07
^{133}Ba	10.5 y	9.48E + 00	2.56E + 02
^{133}Xe	5.245 d	6.92E + 03	1.87E + 05
^{133}I	20.80 h	4.19E + 04	1.13E + 06
^{133}Te	12.45 m	4.20E + 06	1.14E + 08
^{134}Cs	2.062 y	4.79E + 01	1.29E + 03
^{135}I	6.61 h	1.30E + 05	3.51E + 06
^{136}Cs	13.10 d	2.71E + 03	7.33E + 04
^{137}Ce	30 y	3.22E + 00	8.70E + 01
^{137m}Ba	2.552 m	1.99E + 07	5.38E + 08
^{140}Ba	12.74 d	2.71E + 03	7.33E + 04
^{140}La	40.272 h	2.06E + 04	5.57E + 05
^{141}Ce	32.501 d	1.05E + 03	2.84E + 04
^{143}Pr	13.56 d	2.49E + 03	6.73E + 04
^{143}Ce	33.00 h	2.46E + 04	6.65E + 05
^{144}Ce	284.3 d	1.18E + 02	3.19E + 03
^{144}Pr	17.28 m	2.80E+06	7.57E + 07
^{147}Pm	2.62 y	3.43E + 01	9.27E + 02
^{181}W	121.2 d	2.20E + 02	5.95E + 03
^{182}Ta	115 d	2.31E + 02	6.24E + 03
^{185}W	75.10 d	3.48E + 02	9.41E + 03
^{192}Ir	74.02 d	3.40E + 02	9.19E + 03
^{198}Au	2.696 d	9.05E + 03	2.45E + 05
^{199}Au	3.139 d	7.73E + 03	2.09E + 05
^{203}Hg	46.60 d	5.11E + 02	1.38E + 04
^{204}Tl	3.779 y	1.72E + 01	4.65E + 02
^{207}Tl	4.77 m	7.04E + 06	1.90E + 08
^{210}Po	138.38 d	1.66E + 02	4.49E + 03
^{210}Pb	22.3 y	2.83E + 02	7.65E + 03
^{210}Bi	5.012 d	4.59E + 03	1.24E + 05
^{212}Po	3.05E−07 s	6.45E + 15	1.74E + 17
^{224}Ra	3.660 d	5.89E + 03	1.59E + 05
^{226}Ra	1600 y	3.66E − 02	9.89E − 01
^{227}Th	18.718 d	1.14E + 03	3.08E + 04
^{228}Ra	5.75 y	1.01E + 01	2.73E + 02
^{228}Th	1.913 y	3.03E + 01	8.19E + 02
^{228}Ac	6.13 h	8.29E + 04	2.24E + 06
^{229}Th	7340 y	7.87E − 03	2.13E − 01

(*continued*)

Table 1.14 Specific Activity and Half-Life of Select Isotopes by Atomic Weight (*Continued*)

Isotope	Half-Life[a,b]	Specific Activity (TBq/g)[a]	Specific Activity (Ci/g)[a]
^{230}Th	77,000 y	7.47E − 04	2.02E − 02
^{231}Th	25.52 h	1.97E + 04	5.32E + 05
^{232}Th	1.41E + 10 y	4.05E − 09	1.09E − 07
^{232}U	72 y	7.92E − 01	2.14E + 01
^{233}U	1.59E + 05 y	3.57E − 04	9.65E − 03
^{234}U	2.45E+05 y	2.31E − 04	6.24E − 03
^{234}Th	24.10 d	8.56E + 02	2.31E + 04
^{235}U	7.04E + 08 y	8.00E − 08	2.16E − 06
^{236}U	2.34E + 07 y	2.40E − 06	6.49E − 05
^{237}U	6.750 d	3.02E + 03	8.16E + 04
^{238}U	4.47E + 09 y	1.24E − 08	3.35E − 07
^{238}Pu	87.74 y	6.34E − 01	1.71E + 01
^{239}Pu	24,065 y	2.30E − 03	6.22E − 02
^{239}Np	2.355 d	8.58E + 03	2.32E + 05
^{240}Pu	6537 y	8.43E − 03	2.28E − 01
^{241}Am	432.2 y	1.27E − 01	3.43E + 00
^{241}Pu	14.4 y	3.81E + 00	1.03E + 02
^{252}Cf	2.638 y	1.99E + 01	5.38E + 02

Source: Adapted from International Commission on Radiological Protection and CRC Handbook of Radiation Measurement and Protection.

Notes: [a]Values are formatted in exponential units; E + 02 = 10^2 = 100. For example, 1.66E + 02 = 1.66×10^2 = 166. Likewise, 3.66E − 02 = 3.66×10^{-2} = 0.0366.

[b]Half-life units are years (y), days (d), hours (h), minutes (m), and seconds (s).

Labeled specific activity can be used to confirm a substance. For example, a package has caused a police officer's radiation pager to alarm. The package is in the backseat of a car parked on a street. The sender's name matches the name of the vehicle owner, who is a graduate student. The package is labeled to match DOT requirements: nitrates, inorganic, n.o.s., UN 1477, class 5.1, oxidizer hazard label. The package is seized and examined to determine if it is legitimate. The package contains a 500 g container labeled uranyl nitrate hexahydrate [$UO_2(NO_3)_2 \cdot 6H_2O$] with a labeled specific activity of 1.23E+04 Bq/g. Comparing the labeled specific activity to the specific activity of ^{238}U in Table 1.14 (1.24E − 08 TBq/g or 1.24E + 04 Bq/g) shows the labeled value is for pure ^{238}U, not the compound. An accompanying material safety data sheet (MSDS) states the molecular weight is 501.2. The product appears as yellow-green crystals, consistent with the MSDS. A nitrate test strip confirms the presence of nitrate. Is the substance legitimate?

Since nitrate has been confirmed through a chemical test, the focus is on uranium, for which a chemical confirmation test could also be selected. Uranium may also be confirmed through testing with a radiation instrument. First, determine the fraction of the compound that is composed of uranium by dividing the molecular weight of the compound by the molecular weight of uranium:

$$238 \div 501.2 = 0.475$$

Table 1.15 Select Isotopes and Half-Life by Specific Activity

Specific Activity (TBq/g)[a]	Specific Activity (Ci/g)[a]	Isotope	Half-Life[a, b]
4.05E − 09	1.09E − 07	^{232}Th	1.41E + 10 y
1.24E − 08	3.35E − 07	^{238}U	4.47E + 09 y
8.00E − 08	2.16E − 06	^{235}U	7.04E + 08 y
2.40E − 06	6.49E − 05	^{236}U	2.34E + 07 y
6.53E − 06	1.77E − 04	^{129}I	1.57E + 07 y
2.31E − 04	6.24E − 03	^{234}U	2.45E + 05 y
3.57E − 04	9.65E − 03	^{233}U	1.59E + 05 y
7.47E − 04	2.02E − 02	^{230}Th	77,000 y
1.22E − 03	3.30E − 02	^{36}Cl	301,000 y
2.30E − 03	6.22E − 02	^{239}Pu	24,065 y
7.87E − 03	2.13E − 01	^{229}Th	7340 y
8.43E − 03	2.28E − 01	^{240}Pu	6537 y
3.66E − 02	9.89E − 01	^{226}Ra	1600 y
1.27E − 01	3.43E + 00	^{241}Am	432.2 y
1.65E − 01	4.46E + 00	^{14}C	5730 y
2.99E − 01	8.08E + 00	^{59}Ni	75,000 y
6.34E − 01	1.71E + 01	^{238}Pu	87.74 y
7.92E − 01	2.14E + 01	^{232}U	72 y
2.19E + 00	5.92E + 01	^{63}Ni	96 y
3.22E + 00	8.70E + 01	^{137}Ce	30 y
3.81E + 00	1.03E + 02	^{241}Pu	14.4 y
5.05E + 00	1.37E + 02	^{90}Sr	29.12 y
9.48E + 00	2.56E + 02	^{133}Ba	10.5 y
1.01E + 01	2.73E + 02	^{228}Ra	5.75 y
1.72E + 01	4.65E + 02	^{204}Tl	3.779 y
1.99E + 01	5.38E + 02	^{252}Cf	2.638 y
3.03E + 01	8.19E + 02	^{228}Th	1.913 y
3.43E + 01	9.27E + 02	^{147}Pm	2.62 y
4.18E + 01	1.13E + 03	^{60}Co	5.271 y
4.79E + 01	1.29E + 03	^{134}Cs	2.062 y
8.91E + 01	2.41E + 03	^{55}Fe	2.7 y
1.18E + 02	3.19E + 03	^{144}Ce	284.3 d
1.24E + 02	3.35E + 03	^{106}Ru	368.2 d
1.66E + 02	4.49E + 03	^{210}Po	138.38 d
1.76E + 02	4.76E + 03	^{110m}Ag	249.9 d
2.20E + 02	5.95E + 03	^{181}W	121.2 d
2.31E + 02	6.24E + 03	^{22}Na−	2.602 y
2.31E + 02	6.24E + 03	^{182}Ta	115 d
2.83E + 02	7.65E + 03	^{210}Pb	22.3 y
2.86E + 02	7.73E + 03	^{54}Mn	312.5 d
3.05E + 02	8.24E + 03	^{65}Zn	243.9 d
3.13E + 02	8.46E + 03	^{57}Co	270.9 d
3.40E + 02	9.19E + 03	^{192}Ir	74.02 d
3.48E + 02	9.41E + 03	^{185}W	75.10 d
3.57E + 02	9.65E + 03	^{3}H	12.35 y

Table 1.15   Select Isotopes and Half-Life by Specific Activity (*Continued*)

Specific Activity (TBq/g)[a]	Specific Activity (Ci/g)[a]	Isotope	Half-Life[a, b]
5.11E + 02	1.38E + 04	^{203}Hg	46.60 d
6.42E + 02	1.74E + 04	^{125}I	60.14 d
6.58E + 02	1.78E + 04	^{45}Ca	163 d
7.95E + 02	2.15E + 04	^{95}Zr	63.98 h
8.56E + 02	2.31E + 04	^{234}Th	24.10 d
1.05E + 03	2.84E + 04	^{141}Ce	32.501 d
1.07E + 03	2.89E + 04	^{89}Sr	50.50 d
1.14E + 03	3.08E + 04	^{227}Th	18.718 d
1.19E + 03	3.22E + 04	^{103}Ru	39.28 d
1.25E + 03	3.38E + 04	^{46}Sc	83.83 d
1.45E + 03	3.92E + 04	^{95}Nb	35.15 d
1.58E + 03	4.27E + 04	^{35}S	87.44 d
1.84E + 03	4.97E + 04	^{59}Fe	44.529 d
2.23E + 03	6.03E + 04	^{42}K	12.36 h
2.49E + 03	6.73E + 04	^{143}Pr	13.56 d
2.71E + 03	7.33E + 04	^{136}Cs	13.10 d
2.71E + 03	7.33E + 04	^{140}Ba	12.74 d
3.01E + 03	8.14E + 04	^{86}Rb	18.66 d
3.02E + 03	8.16E + 04	^{237}U	6.750 d
3.42E + 03	9.24E + 04	^{51}Cr	27.704 d
4.59E + 03	1.24E + 05	^{131}I	8.040 d
4.59E + 03	1.24E + 05	^{210}Bi	5.012 d
5.89E + 03	1.59E + 05	^{224}Ra	3.660 d
6.92E + 03	1.87E + 05	^{133}Xe	5.245 d
7.73E + 03	2.09E + 05	^{199}Au	3.139 d
8.58E + 03	2.32E + 05	^{239}Np	2.355 d
9.05E + 03	2.45E + 05	^{198}Au	2.696 d
1.06E + 04	2.87E + 05	^{32}P	14.29 d
1.12E + 04	3.03E + 05	^{132}Te	78.20 h
1.29E + 04	3.49E + 05	^{7}Be	53.44 d
1.77E + 04	4.78E + 05	^{99}Mo	66.00 h
1.97E + 04	5.32E + 05	^{231}Th	25.52 h
2.06E + 04	5.57E + 05	^{140}La	40.272 h
2.46E + 04	6.65E + 05	^{143}Ce	33.00 h
4.00E + 04	1.08E + 06	^{82}Br	35.30 h
4.19E + 04	1.13E + 06	^{133}I	20.80 h
5.79E + 04	1.57E + 06	^{76}As	26.32 h
7.14E + 04	1.93E + 06	^{123}I	13.20 h
8.29E + 04	2.24E + 06	^{228}Ac	6.13 h
1.14E + 05	3.08E + 06	^{72}Ga	14.10 h
1.30E + 05	3.51E + 06	^{135}I	6.61 h
1.43E + 05	3.87E + 06	^{64}Cu	12.70 h
1.94E + 05	5.24E + 06	^{99m}Tc	6.02 h
3.22E + 05	8.70E + 06	^{24}Na	15.00 h
3.56E + 05	9.62E + 06	^{92}Y	3.54 h

(*continued*)

Table 1.15   Select Isotopes and Half-Life by Specific Activity (*Continued*)

Specific Activity (TBq/g)[a]	Specific Activity (Ci/g)[a]	Isotope	Half-Life[a, b]
3.82E + 05	1.03E + 07	^{132}I	2.30 h
4.65E + 05	1.26E + 07	^{92}Sr	2.71 h
7.08E + 05	1.91E + 07	^{65}Ni	2.52 h
8.03E + 05	2.17E + 07	^{56}Mn	2.579 h
1.55E + 06	4.19E + 07	^{41}Ar	1.827 h
2.01E + 06	5.43E + 07	^{90}Y	64.00 h
2.80E + 06	7.57E + 07	^{144}Pr	17.28 m
3.52E + 06	9.51E + 07	^{18}F	109.74 m
4.20E + 06	1.14E + 08	^{133}Te	12.45 m
7.04E + 06	1.90E + 08	^{207}Tl	4.77 m
1.99E + 07	5.38E + 08	^{137m}Ba	2.552 m
3.10E + 07	8.38E + 08	^{11}C	20.38 m
5.37E + 07	1.45E + 09	^{13}N	9.97 m
1.54E + 08	4.16E + 09	^{110}Ag	24.60 s
3.62E + 09	9.78E + 10	^{16}N	7.20 s
6.45E + 15	1.74E + 17	^{212}Po	3.05E–07 s

Source: Adapted from International Commission on Radiological Protection and CRC Handbook of Radiation Measurement and Protection.

Notes: [a]Values are formatted in exponential units; E + 02 = 10^2 = 100. For example, 1.66E + 02 = 1.66×10^2 = 166. Likewise, 3.66E–02 = 3.66×10^{-2} = 0.0366.

[b]Half-life units are years (y), days (d), hours (h), minutes (m), and seconds (s).

Next, determine the mass of uranium in the compound by multiplying 500 g of compound by the fraction of mass attributed to uranium:

$$500 \text{ g} \times 0.475 = 237.5 \text{ g}$$

Finally, determine the expected activity by multiplying the mass of the uranium by the specific activity:

$$237.5 \text{ g} \times 1.23\text{E} + 04 \text{ Bq/g} = 2{,}921{,}250 \text{ Bq}$$

Compare the activity measured with an appropriate instrument to the calculated activity. Since some of the material may have been removed from the labeled quantity, the measured activity is expected to be equal to or less than about 3 million Bq. A measured result greater than about 3 million Bq is cause for concern. Compare this result to the previous example; 2,921,250 Bq is equivalent to 2.92E – 06 TBq.

Solubility

Solubility can be estimated by simple field tests. Solubility can also be manipulated to confirm identity of the product's constituents.

Solubility of solids in water will increase with increasing temperature. Solubility of gases in water will increase with decreasing temperature and increasing pressure.

Expect increased heat to increase the solubility of solids in water and to drive gases out of water solutions. Increased heat will also drive higher-vapor pressure (organic) liquids out of water solutions. This principle may be useful when attempting to detect a gas dissolved in a liquid. A test tube of liquid can be gently heated with a torch and the head space of the tube can be sampled with an appropriate gas or vapor test selected from Chapter 2, "Chemical Tests."

Confirming Water Solubility

To estimate water solubility of a solid, add about ½ inch of water to a test tube and add a gram of solid sample to the water. Look for any dissolving as evidenced by a reduced volume of solid, softening of sharp corners on solid particles, change in color of the liquid, and so on. This method can determine if the solid is water soluble, but it does not determine the percentage of solubility.

To determine the percent solubility of a solid, place a precise volume of water in a container. Weigh a small amount of the solid and dissolve it into the water. When it completely dissolves, weigh another small amount and dissolve it completely. Repeat this process until the solution is saturated. Determine the solubility by adding the weights of all solid material added to the water and dividing by the volume of water. The result will be a weight per volume (w/v) ratio.

A practical guide to predicting the water solubility of ionic compounds is described in the following statements, which are intended as general guidelines. As always during an analysis, consult a reliable resource such as *Handbook of Chemistry and Physics* (CRC Press).

- All compounds of ammonium and of alkali metals are water soluble.
- All nitrates and acetates are water soluble.
- All chlorides, bromides, and iodides are water soluble except those of silver, lead, and mercury(I).
- All sulfates are water soluble except those of silver, lead, mercury(I), barium, strontium, and calcium.
- All carbonates, sulfites, and phosphates are water insoluble except those of ammonium and alkali metals.
- All hydroxides are water insoluble except those of ammonium, barium, and alkali metals.
- All sulfides are water insoluble except those of ammonium, alkali metals, and alkali earth cations.
- All oxides are water insoluble except those of calcium, barium, and alkali metals, which hydrolyze to form water-soluble hydroxides.

To determine water solubility of a liquid, add about ½ inch of water to a test tube and mark the test tube at the surface of the water. Next, add an equal volume of liquid sample with a pipette by slowly running it down the side of the test tube. Be careful to point the tube in a safe direction as contents of the tube may be ejected. Next, gently agitate the contents with a pipette but not so forcefully as to inject air into the liquid. Allow the liquid to settle and determine the solubility by observing the visible division between liquids and comparing it to the original mark. No division indicates complete solubility or miscibility. No change in the division between layers indicated insolubility. If the division exists but is no longer at the mark, estimate the percentage of solubility as shown in Figure 1.38.

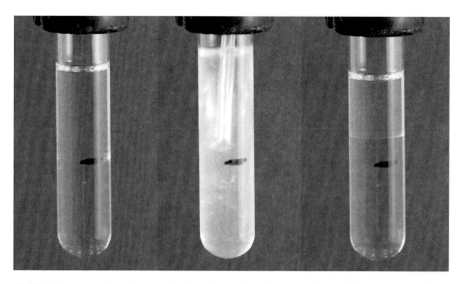

Figure 1.38 The water level is marked on the side of a test tube and then an equal volume of sample is added. In this case, the lighter-than-water liquid sample floats on the water with a barely visible boundary formed at the black mark (left). The contents of the test tube are mixed gently with a pipette (center). After the contents settle, compare the mark on the test tube to the new dividing line formed at the boundaries of the liquids to estimate solubility, which is about 30% in the case of this lighter-than-water liquid (right).

Table 1.16 displays the water solubility of several inorganic compounds. As is evident in the table, water solubility must be determined for each compound.

Confirming Vapor Solubility

You can estimate the relative water solubility of a vapor or gas by using the same air flow "smoke" tubes described in the section on estimating vapor density. Prepare a sample as described for estimating vapor density and observe the visual density of the fume in the sample while it is covered and still resting.

A sample with higher moisture content will produce a "thicker" or more opaque cloud when either the Sensidyne or Drager air current tube is used. For example, one pump of fume injected into trichloroethylene vapor will appear to be "thin" while one pump of fume into a sample of 91% isopropyl alcohol (9% water) vapor will appear to be "thick." Low humidity samples, such as hexane vapor, may contain fume that is not visible until making contact with atmospheric moisture when the sample is spilled from the beaker.

The relative visual density of the fume inside the sample beaker can give you a relative idea of the water solubility of the vapor or gas. Samples that immediately produce a thick white fume in the beaker most likely already contain water and are likely to be fairly water soluble. Samples that hold a thin, lightly visible white fume contain little water vapor and the vapor or gas is displacing humid air from the beaker. The visual fume density of anhydrous samples may increase once the vapor is released from the beaker and mixes with humid air. You can use this estimate of the water solubility of the vapor or gas to determine how effective water spray might be in absorbing or scrubbing vapors from the air. Water-soluble gases and vapors tend to cause injury to the upper respiratory system due to early entrapment by moist membranes in the upper airway. Water-insoluble gases and vapors

Table 1.16 Water Solubility of Various Inorganic Compounds

Cation	Acetate	Borate	Bromide	Carbonate	Chlorate	Chloride	Chromate	Hydroxide	Iodide	Nitrate	Nitrite	Oxalate	Perchlorate	Permanganate	Phosphate	Sulfate	Sulfide	Sulfite	Thiocyanate
Aluminum	S	S	S	--	S	S	--	I	S	S	--	I	S	--	I	S	D	--	--
Ammonium	S	S	S	S	S	S	S	S	S	S	S	S	S	S	S	S	S	S	S
Barium	S	--	S	I	S	S	I	S	S	S	S	I	S	S	I	I	D	I	S
Beryllium	I	--	S	I	--	S	--	--	D	S	--	S	--	--	S	I	D	--	--
Cadmium	S	S	S	I	S	S	--	I	S	S	--	--	S	S	I	S	I	--	--
Calcium	S	M	S	I	S	S	I	M	S	S	S	I	S	S	I	M	I	I	S
Cesium	S	--	S	S	S	S	S	S	S	S	--	S	M	I	I	S	S	--	--
Chromium	S	--	S	--	--	S	--	D	S	S	--	S	--	--	M	S	I	--	--
Cobalt	S	S	S	I	S	S	I	I	S	S	--	I	S	--	I	S	I	I	S
Copper	S	--	S	I	S	S	I	I	I	S	I	I	S	--	I	S	I	I	I
Gold	--	--	I	--	--	S	--	--	M	--	--	--	--	--	--	--	I	--	--
Hydrogen	--	S	S	S	S	S	S	S	S	S	S	S	--	S	S	S	S	S	--
Iron	S	--	S	I	--	S	--	I	S	S	--	S	S	--	I	M	I	I	S
Lead	S	I	M	I	S	M	I	I	I	S	S	I	S	--	I	I	I	I	I
Lithium	S	S	S	S	S	S	--	S	S	S	S	S	S	S	I	S	S	S	S
Magnesium	S	I	S	I	S	S	S	I	S	S	S	I	S	S	I	S	D	S	--
Manganese	S	--	S	I	--	S	--	I	S	S	--	I	--	--	I	S	I	--	S
Molybdenum	--	--	I	--	--	I	--	M	I	--	--	--	--	--	I	--	--	I	--
Nickel	S	I	S	I	M	S	--	--	S	--	--	I	S	--	I	S	I	I	--
Palladium	--	--	I	--	--	S	--	--	I	S	--	--	--	--	--	S	I	--	--
Platinum	--	--	I	--	--	S	--	I	I	--	--	--	--	--	--	S	I	--	--
Potassium	S	S	S	S	S	S	S	S	S	S	S	S	S	S	S	S	S	S	S
Silver	M	--	I	I	S	I	I	I	I	S	I	I	S	M	I	M	I	I	I
Sodium	S	S	S	S	S	S	S	S	S	S	S	S	S	S	S	S	S	S	S
Strontium	S	S	S	I	S	S	I	M	S	S	S	I	S	S	I	I	I	I	S
Titanium	--	--	S	--	--	S	--	--	D	--	--	S	S	--	--	I	I	--	--
Vanadium	--	--	S	--	--	S	--	--	S	--	--	--	--	--	--	I	--	--	--

Notes: S = Soluble in water (> 0.1 M)

M = Moderately soluble in water (0.01 – 0.1 M)

I = Insoluble (< 0.01 M)

D = Decomposes in water

A double-dash (--) denotes compounds lacking data or that do not exist.

tend to cause injury deep in the lungs due to the inability of the moist upper airway membranes to absorb the gas or vapor. Obviously, an unknown material can cause other types of injury by being absorbed and transported throughout the body.

It might be possible that an unknown anhydrous sample could react with the smoke tube reagents to form a fume, so do not use this test as a firm indicator of the water content. In fact, you should never rely on a single test as a sole source of information on which to plan a response. Figure 1.39 shows the density of fume produced by a Sensidyne Tube 5100

Figure 1.39 The relative water vapor content and the relative water solubility of the vapor or gas can be estimated by the visual density of the fume. Samples are described from left to right with water solubility values expressed in percent. Top row: room air and ambient humidity. Middle row: hexane (0.002%), propane (0.01%), trichloroethylene (0.1%), 1,1,1,2-tetrafluoroethane (0.1%). Bottom row: 91% isopropyl alcohol, (miscible), acetone (miscible), ammonium hydroxide (soluble), glacial acetic acid (miscible). The bottom row illustrates the higher water content of water-soluble vapors. The middle row has had much of the room air displaced by insoluble vapors or gas. An exception is propane (second from left), which was injected into the beaker with a propane torch head. The torch head mixes room air with the propane in a proper proportion to support the torch flame; therefore, the propane sample contains a higher proportion of room air and humidity than the other samples in the middle row.

Smoke Tube (Air Flow Indicator Tube) for several samples along with the solubility of the liquid in water.

Vapor Pressure, Volatility, and Evaporation Rate

Vapor pressure is influenced significantly by temperature. As the temperature rises, vapor pressure rises. When the vapor pressure equals atmospheric pressure, the liquid is at its boiling point. Most references provide a vapor pressure measured at room temperature unless otherwise specified. The vapor pressure of water at standard temperature is about

18 mmHg and is useful as a benchmark. Vapor pressures of other common substances are shown in Table 1.17.

Some sources, usually MSDSs, list the evaporation rate at which a particular material will vaporize when compared to the rate of vaporization of a known material. The evaporation is a ratio and must be related to a standard evaporation rate. The standard, among others, can be n-butyl acetate (NBUAC or n-BuAc) with a vaporization rate designated as 1.0 and a vapor pressure of about 10 mmHg. Vaporization rates of other solvents or materials are then classified as fast evaporating (>3.0), medium evaporating (0.8–3.0), or slow evaporating (<0.8). Ethyl ether is used as a standard, but it is difficult to store and deploy for field use since it can form explosive peroxides over time.

The evaporation rate test will provide a more confident confirmation test since it describes the actual evaporation. Vapor pressure is an estimate of the rate of evaporation, and evaporation may be affected by other, unknown factors at the time of the test. However, evaporation data are not always available for materials and vapor pressure seems to be a more commonly provided value. Table 1.17, "Approximate Vapor Pressure and Evaporation Rates of Common Substances," shows that vapor pressure and evaporation rates do not correlate precisely.

Confirming Vapor Pressure or Evaporation Rate

Estimate vapor pressure by using water as a comparison (VP 18 mmHg). Place a drop of water next to a drop of liquid sample on an inert surface and observe the evaporation rate. If you have other known materials you can use them as a side-by-side guide to estimate vapor pressure (Table 1.17). Other liquids with known vapor pressures could easily be adapted.

Estimate the evaporation rate in a similar manner. Place a drop of the liquid sample next to a drop of a liquid with a known evaporation rate on an inert surface and estimate the relative evaporation rates. This is not a highly precise method but is adequate for field use.

A more precise method involves weighing a liquid sample as it evaporates and then comparing the rate of loss to a standard liquid. When variables affecting evaporation are tightly controlled, this gravimetric method has been shown to be accurate within a few percent of absolute liquid evaporation rates.

Viscosity

Viscosity is a term used to describe the "thickness" of a liquid or its resistance to flow. High-viscosity liquids are thick and slow to pour. Viscosity is most often measured in a unit termed *poise*.

Viscosity is affected mostly by temperature, solvents, or polymers. Viscosity can be decreased by mixing with a less viscous liquid or by heating. Viscosity can be increased by adding a solid or more viscous liquid, adding a polymer such as starch, or cooling the material. A pure substance can be verified by comparing its estimated viscosity to that listed in a reference source at a specified temperature. This can be done by comparing the sample to a known liquid in a side-by-side flow test. Liquids that are less viscous than water have an organic component.

Table 1.17   Approximate Vapor Pressure and Evaporation Rates of Common Substances

Substance	Evaporation Rate, n-BuAcᵃ = 1	VP, mmHg @ 20°C
Water	0.3	18
n-Butyl acetateᵃ	1	10
Hexanone	1	10
Octane	1.4	11
Ethanol	1.4	45
Isopropanol	2.8	33
Butanone (MEK)	3.8	71
Acetone	5.6	184
Methanol	5.9	98
Hexane	8.3	132
Pentane	28.6	426
Ether	37.5	440

Note: ᵃ n-BuAc or NBUAC is n-butyl acetate (CAS 123-86-4)

Estimating Viscosity

One method of estimating viscosity involves filling a test tube one-third of the way with the sample and capping it. A second test tube is filled with a liquid of known viscosity and capped. After assuring both liquids are at equal temperature, place a rubber band around the tubes. Place the base of the tubes on a notepad and let them fall horizontally (Figure 1.40). Watch the leading edge of the liquids as they flow toward the caps. Use the relative

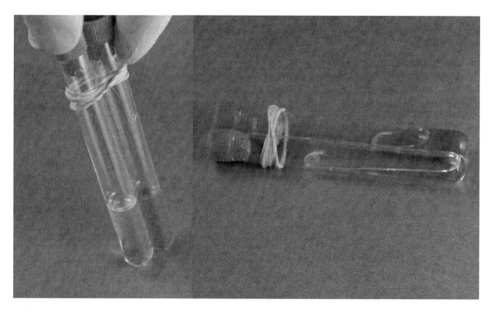

Figure 1.40 Viscosity may be estimated by comparing the flow of the sample liquid to that of a known standard (in this case glycerin, 757 cSt) at approximately room temperature. The sample and standard are added to separate test tubes at equal volumes and temperatures. It is helpful to hold the stoppered test tubes with a rubber band (left). Lay the tubes horizontally and compare the movement of the sample to the standard (right). The sample was found to have viscosity greater than 757 cSt at 25°C.

Table 1.18 Measured Viscosity of Common Substances[a]

Substance at 65°F (25°C)	Viscosity (cSt)
Water	1.0
Brake fluid	12.7
Automotive antifreeze	17.6
Vegetable oil	65
Automatic transmission fluid	78
Motor oil, SAE 30	270
Glycerin	757
Corn syrup	5180

Note: [a]These substances represent a range of viscosity useful for field measurements. Additional materials may be added for greater specificity.

speeds to estimate viscosity. It will be helpful to have several liquids of known viscosity available to use as a "viscosity library" (Table 1.18).

Confirming Viscosity

A laboratory method of measuring viscosity involves the use of glassware. The liquid is placed in the apparatus and is allowed to flow from an upper mark to a lower mark (Figure 1.41). The flow is timed, and the viscosity is extrapolated from a chart. The glass is fragile and difficult to protect in field use.

Water Reactivity

Water reactive compounds react spontaneously with water to release energy or produce a toxic substance.

The term *water reactive* has certain definitions based on transportation, manufacturing, or other concerns. Placing the unknown substance in contact with water can produce many reactions, but this section is concerned with the hazardous reactions formed from an unknown substance in contact with water.

Water reactive materials may cause formation of a gas or vapor that is flammable, corrosive, toxic, oxidizing, or asphyxiating, many of which may be verified using the appropriate test. Water reactive materials may also produce a fire or leave a corrosive by-product.

Certain materials classified as water reactive by DOT produce one of several toxic gases. Using the "Table of Initial Isolation and Protective Action Distances" from the DOT's *2004 Emergency Response Guidebook,* it is possible to group water reactive materials by the toxic gas produced when the material contacts water. The following information is from a previous work, "Emergency Characterization of Unknown Materials" (CRC Press, 2007), and may be helpful in selecting chemical tests from Chapter 2.

Confirming Water Reactivity

The simplest way to determine water reactivity is to add a small amount of the sample to water and observe. Be concerned with observing any chemical or physical reaction of the

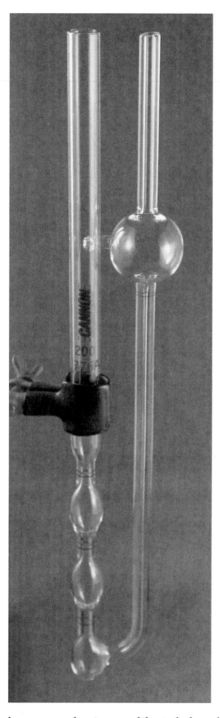

Figure 1.41 Viscosity may be measured using a calibrated glass device, such as the pictured viscometer, at a defined temperature. The flow of the liquid is timed as the leading edge passes set lines on the glass tube and viscosity is extrapolated from a chart provided with the apparatus. Other methods are available, such as timing a sphere as it falls through a cylinder of liquid.

unknown substance with water so its subsequent behavior may be predicted if a larger amount makes contact with water. For example, tetrachlorosilane will evolve a toxic gas and fit the definition of a Department of Transportation water reactive, toxic inhalation hazard. Concentrated hydrochloric acid will spontaneously emit white fumes; however, as a fuming acid, it does not fit the criteria as a Department of Transportation–Water Reactive Toxic Inhalation Hazard (DOT WRTIH).

Substances which are water reactive are dangerous, but the by-products of these reactions are often more hazardous than the original substance. Water reactive substances often turn the water solution strongly acidic or basic or may begin burning spontaneously. The most severe hazard involves those water reactive substances which produce a toxic gas or vapor.

The U.S. Department of Transportation has done extensive research involving toxic inhalation hazards (TIH) resulting from a transportation accident of any type of hazardous material. Within the TIH group are the water reactive, toxic inhalation hazard (WRTIH) substances. All of these water reactive chemicals produce a significantly toxic gas or vapor. Of the 70 water reactive materials researched for transportation, several are also TIH compounds that produce a secondary, sometimes more toxic TIH gas upon reaction with water.

DOT's 2004 Emergency Response Guidebook lists response guides for 70 water reactive materials, defined here as materials that emit a TIH gas upon contact with water. The TIH gases produced by these 70 water reactive materials are listed in Table 1.19. These 70 water reactive materials are allowed to be transported. Other water reactive material exists that is forbidden from transport; these materials are likely to be found engineered into fixed sites or possibly used in an illegal manner.

The researchers excluded from their definition water reactive substances that produce flammable gases that do not otherwise pose a toxic hazard. Examples would include hydrogen gas from wet calcium hydride and acetylene gas from wet calcium carbide.

Alkali metals react vigorously with water to release flammable hydrogen gas and form the corresponding hydroxide. For example, potassium metal in water would produce

Table 1.19 Toxic Gases Produced By Water Reactive Materials When Spilled In Water

Symbol	Chemical	Immediately Dangerous to Life or Health (IDLH), ppm	CAS Number
Br_2	Bromine	3	7726-95-6
Cl_2	Chlorine	10	7782-50-5
HBr	Hydrogen bromide	30	10035-10-6
HCl	Hydrogen chloride	50	7647-01-0
HCN	Hydrogen cyanide	50	74-90-8
HF	Hydrogen fluoride	30	7664-39-3
H_2S	Hydrogen sulfide	100	7783-06-4
NH_3	Ammonia	300	7664-41-7
PH_3	Phosphine	50	7803-51-2
SO_2	Sulfur dioxide	100	7446-09-5
SO_3	Sulfur trioxide	15 mg/m^3 as sulfuric acid	7446-11-9

hydrogen gas and potassium hydroxide. The resulting solution is strongly basic. The rate of reaction increases as atomic weight increases. Therefore, lithium metal reacts more slowly than sodium metal, which reacts more slowly than potassium metal. Considerable heat is generated quickly from these reactions and may be sufficient to ignite the hydrogen gas.

References

2004 Emergency Response Guidebook. United States Department of Transportation, Washington, DC, 2005.

29 CFR 1910.109. United States Department of Transportation, Code of Federal Regulations, Washington, DC.

49 CFR 173.124. United States Department of Transportation, Code of Federal Regulations, Washington, DC.

A Basic Guide to RTR Radioactive Materials, Revision 3, Sandia National Laboratories, Albuquerque, New Mexico, July 2004.

Aarino, P. *Expert systems in radiation source identification,* www.tkk.fi/Units/AES/projects/radphys/shaman.htm, Helsinki University of Technology, Finland, accessed December 7, 2006.

Bell, Harold M. *The Organic Compounds Database,* Virginia Polytechnic Institute and State University, www.colby.edu/chemistry/cmp/cmp.html, accessed September 6, 2007.

Beverley, K, Clint, J, and Fletcher, P. Evaporation rates of pure liquids measured using a gravimetric technique, *Phys. Chem. Chem. Phys.,* 1999, 1, 149È153, October 13, 1998.

Brown, DF, Freeman, WA, Carhart, RA, and Krumpolc, M. *Development of the Table of Initial Isolation and Protective Action Distances for the 2004 Emergency Response Guidebook,* ANL/DIS-05-2, Argonne National Laboratory, U.S. DOE, Argonne, Illinois, May 2005.

Chapman, S. *Police Dogs in North America,* Charles C Thomas, Springfield, Illinois, 1990.

Chemical Hazards Response Information System (CHRIS). United States Coast Guard, U.S. Department of Transportation, Washington, DC, March 1999.

CRC Handbook of Radiation Measurement and Protection, Sec. A, Vol. II: Biological and Mathematical Information. CRC Press, Boca Raton, Florida, 1982.

Dräger Safety AG & Co. *Flow Check Air Flow Indicator 6400761,* KGaA, Lübeck, Germany, October 2004.

Eisenbud, M, and Gesell, T. *Environmental Radioactivity from Natural, Industrial and Military Sources,* 4th ed., Academic Press, San Diego, 1997, 135.

Emergency Response to Terrorism—Weapons of Mass Destruction (WMD) Radiation/Nuclear Course for Hazardous Materials Technicians. Office of Justice Programs, Office for Domestic Preparedness, Washington, DC, June 2004.

Engelder, C, Dunkelberger, T, and Schiller, W. *Semi-Micro Qualitative Analysis*, 5th ed., John Wiley & Sons, New York, May 1946.

Feigl, F. *Spot Tests in Inorganic Analysis,* 5th ed., Elsevier Publishing Company, New York, 1958.

Goodman, JM, Kirby, PD, and Haustedt, LO. *Tetrahedron Letters,* 2000, 41, 9879–9882.

Houghton, R. *Emergency Characterization of Unknown Materials,* CRC Press, Boca Raton, Florida, October 2007.

How Can You Detect Radiation? Health Physics Society, http://hps.org/publicinformation/ate/faqs/radiationdetection.html, accessed August 29, 2005.

ICRP Publication 38: Radionuclide Transformations: Energy and Intensity of Emissions. Elsevier, New York, 1983.

Irritant Smoke (Stannic Chloride) Protocol, 29 CFR 1910.134. Occupational Safety and Health Administration Department of Labor, Washington, DC, July 1, 2004.

K-9 SOS website, http://www.k9sos.com/, K-9 SOS, Oak Ridge, Tennessee, 2007.

Kirby, P, *The Variation of Boiling Point with Pressure,* www.ch.cam.ac.uk/magnus/boil.html, Goodman Group, 2000.

Kristofeck, W. *A Study of Attitudes, Knowledge and Utilization of Canine Teams by the Louisville Division of Police,* University of Louisville, Louisville, Kentucky, 1991.

Kurtz, M, et al. Evaluation of Canines for Accelerant Detection at Fire Scenes, *Journal of Forensic Science,* 1994, 39 (6), 1528–1536.

Material Safety Data Sheet 6400812. Dräger Safety AG & Co., KGaA, Lübeck, Germany, May 30, 2000.

Material Safety Data Sheet 9030166. Dräger Safety AG & Co., KGaA, Lübeck, Germany, January 21, 2005.

Mesloh, C. *Scent as Forensic Evidence and its Relationship to the Law Enforcement Canine,* University of Central Florida, Orlando, Florida, November 2, 2000.

National Detector Dog Association. *Explosive Detection Certification,* NNDDA, Carthage, Texas, June 22, 2006.

National Detector Dog Association. *Narcotics Detection Standard,* NNDDA, Carthage, Texas, June 22, 2006.

National Fire Protection Association. *NFPA 30 Flammable and Combustible Liquids Code,* 2002 ed., NFPA, Quincy, 2003.

National Police Canine Association. www.npca.net/standards.html, NPCA, Waddell, Arizona, September 29, 2007.

NIOSH Pocket Guide to Chemical Hazards, US Department of Health and Human Services, National Institute of Occupational Safety and Health, Centers for Disease Control and Prevention, Atlanta, Georgia, September 2005.

North American Police Work Dog Association. *Bylaws and Certification Rules,* NAPWDA, Perry, Ohio, June 29, 2007.

Notation for States and Processes, Significance of the Word Standard in Chemical Thermodynamics, and Remarks on Commonly Tabulated Forms of Thermodynamic Functions. *Pure & Appl. Chem.,* 1982, 54 (6), 1239–1250, Pergamon Press, Great Britain.

Notes from Conversation. National Weather Service Weather Forecast Office, National Oceanic and Atmospheric Administration, Grand Rapids, Michigan, June 10, 2007.

O'Block, R, Doeren, S, and True, N. The Benefits of Canine Squads, *Journal of Police Science and Administration,* 1979, 7 (2), 155–160, Arlington, Virginia.

Occupational Safety and Health Guideline for Heptane. www.osha.gov/SLTC/healthguidelines/ heptane/recognition.html, Occupational Safety and Health Administration, United States Department of Labor, Washington, DC, October 9, 2007.

Occupational Safety and Health Guideline for Isopropyl Alcohol. www.osha.gov/SLTC/healthguide-lines/isopropylalcohol/recognition.html, Occupational Safety and Health Administration, United States Department of Labor, Washington, DC, October 9, 2007.

Primer on Spontaneous Heating and Pyrophoricity. United States Department of Energy, Washington, DC, December 1994.

Radionuclides. US Environmental Protection Agency, Washington, DC, December 14, 2005.

Ramsey, M. *Notes from Conversation and Demonstration,* Lansing Police Department, Lansing, Michigan, September 27, 2007.

Royal Oak Enterprises. *Material Safety Data Sheet 179267,* Roswell, Georgia, February 28, 2007.

Safe Handling of Alkali Metals and Their Reactive Compounds. Environmental Safety and Health Manual, Document 14.7, Lawrence Livermore National Laboratory, Livermore, California, revised October 13, 2005.

Scientific Working Group on Dog and Orthogonal Detector Guidelines (SWGDOG). *SWGDOG Guidelines,* Florida International University, Miami, Florida, August 1, 2007.

Seager, S, and Slabaugh, M. *Chemistry for Today: General, Organic, and Biochemistry,* 4th ed., Thomson Brooks/Cole, 1999.

Sensidyne. *Material Safety Data Sheet #5100,* Clearwater, Florida, February 2007.

Settle, R, et al. Human Scent Matching Using Specially Trained Dogs, *Animal Behavior,* 1994, 48, 1443–1448.

The Engineering ToolBox. *Dynamic, Absolute and Kinematic Viscosity,* www.engineeringtoolbox.com/dynamic-absolute-kinematic-viscosity-d_412.html, accessed July 16, 2008.

Trouton, F. *Nature* 1883, 27, 292.

United States Police Canine Association. *Certification Rules and Regulations,* USPCA, Springboro, Ohio, March 19, 2007.

Weast, RC, *Handbook of Chemistry and Physics,* 57th ed., CRC Press, Boca Raton, 1976.

Williams, M, et al. *Canine Detection Odor Signatures for Explosives,* Paper presented at the Society of Photographic Instrumentation Engineers Conference on Enforcement and Security Technologies Conference, Boston, Massachusetts, 1998.

Chemical Confirmation Tests

2

Introduction

This chapter contains over 400 chemical tests suitable for various uses in field testing of suspicious substances. The tests were selected from several sources from the late 1800s through the early 2000s. Chemical tests have been used and modified for specialized purposes as needed, and this book is no different.

The chemical tests were chosen for their usefulness under field testing conditions, specificity, sensitivity, and simplicity. As with other technologies, a single test result should not be considered definitive. Consider any result as suggestive and consider ways to increase confidence by physically manipulating the sample, using additional tests, or integrating other technologies. Finally, maintain a strong situational awareness and use clues related to a particular circumstance.

Each chemical test in this chapter is formatted uniformly. Table 2.1 describes the test presentation format. A 16-page insert following page 234 contains color images of certain test details.

<table>
<tr><td colspan="2">Table 2.1   Description of Test Format</td></tr>
<tr><td colspan="2">Test Name: Each test has a unique name based on the traditional reagent name or test purpose. Prefixes are deleted in the test name to simplify alphabetical order. For example, a test using a reagent called p-aminoacetophenone is named "aminoacetophenone solution."</td></tr>
<tr><td>Detects</td><td>Test Principle</td></tr>
<tr>
<td rowspan="5">
<ul>
<li>An alphabetical list of target substances that correspond to the Index of Target Substances.</li>
<li>The Index of Target Substances follows this section.</li>
</ul>
</td>
<td>A brief description of the chemistry of the test.</td>
</tr>
<tr><td>Test Phase</td></tr>
<tr><td>

	Solid	Test phase describes whether the sample should be solid, liquid, and/or gas. Some manipulation may be necessary, for example, solids that can be dissolved in a solvent may be tested as liquids.
✓	Liquid	
	Gas/Vapor	

</td></tr>
<tr><td>Sensitivity</td></tr>
<tr><td>Sensitivity describes the detection limit of the test under optimal conditions as described by the test directions.</td></tr>
<tr><td colspan="2">Directions and Comments</td></tr>
<tr><td colspan="2">Reagent chemical names, CAS numbers, mixing instructions, and test procedures are described in this section. Comments regarding similar tests, additional methods, and so on, are included.</td></tr>
<tr><td>References</td><td>Manufacturer</td></tr>
<tr><td>References are provided for all pertinent information used to formulate the test. Consult the reference directly if more detailed information is necessary for a particular use.</td><td>One or more manufacturers are listed if the test is provided commercially. Manufacturers of chemical reagents are not listed as nearly all reagents are commonly available through several commercial sources.</td></tr>
</table>

How to Select Tests

Tests may be selected by analyte (target substance) from the Index of Target Substances, and then the test may be located by the corresponding alphabetical listing in the Index of Tests. By using the Index of Tests a list of possible tests for the target substance immediately becomes available.

When searching for a target substance, consider synonyms or chemical "families" that might also apply. For example, a label states a substance contains acetaldehyde. A search for the target substance in the Index of Target Substances for "acetaldehyde" produces two tests:

- Sensidyne Tube #133A
- Sensidyne Tube #171SB

Next, find these two tests and determine if the test is appropriate for the situation. If more tests are desired, a search for the more general term "aldehyde" will yield several more test options:

- Brady's Reagent
- Drager Acetaldehyde 100/a
- Drager Acetone 100/b
- Drager Simultaneous Test Set III—Tube 1
- Drager Simultaneous Test Set III—Tube 3
- Fehling's Solution
- Merckoquant® Formaldehyde Test
- Nitrophenylhydrazine Solution
- Phenylenediamine Reagent
- Reducing Substance Test
- Schiff's Reagent
- Sensidyne Tube #102SC
- Sensidyne Tube #232SA

Tests may also be selected by reagent or by test name directly from the Index of Tests, which are listed alphabetically with the page number. The Index of Tests follows the Index of Target Substances.

The appendix contains detailed information on a drug screening test and explosive material tests.

Index of Target Substances

(continued)

Substance	Test
Alcohol	Drager Simultaneous Test Set III - Tube 3
Alcohol	Reducing Substance Test
Alcohol	Sensidyne Tube #102SD
Alcohol	Sensidyne Tube #111U
Alcohol	Sensidyne Tube #137U
Alcohol	Sensidyne Tube #186
Alcohol	Sensidyne Tube #190U
Alcohol	Sensidyne Tube #214S
Aldehyde	Brady's Reagent
Aldehyde	Drager Acetaldehyde 100/a
Aldehyde	Drager Acetone 100/b
Aldehyde	Drager Simultaneous Test Set III - Tube 1
Aldehyde	Drager Simultaneous Test Set III - Tube 3
Aldehyde	Fehling's Solution
Aldehyde	Merckoquant® Formaldehyde Test
Aldehyde	Nitrophenylhydrazine Solution
Aldehyde	Phenylenediamine Reagent
Aldehyde	Reducing Substance Test
Aldehyde	Schiff's Reagent
Aldehyde	Sensidyne Tube #102SC
Aldehyde	Sensidyne Tube #232SA
Aliphatic	Drager Simultaneous Test Set III - Tube 4
Aliphatic amine	Unsaturated Hydrocarbon Test
Alkali peroxide	Lead Sulfide Paper
Alkaline earth sulfate	Arsine From Elemental Arsenic Test
Alkaloid	Aloy's Reagent
Alkaloid	Cobalt Thiocyanate Reagent
Alkaloid	Dragendorff's Reagent (Alkaloids)
Alkaloid	Froede Reagent
Alkaloid	Mandelin Reagent
Alkane	Drager Simultaneous Test Set III - Tube 4
Alkane	Sensidyne Tube #102SD
Alkane	Sensidyne Tube #214S
Alkene	Bromine Water
Alkene	Drager Ethylene 0.1/a
Alkene	Drager Olefine 0.05%/a
Alkene	Drager Simultaneous Test Set III - Tube 4
Alkene	Potassium Permanganate Solution
Alkene	Unsaturated Hydrocarbon Test
Alkyl chloride	Clor-N-Oil 50

(continued)

Substance	Test
Alkyl halide	Beilstein Test
Alkyl nitrate	Ferrous Hydroxide Reagent
Alkyl nitrite	Ferrous Hydroxide Reagent
Alkyne	Bromine Water
Alkyne	Drager Simultaneous Test Set III - Tube 4
Alkyne	Potassium Permanganate Solution
Alkyne	Unsaturated Hydrocarbon Test
Alloys	Ferric Ferricyanide Test
Aluminum	Alizarin S Reagent
Aluminum	Aluminon Reagent
Aluminum	Aluminon Reagent on Paper
Aluminum	Ammonium Dithiocarbamate Solution
Aluminum	Eriochromocyanin R
Aluminum	Merckoquant® Aluminum Test
Aluminum	Morin Reagent
Aluminum	Quinalizarin Reagent (Aluminum)
Aluminum hydroxide	Alizarin Reagent (Metal)
Aluminum metal	Dinitrobenzene Reagent
Aluminum salt	Quinalizarin Reagent (Aluminum)
Amalgam	Cuprous Iodide Paper
Amalgam	Palladium Chloride Paper
Amide	Hydroxylamine Hydrochloride Reagent
Amine	Ammonium Test (Gutzeit Scheme)
Amine	Biuret Protein Test
Amine	Dragendorff's Reagent (Alkaloids)
Amine	Drager Clan Lab Simultaneous Test Set - Tube 3
Amine	Drager Organic Basic Nitrogen Compounds
Amine	Drager Simultaneous Test Set I - Tube 4
Amine	Gastec Tube #180
Amine	Sensidyne Deluxe Haz Mat III Tube 131
Amine	Sensidyne Tube #105SB
Amine	Sensidyne Tube 183U
Amino acid	Biuret Paper
Amino acid	Ninhydrin Solution
Ammonia	Drager Acetone 100/b
Ammonia	Drager Clan Lab Simultaneous Test Set - Tube 3
Ammonia	Drager Simultaneous Test Set I - Tube 4
Ammonia	Merckoquant® Ammonium Test
Ammonia	Sensidyne Deluxe Haz Mat III Tube 131
Ammonia	Sensidyne Tube #101S

(continued)

Substance	Test
Ammonia	Sensidyne Tube #105SB
Ammonia	Sensidyne Tube #105SH
Ammonia	Sensidyne Tube 183U
Ammonium	Ammonium Test
Ammonium	Ammonium Test (Gutzeit Scheme)
Ammonium	pH Test For Solids
Ammonium salt	Merckoquant® Ammonium Test
Amobarbital	Dille-Koppanyi Reagent, Modified
Amphetamine	Marquis Reagent
Amphetamine	Simon's Reagent
Aniline	Drager Aniline 0.5/a
Aniline	Sensidyne Tube #181S
Antimony	Firearm Discharge Residue Test
Antimony	Jaffe's Reagent
Antimony	Rhodamine B Reagent (Antimony)
Antimony	Rhodamine B Reagent (Gallium)
Antimony	Triphenylmethylarsonium Reagent
Antimony hydride	Drager Arsine 0.05/a
Antimony hydride	Drager Phosphine 0.1/a
Antimony hydride	Drager Phosphine 0.1/a
Aromatic	Bromine Water
Aromatic	Drager Aniline 0.5/a
Aromatic	Drager Benzene 0.5/a
Aromatic	Drager Simultaneous Test Set III - Tube 2
Aromatic	Sensidyne Tube #102SD
Aromatic	Sensidyne Tube #186
Aromatic	Sensidyne Tube #214S
Aromatic alcohol	Drager Phenol 1/b
Arsenate	Ammonium Molybdate Solution (Deniges)
Arsenate	Bettendorf's Test
Arsenic	Ammonium Molybdate Reagent
Arsenic	Arsine From Elemental Arsenic Test
Arsenic	Bettendorf's Test
Arsenic	ITS SenSafe™ Water Metal Test
Arsenic	Merckoquant® Arsenic Test
Arsenic	Silver Nitrate Paper
Arsenic(III)	Bettendorf's Test
Arsenic(III)	Ethyl-8-Hydroxytetrahydroquinoline HCl Reagent
Arsenic(III)	Merckoquant® Arsenic Test

(continued)

Substance	Test
Arsenic(IV)	Bettendorf's Test
Arsenic(V)	Merckoquant® Arsenic Test
Arsenic, organic	Drager Organic Arsenic Compounds and Arsine
Arsenite	Bettendorf's Test
Arsenite	Ethyl-8-Hydroxytetrahydroquinoline HCl Reagent
Arsenite	Silver Nitrate Paper
Arsine	Drager Arsine 0.05/a
Arsine	Drager CDS Simultaneous Test Set I - Tube 4
Arsine	Drager Clan Lab Simultaneous Test Set - Tube 1
Arsine	Drager Hydrogen Sulfide 0.2/b
Arsine	Drager Natural Gas Odorization Tertiary Butylmercaptan
Arsine	Drager Organic Arsenic Compounds and Arsine
Arsine	Drager Phosphine 0.1/a
Arsine	Drager Phosphine 0.1/a
Arsine	Sensidyne Deluxe Haz Mat III Tube 131
Arsine	Sensidyne Tube #121U
Arsine	Sensidyne Tube #165SB
Arsine	Sensidyne Tube #167S
Ascorbic acid	Merckoquant® Ascorbic Acid Test
Aspirin	Froede Reagent
Aspirin	Mandelin Reagent
Aspirin	Marquis Reagent
Auric	Pyridine Reagent
Azide	Gas Production With Sulfuric Acid
Azide	Iodine-Sulfide Reagent
Azide	Nitric Acid - Silver Nitrate Solution
Barbiturate	Dille-Koppanyi Reagent, Modified
Barbiturate	Zwikker Reagent
Barium	Firearm Discharge Residue Test
Barium	Lead(II) Test
Barium	Sodium Rhodizonate Solution (Barium)
Barium	Triphenylmethylarsonium Reagent
Base	Drager Amine Test
Base	Drager Ammonia 2/a
Base	Drager Clan Lab Simultaneous Test Set - Tube 3
Base	Drager Simultaneous Test Set I - Tube 4
Base	Universal pH Test Strip
Bean	Melibiose Test

(continued)

Substance	Test
Benzene	Drager Benzene 0.5/a
Benzene	Sensidyne Tube #118SB
Benzoyl peroxide	Aminodiphenylamine Sulfate Reagent
Beryllium	Aluminon Reagent
Beryllium	Curcumin Solution
Beryllium	Morin Reagent
Beryllium	Quinalizarin Reagent (Aluminum)
Beryllium metal	Dinitrobenzene Reagent
Beryllium salt	Quinalizarin Reagent (Magnesium)
Bicarbonate	Fizz Test
Bismuth	Cinchonine - Potassium Iodide Solution
Bismuth	Dimethylglyoxime Reagent
Bismuth	Jaffe's Reagent
Bismuth chloride	Rhodamine B Reagent (Antimony)
Bisulfite	Bisulfite Test
Blister agent	M256 Series Chemical Agent Detector Kit
Blister agent	M256A1 Blister Agent Test
Blood agent	M256 Series Chemical Agent Detector Kit
Blood agent	M256A1 Blood Agent Test
Borate	Borate Test (Gutzeit Scheme)
Boric acid	Benzoin Reagent
Boric acid	Borate Test (Gutzeit Scheme)
Boric acid	Carmine Red Reagent
Boric salt	Benzoin Reagent
Bromate	Bromic Acid Test
Bromate	Drop-Ex Plus™
Bromate	Sulfanilic Acid Reagent
Bromic acid	Bromic Acid Test
Bromic acid	Sulfanilic Acid Reagent
Bromide	Fluorescein Reagent (Bromides)
Bromide	Gold Chloride Reagent
Bromide	Nitric Acid - Silver Nitrate Solution
Bromide	Silver Ferrocyanide Paper
Bromine	Aminodimethylaniline Reagent
Bromine	Drager Chlorine 0.2/a
Bromine	Drager Perchloroethylene 0.1/a
Bromine	Drager Simultaneous Test Set II - Tube 2
Bromine	Fluorescein Reagent (Halogens)
Bromine	Michler's Thioketone
Bromine	Sensidyne Deluxe Haz Mat III Tube 131

(continued)

Substance	Test
Bromine	Sensidyne Tube #109SB
Bromine	Sensidyne Tube #114
Bromite	Nickel Hydroxide Test
Butadiene	Sensidyne Tube #101S
Butanol	Sensidyne Tube #190U
Butylmercaptan	Drager Natural Gas Odorization Tertiary Butylmercaptan
C-4	RDX Test (J-Acid)
C-4	RDX Test (Pyrolytic Oxidation)
C-4	RDX Test (Thymol)
Cadmium	Allylthiourea Reagent
Cadmium	ITS SenSafe™ Water Metal Test
Cadmium	Lead(II) Test
Cadmium	Merckoquant® Calcium Test
Cadmium	Merckoquant® Zinc Test
Calcium	Merckoquant® Calcium Test
Calcium salt	Sodium Rhodizonate Solution (Calcium)
Cannabis	Duquenois-Levine Reagent, Modified
Carbamate pesticide	Agri-Screen Ticket
Carbamate pesticide	M256A1 Nerve Agent Test
Carbohydrate	Reducing Substance Test
Carbon dioxide	Drager Carbon Dioxide 5%/A
Carbon dioxide	Drager Simultaneous Test Set II - Tube 4
Carbon dioxide	Sensidyne Tube #126SA
Carbon dioxide	Sensidyne Tube #126UH
Carbon disulfide	Sensidyne Tube #101S
Carbon disulfide	Sensidyne Tube #112SA
Carbon disulfide	Sensidyne Tube #141SB
Carbon disulfide	Sensidyne Tube #239S
Carbon monoxide	Drager Carbon Monoxide 2/a
Carbon monoxide	Drager Ethylene 0.1/a
Carbon monoxide	Drager Simultaneous Test Set I - Tube 3
Carbon tetrachloride	Drager Carbon Tetrachloride 5/c
Carbon tetrachloride	Sensidyne Tube #147S
Carbonate	Fizz Test
Carbonate	pH Test For Solids
Carbonates	Nickel Dimethylglyoxime Equilibrium Solution
Carbonyl	Nitrophenylhydrazine Solution
Carbonyl bromide	Drager CDS Simultaneous Test Set I - Tube 2
Carbonyl bromide	Drager CDS Simultaneous Test Set V - Tube 3

(continued)

Substance	Test
Carbonyl bromide	Drager Phosgene 0.05/a
Carbonyl group	Benedict's Solution (Carbonyl)
Carbonyl group	Brady's Reagent
Carbonyl group	Phenylenediamine Reagent
Carbonyl sulfide	Sensidyne Tube #239S
Caustic	Universal pH Test Strip
Cerium	Aluminon Reagent
Cerium salt	Quinalizarin Reagent (Magnesium)
Cerium(III)	Ammoniacal Silver Oxide Solution
Cerium(IV)	Aminoacetophenone Reagent
Cesium	Ball's Reagent
Cesium	Potassium Bismuth Iodide Reagent
Cheese	Catalase Test
Chemical warfare agent	M256 Series Chemical Agent Detector Kit
Chlorate	Aniline Sulfate Reagent
Chlorate	Chloric Acid Test
Chlorate	Chlorine Dioxide Test
Chlorate	Drop-Ex Plus™
Chlorate	Indigo Carmine Reagent (Chlorate)
Chlorate	Offord's Reagent
Chlorate	Phenylanthranilic Acid Test
Chloric acid	Chloric Acid Test
Chloride	Merckoquant® Chloride Test
Chloride	Nitric Acid - Silver Nitrate Solution
Chloride	Phenylanthranilic Acid Test
Chloride	Silver Ferrocyanide Paper
Chloride	Silver Nitrate Reagent
Chlorinated hydrocarbon	Beilstein Test
Chlorinated hydrocarbon	Clor-N-Oil 50
Chlorine	Aminodimethylaniline Reagent
Chlorine	Drager CDS Simultaneous Test Set V - Tube 4
Chlorine	Drager Chlorine 0.2/a
Chlorine	Drager Nitrogen Dioxide 0.5/c
Chlorine	Drager Nitrous Fumes 0.5/a
Chlorine	Drager Nitrous Fumes 20/a
Chlorine	Drager Ozone 0.05/b
Chlorine	Drager Perchloroethylene 0.1/a
Chlorine	Drager Phosgene 0.02/a
Chlorine	Drager Simultaneous Test Set I - Tube 5
Chlorine	Drager Simultaneous Test Set II - Tube 2

(*continued*)

Substance	Test
Chlorine	Fluorescein Reagent (Halogens)
Chlorine	Merckoquant® Chlorine Tests 1.17924.0001, 1.17925.0001
Chlorine	Michler's Thioketone
Chlorine	Potassium Iodide Starch Test Strip (JD260)
Chlorine	Sensidyne Deluxe Haz Mat III Tube 131
Chlorine	Sensidyne Tube #101S
Chlorine	Sensidyne Tube #109SB
Chlorine	Sensidyne Tube #114
Chlorine	Sensidyne Tube #156S
Chlorine	Sensidyne Tube #174A
Chlorine	Sensidyne Tube #216S
Chlorine	Villiers-Fayolle's Reagent
Chlorine dioxide	Chlorine Dioxide Test
Chlorine dioxide	Drager Chlorine 0.2/a
Chlorine dioxide	Drager Simultaneous Test Set II - Tube 2
Chlorine dioxide	Sensidyne Tube #109SB
Chlorine dioxide	Sensidyne Tube #114
Chlorite	Nickel Hydroxide Test
Chlorohydrocarbon	Drager Carbon Tetrachloride 5/c
Chlorohydrocarbon	Drager Simultaneous Test Set III - Tube 5
Chlorohydrocarbons	Sensidyne Tube #147S
Chloropicrin	Sensidyne Tube #147S
Chlorpromazine HCl	Ferric Chloride Reagent
Chromate	Diphenylcarbazide Reagent
Chromate	Diphenylcarbazide Test
Chromate	Hydrogen Peroxide Reagent
Chromate	Indigo Carmine Reagent (Chromate)
Chromate	Merckoquant® Chromate Test
Chromate	Methylene Blue Reagent
Chromate	Silver Nitrate Paper
Chromate	Zwikker's Reagent
Chromic acid	Diphenylcarbazide Test
Chromic acid	Van Eck's Reagent
Chromic hydroxide	Alizarin Reagent (Metal)
Chromium	ITS SenSafe™ Water Metal Test
Chromium(III)	Ammoniacal Silver Oxide Solution
Chromium(III)	Diphenylcarbazide Test
Chromium(III)	Merckoquant® Chromate Test
Chromium(VI)	Diphenylcarbazide Test

(continued)

Substance	Test
Chromium(VI)	Merckoquant® Chromate Test
Chromium(VI)	Van Eck's Reagent
Chromium, bound	Merckoquant® Chromate Test
CK	M256A1 Blood Agent Test
Cobalt	Ammonium Dithiocarbamate Solution
Cobalt	Copper Metal Test
Cobalt	Dimethylglyoxime Reagent
Cobalt	Formaldoxime Reagent
Cobalt	ITS SenSafe™ Water Metal Test
Cobalt	Merckoquant® Cobalt Test
Cobalt	Merckoquant® Nickel Test
Cobalt	Nitroso-R-Salt Solution
Cobalt	Offord's Reagent
Cobalt	Rubeanic Acid (Cobalt)
Cobalt	Sodium Thiosulfate Reagent
Cobalt	Thiocyanate Solution Test (Cobalt)
Cobalt(II)	Ammoniacal Silver Oxide Solution
Cocaine	Cobalt Thiocyanate Reagent
Codeine	Froede Reagent
Codeine	Mandelin Reagent
Codeine	Marquis Reagent
Codeine	Mecke Reagent
Copper	Aminophenol Hydrochloride Solution
Copper	Cinchonine - Potassium Iodide Solution
Copper	Cuprion Reagent
Copper	Dimethylglyoxime Reagent
Copper	ITS SenSafe™ Water Metal Test
Copper	Merckoquant® Calcium Test
Copper	Merckoquant® Copper Test
Copper	Merckoquant® Nickel Test
Copper	Merckoquant® Zinc Test
Copper	Rubeanic Acid (Cobalt)
Copper	Rubeanic Acid Reagent
Copper metal	Copper Metal Test
Copper(I)	Cuprion Reagent
Copper(I)	Merckoquant® Copper Test
Copper(II)	Cuprion Reagent
Copper(II)	Merckoquant® Cobalt Test
Copper(II)	Merckoquant® Copper Test
Copper(II)	Thiocyanate Solution Test (Cobalt)

(continued)

Substance	Test
Corrosive	Universal pH Test Strip
Cresol	Drager Phenol 1/b
Cresol	Sensidyne Tube 183U
Cupric	Formaldoxime Reagent
Cupric	Offord's Reagent
CX	M256 Series Chemical Agent Detector Kit
CX	M256A1 Blister Agent Test
Cyanate	Hydroxylamine Hydrochloride Reagent
Cyanic acid	Hydroxylamine Hydrochloride Reagent
Cyanide	Drager Cyanide 2/a
Cyanide	Gas Production With Sulfuric Acid
Cyanide	ITS Cyanide ReagentStrip™
Cyanide	Merckoquant® Cyanide Test
Cyanide	Nitric Acid - Silver Nitrate Solution
Cyanide	Prussian Blue Test
Cyanogen bromide	Drager CDS Simultaneous Test Set V - Tube 1
Cyanogen bromide	Drager Cyanogen Chloride 0.25/a
Cyanogen chloride	Drager CDS Simultaneous Test Set V - Tube 1
Cyanogen chloride	Drager Cyanogen Chloride 0.25/a
Cyanogen chloride	ITS Cyanide ReagentStrip™
Cyanogen chloride	M256A1 Blood Agent Test
Cyclonite	RDX Test (J-Acid)
Cyclonite	RDX Test (Pyrolytic Oxidation)
Cyclonite	RDX Test (Thymol)
Cyclonite	Trinitrophenylmethylnitramine Test
Cyclotrimethylenetrinitramine	RDX Test (J-Acid)
Cyclotrimethylenetrinitramine	RDX Test (J-Acid)
Cyclotrimethylenetrinitramine	RDX Test (Pyrolytic Oxidation)
Cyclotrimethylenetrinitramine	RDX Test (Thymol)
Dairy	Catalase Test
Dairy	Lactase Test
Decaborane	Diethylnicotinamide Reagent
Dialkyl-lead salt	Dithizone Reagent (Lead)
Diborane	Sensidyne Tube #121U
Dichlorobenzene	Sensidyne Tube #214S
Dichlorobenzene	Sensidyne Tube #215S
Dichloroethane	Sensidyne Tube #235S
Dichloroethylene	Sensidyne Tube #134SB
Dichromate	Diphenylamine Indicator Solution
Dichromate	Diphenylcarbazide Test

(continued)

Substance	Test
Dichromate	Merckoquant® Chromate Test
Dicyanide	Sensidyne Tube #112SA
Diethyl ether	Sensidyne Tube 107SA
Dimethylsulfoxide	Reducing Substance Test
Dinitrotoluene	Dinitrotoluene Test
Dinitrotoluene	Janovsky Test
Dinitrotoluene	Nitrotoluene Test
Dinitrotoluene	Trinitrotoluene Test
Dioxane	Ceric Nitrate Reagent
DMSO	Reducing Substance Test
DNT	Dinitrotoluene Test
DNT	Janovsky Test
DNT	Nitrotoluene Test
DNT	Trinitrotoluene Test
Dysprosium	Arsenazo I Reagent
E85	Ceric Nitrate Reagent
Ecstasy	Simon's Reagent
Elemental metal	Ferric Ferricyanide Test
Ephedrine	Cobalt Thiocyanate Reagent
Ester	Drager Acetaldehyde 100/a
Ester	Drager Aniline 0.5/a
Ester	Drager Simultaneous Test Set III - Tube 3
Ester	Sensidyne Tube #102SD
Ester	Sensidyne Tube #111U
Ester	Sensidyne Tube #214S
Ether	Drager Acetaldehyde 100/a
Ether	Drager Aniline 0.5/a
Ether	Drager Simultaneous Test Set III - Tube 3
Ethyl acetate	Sensidyne Tube #111U
Ethyl benzene	Drager Benzene 0.5/a
Ethyl mercaptan	Sensidyne Test #165SA
Ethyl mercaptan	Sensidyne Tube #165SB
Ethylene	Drager Ethylene 0.1/a
Ethylene glycol	Sensidyne Tube #232SA
Ethylene oxide	Sensidyne Tube #232SA
Europium	Arsenazo I Reagent
Explosive	Dichlorofluorescein Reagent
Explosive	Dinitrotoluene Test
Explosive	Ferrous Hydroxide Reagent
Explosive	Nitrotoluene Test

(continued)

Substance	Test
Explosive	Polynitroaromatic Screening Test
Explosive	TATB Test
Explosive	Trinitrotoluene Test
Ferric	Chromotropic Acid Reagent
Ferric	Dihydroxyacetophenone Solution
Ferric	Dipyridyl Reagent
Ferric	Formaldoxime Reagent
Ferric	Offord's Reagent
Ferric	Rubeanic Acid Reagent
Ferric hydroxide	Alizarin Reagent (Metal)
Ferricyanate	Methylene Blue Reagent
Ferricyanide	Ferricyanide Test
Ferrocyanate	Methylene Blue Reagent
Ferrocyanide	Ferric Chloride Reagent
Ferrocyanide	Ferrocyanide Test (Ferric Chloride)
Ferrocyanide	Ferrocyanide Test (Gutzeit Scheme)
Ferrocyanide	Prussian Blue Test
Ferrous	Dimethylglyoxime Reagent
Ferrous	Dipyridyl Reagent
Fluoride	Fluoride Test Strip
Fluoride	Zirconium-Alizarin Reagent
Fluorine	Drager Perchloroethylene 0.1/a
Formaldehyde	Diphenylamine Reagent
Formaldehyde	Merckoquant® Formaldehyde Test
Formaldehyde	RDX Test (J-Acid)
Formaldehyde	Sensidyne Tube #171SB
Fructose	Fehling's Solution
Fructose	Luff's Reagent
Gadolinium	Arsenazo I Reagent
Gallium	Morin Reagent
Gallium	Rhodamine B Reagent (Gallium)
Gallium(III)	Quinizarin Reagent
Gasoline (10% ethanol)	Ceric Nitrate Reagent
Germanium	Diphenylcarbazone Reagent (Germanium)
Glucose	Fehling's Solution
Glucose	Luff's Reagent
Glucose	Merckoquant® Glucose Test
G-nerve agent	Drager CDS Simultaneous Test Set V - Tube 5
G-nerve agent	M256A1 Nerve Agent Test
Gold	Aqua Regia Reagent

(continued)

Substance	Test
Gold	Arnold's Base
Gold	Pyridine Reagent
Gold	Rhodamine B Reagent (Antimony)
Gold	Rhodamine B Reagent (Gallium)
Gold(III)	Potassium Iodide Reagent (Gold)
G-verve agent	Drager Phosphoric Acid Esters 0.05/a
Hafnium	Neothorin Reagent
Halide	Gas Production With Sulfuric Acid
Halogen	Aminodimethylaniline Reagent
Halogen	Drager Halogenated Hydrocarbon 100/a
Halogen	Drager Perchloroethylene 0.1/a
Halogen	Sensidyne Tube #157SB
Halogen	Sensidyne Tube #235S
Halogenated hydrocarbon	Beilstein Test
Halohydrocarbon	Drager Halogenated Hydrocarbon 100/a
Halohydrocarbon	Drager Methyl Bromide 0.5/a
Halohydrocarbon	Drager Perchloroethylene 0.1/a
Halohydrocarbon	Sensidyne Tube #134SB
Halohydrocarbon	Sensidyne Tube #157SB
Halohydrocarbon	Sensidyne Tube #214S
Halohydrocarbon	Sensidyne Tube #235S
Hashish	Duquenois-Levine Reagent, Modified
Heavy metal	ITS SenSafe™ Water Metal Test
Heroin	Nitric Acid Reagent
Hexogen	RDX Test (J-Acid)
Hexogen	RDX Test (Pyrolytic Oxidation)
Hexogen	RDX Test (Thymol)
HMX	Dichlorofluorescein Reagent
HMX	RDX Test (Thymol)
H-Series blister agent	M256A1 Blister Agent Test
Hydrazine	Drager Hydrazine 0.2/a
Hydrazine	Sensidyne Deluxe Haz Mat III Tube 131
Hydrazine	Sensidyne Tube #105SB
Hydrazine, organic	Drager Hydrazine 0.2/a
Hydrazoic acid	Iodine-Sulfide Reagent
Hydrobromic acid	Fluorescein Reagent (Bromides)
Hydrocarbon	Ferric Thiocyanate Reagent
Hydrocarbon	Sensidyne Tube #186
Hydrocarbon	Sulfur Reagent
Hydrocarbon (larger aliphatic)	Sensidyne Tube #111U

(continued)

Substance	Test
Hydrocarbon (petroleum)	Drager Acetaldehyde 100/a
Hydrocarbon (petroleum)	Drager Aniline 0.5/a
Hydrochloric acid	Drager Clan Lab Simultaneous Test Set - Tube 4
Hydrochloric acid	Drager Phosgene 0.02/a
Hydrocodone tartrate	Mecke Reagent
Hydrocyanic acid	Drager CDS Simultaneous Test Set I - Tube 3
Hydrocyanic acid	Drager Hydrocyanic Acid 2/a
Hydrocyanic acid	Drager Simultaneous Test Set I - Tube 2
Hydrofluoric acid	Zirconium-Alizarin Reagent
Hydrogen	Sensidyne Tube #101S
Hydrogen	Sensidyne Tube #137U
Hydrogen (organic)	Sulfur Reagent
Hydrogen bromide	Sensidyne Deluxe Haz Mat III Tube 131
Hydrogen chloride	Sensidyne Deluxe Haz Mat III Tube 131
Hydrogen chloride	Sensidyne Tube #156S
Hydrogen chloride	Sensidyne Tube #216S
Hydrogen cyanide	Drager Hydrogen Sulfide 0.2/a
Hydrogen cyanide	Drager Hydrogen Sulfide 0.2/b
Hydrogen cyanide	M256A1 Blood Agent Test
Hydrogen cyanide	Sensidyne Deluxe Haz Mat III Tube 131
Hydrogen cyanide	Sensidyne Tube #112SA
Hydrogen cyanide	Sensidyne Tube #112SB
Hydrogen cyanide	Sensidyne Tube #121U
Hydrogen cyanide	Sensidyne Tube #126SA
Hydrogen cyanide	Sensidyne Tube #128SC
Hydrogen cyanide	Sensidyne Tube #165SB
Hydrogen fluoride	Drager Hydrogen Fluoride 1.5/b
Hydrogen fluoride	Drager Hydrogen Fluoride 1.5/b
Hydrogen fluoride	Sensidyne Tube #156S
Hydrogen halide	Drager Halogenated Hydrocarbon 100/a
Hydrogen peroxide	Arnold-Mentzel's Solution
Hydrogen peroxide	Drager Hydrogen Peroxide 0.1/a
Hydrogen peroxide	Lead Sulfide Paper
Hydrogen peroxide	Merckoquant® Peroxide Test 1.10337.0001
Hydrogen peroxide	Merckoquant® Peroxide Tests 1.10011.0001, 1.10011.0002, 1.10081.0001
Hydrogen peroxide	Peroxide (10%) Test Strip
Hydrogen peroxide	Peroxide Test Strip (PER 400)
Hydrogen peroxide	Potassium Iodide Starch Test Strip (JD260)
Hydrogen peroxide	Sensidyne Tube #247S

(continued)

Substance	Test
Hydrogen selenide	Sensidyne Tube #121U
Hydrogen selenide	Sensidyne Tube #165SB
Hydrogen selenide	Sensidyne Tube #167S
Hydrogen sulfide	Drager Hydrazine 0.2/a
Hydrogen sulfide	Drager Hydrogen Sulfide 0.2/a
Hydrogen sulfide	Drager Hydrogen Sulfide 0.2/b
Hydrogen sulfide	Drager Natural Gas Odorization Tertiary Butylmercaptan
Hydrogen sulfide	Drager Simultaneous Test Set II - Tube 1
Hydrogen sulfide	Drager Simultaneous Test Set II - Tube 3
Hydrogen sulfide	Lead Acetate Test Strip (JD175)
Hydrogen sulfide	Sensidyne Deluxe Haz Mat III Tube 131
Hydrogen sulfide	Sensidyne Test #165SA
Hydrogen sulfide	Sensidyne Tube #112SA
Hydrogen sulfide	Sensidyne Tube #112SB
Hydrogen sulfide	Sensidyne Tube #120SB
Hydrogen sulfide	Sensidyne Tube #126SA
Hydrogen sulfide	Sensidyne Tube #141SB
Hydrogen sulfide	Sensidyne Tube #142S
Hydrogen sulfide	Sensidyne Tube #167S
Hydrogen sulfide	Sensidyne Tube #186
Hydrogen sulfide	Sensidyne Tube #232SA
Hydrogen sulfide	Sensidyne Tube #239S
Hydrogen sulfide	Sensidyne Tube 282S
Hydrogen sulfide	Sensidyne Tube 282S
Hydrogen sulfide	Sodium Nitroprusside Paper
Hydroiodic acid	Fluorescein Reagent (Bromides)
Hydroiodic acid	Palladous Chloride Reagent
Hydroperoxide	Merckoquant® Peroxide Test 1.10337.0001
Hydroperoxide	Merckoquant® Peroxide Tests 1.10011.0001, 1.10011.0002, 1.10081.0001
Hydroperoxide	Peroxide Test Strip (PER 400)
Hydroxide	pH Test For Solids
Hydroxy aldehyde	Benedict's Solution
Hydroxy ketone	Benedict's Solution
Hydroxylamine	Ferrous Hydroxide Reagent
Hypobromite	Hypohalogenite Differentiation Test
Hypobromite	Hypohalogenite Test
Hypobromite	Safranin O Solution
Hypobromous acid	Safranin O Solution

(continued)

Substance	Test
Hypochloric acid	Safranin O Solution
Hypochlorite	Diphenylamine Indicator Solution
Hypochlorite	Hypohalogenite Differentiation Test
Hypochlorite	Hypohalogenite Test
Hypochlorite	Safranin O Solution
Hypohalogenite	Hypohalogenite Differentiation Test
Hypohalogenite	Hypohalogenite Test
Hypohalogenous acid	Safranin O Solution
Hypoiodite	Hypohalogenite Differentiation Test
Hypoiodite	Hypohalogenite Test
Hypoiodite	Safranin O Solution
Hypoiodous acid	Safranin O Solution
Illegal drug	Ferric Chloride Reagent
Illegal drug	Mandelin Reagent
Illegal drug	Marquis Reagent
Illegal drug	Mecke Reagent
Illegal drug	Nitric Acid Reagent
Indium	Morin Reagent
Indium(III)	Quinizarin Reagent
Iodate	Iodic Acid Test 1
Iodate	Iodic Acid Test 2
Iodic acid	Iodic Acid Test 1
Iodic acid	Iodic Acid Test 2
Iodide	Fluorescein Reagent (Bromides)
Iodide	Gold Chloride Reagent
Iodide	Nitric Acid - Silver Nitrate Solution
Iodide	Palladous Chloride Reagent
Iodide	Silver Ferrocyanide Paper
Iodine	Aminodimethylaniline Reagent
Iodine	Drager Clan Lab Simultaneous Test Set - Tube 5
Iodine	Fluorescein Reagent (Halogens)
Iodine	Michler's Thioketone
Iodine	Potassium Iodide Starch Test Strip (JD260)
Iodite	Nickel Hydroxide Test
Iron	Aluminon Reagent
Iron	Aminophenol Hydrochloride Solution
Iron	Chromotropic Acid Reagent
Iron	Dihydroxyacetophenone Solution
Iron	Merckoquant® Iron Test
Iron salt	Hydrogen Peroxide Reagent

(continued)

Substance	Test
Iron tetracarbonyl	Drager Nickel Tetracarbonyl 0.1/a
Iron(II)	Dipyridyl Reagent
Iron(II)	Merckoquant® Iron Test
Iron(III)	Ammoniacal Silver Oxide Solution
Iron(III)	Dipyridyl Reagent
Iron(III)	Merckoquant® Cobalt Test
Iron(III)	Merckoquant® Iron Test
Iron(III)	Thiocyanate Solution Test (Cobalt)
Keto aldehyde	Benedict's Solution
Ketone	Brady's Reagent
Ketone	Drager Acetaldehyde 100/a
Ketone	Drager Acetone 100/b
Ketone	Drager Aniline 0.5/a
Ketone	Drager Simultaneous Test Set III - Tube 1
Ketone	Drager Simultaneous Test Set III - Tube 3
Ketone	Nitrophenylhydrazine Solution
Ketone	Phenylenediamine Reagent
Ketone	Schiff's Reagent
Ketone	Sensidyne Tube #102SC
Ketone	Sensidyne Tube #102SD
Ketone	Sensidyne Tube #111U
Ketone	Sensidyne Tube #186
Lactose	Lactase Test
Lanthanum salt	Quinalizarin Reagent (Magnesium)
Lead	Ammonium Dithiocarbamate Solution
Lead	Cinchonine - Potassium Iodide Solution
Lead	Dithizone Reagent (Lead)
Lead	Firearm Discharge Residue Test
Lead	ITS SenSafe™ Water Metal Test
Lead	Lead(II) Test
Lead	Merckoquant® Calcium Test
Lead	Merckoquant® Lead Test
Lead	Merckoquant® Zinc Test
Lead	Potassium Iodide Reagent (Thallium)
Lead	Triphenylmethylarsonium Reagent
Lead	Watersafe® Lead Test
Lead metal	Dinitrobenzene Reagent
Lewisite	Drager CDS Simultaneous Test Set I - Tube 4
Lewisite	Drager Simultaneous Test Set I - Tube 3
Lewisite	M256 Series Chemical Agent Detector Kit

(continued)

Substance	Test
Lewisite	M256A1 Lewisite Test
Lewisite	Michler's Thioketone
Lithium	Procke-Uzel's Reagent
Lithium	Sodium Arsenite Reagent
LSD	Dimethylaminobenzaldehyde (p-DMAB)
LSD	Froede Reagent
LSD	Mecke Reagent
Lutetium	Arsenazo I Reagent
Lysergic acid diethylamide	Dimethylaminobenzaldehyde (p-DMAB)
Mace	Froede Reagent
Mace	Mandelin Reagent
Mace	Marquis Reagent
Magnesium	Diaminobenzene Solution
Magnesium	Quinalizarin Reagent (Magnesium)
Magnesium metal	Dinitrobenzene Reagent
Magnesium salt	Quinalizarin Reagent (Magnesium)
Manganese	Merckoquant® Manganese Test
Manganese hydroxide	Alizarin Reagent (Metal)
Manganese(II)	Ammoniacal Silver Oxide Solution
Manganous	Formaldoxime Reagent
Marijuana	Duquenois-Levine Reagent, Modified
MDMA HCl	Simon's Reagent
Melibiose	Melibiose Test
Mercaptan	Benedict's Solution (Sulfur)
Mercaptan	Bromine Water
Mercaptan	Drager Hydrogen Sulfide 0.2/b
Mercaptan	Drager Mercaptan 0.5/a
Mercaptan	Drager Natural Gas Odorization Tertiary Butylmercaptan
Mercaptan	Gold Chloride Reagent (Thiol)
Mercaptan	Isatin Reagent (Mercaptans)
Mercaptan	Raikow's Reagent
Mercaptan	Sensidyne Tube #121U
Mercaptan	Sensidyne Tube #165SB
Mercaptan	Sensidyne Tube 282S
Mercaptan	Sensidyne Tube 282S
Mercaptan	Sodium Nitroprusside Paper
Mercuric	Benzopurpurin 4 B Reagent
Mercurous	Benzopurpurin 4 B Reagent
Mercurous	Rubeanic Acid Reagent

(continued)

Substance	Test
Mercurous chloride	Methylene Blue Reagent
Mercury	Arnold's Base
Mercury	Benzopurpurin 4 B Reagent
Mercury	Cinchonine - Potassium Iodide Solution
Mercury	Cuprous Iodide Paper
Mercury	Drager Mercury Vapour 0.1/b
Mercury	ITS SenSafe™ Water Metal Test
Mercury	Merckoquant® Nickel Test
Mercury	Merckoquant® Zinc Test
Mercury	Potassium Iodide Reagent (Thallium)
Mercury	Sensidyne Tube #142S
Mercury	Sensidyne Tube #167S
Mercury chlorides	Rhodamine B Reagent (Antimony)
Mercury salt	Cuprous Iodide Paper
Mercury salt	Diphenylcarbazone Reagent (Mercury)
Mercury salt	Dithizone Reagent (Lead)
Mercury vapor	Palladium Chloride Paper
Mercury(I)	Merckoquant® Cobalt Test
Mercury, organic	Diphenylcarbazone Reagent (Mercury)
Mescaline	Nitric Acid Reagent
Metal	Ammonium Dithiocarbamate Solution
Metal	Diphenylcarbazide Reagent
Metal	Ferric Ferricyanide Test
Metal	ITS SenSafe™ Water Metal Test
Methadone	Cobalt Thiocyanate Reagent
Methamphetamine HCl	Simon's Reagent
Methyl aniline	Drager Aniline 0.5/a
Methyl bromide	Drager Methyl Bromide 0.5/a
Methyl bromide	Sensidyne Tube #157SB
Methyl ethyl ketone	Sensidyne Tube #133A
Methyl ethyl ketone	Unsaturated Hydrocarbon Test
Methyl isobutyl ketone	Sensidyne Tube #133A
Methyl mercaptan	Sensidyne Test #165SA
Methylenedioxy-N-methamphetamine	Simon's Reagent
Methylphenidate HCl	Simon's Reagent
Milk	Catalase Test
Milk	Lactase Test
Molybdate	Dipyridyl Reagent
Molybdate	Hydrogen Peroxide Reagent
Molybdate	Potassium Ethyl Xanthate Reagent

(*continued*)

Substance	Test
Molybdate	Stannous Chloride Reagent (Molybdate)
Molybdate salt	Rhodamine B Reagent (Antimony)
Molybdenum	Merckoquant® Molybdenum Test
Molybdenum	Merckoquant® Nickel Test
Molybdenum metal	Molybdenum Metal Test
Monomer	Sensidyne Tube #171SB
Morphine	Aloy's Reagent
Morphine	Ferric Chloride Reagent
Morphine	Froede Reagent
Morphine	Mecke Reagent
Morphine	Nitric Acid Reagent
Mustard	Gold Chloride Reagent (Thiol)
Neodymium	Arsenazo I Reagent
Neodymium salt	Quinalizarin Reagent (Magnesium)
Nerve agent	Agri-Screen Ticket
Nerve agent	M256 Series Chemical Agent Detector Kit
Nerve agent	M256A1 Nerve Agent Test
Nickel	Copper Metal Test
Nickel	Dimethylglyoxime Reagent
Nickel	Formaldoxime Reagent
Nickel	ITS SenSafe™ Water Metal Test
Nickel	Merckoquant® Nickel Test
Nickel	Nickel Dimethylglyoxime Equilibrium Solution
Nickel	Offord's Reagent
Nickel	Rubeanic Acid (Cobalt)
Nickel carbonyl	Sensidyne Tube #167S
Nickel tetracarbonyl	Drager Nickel Tetracarbonyl 0.1/a
Niobium	Pyridylazo Resorcinol (PAR) Reagent
Nitramine	Drop-Ex Plus™
Nitrate	Alvarez's Reagent
Nitrate	Diphenylamine Indicator Solution
Nitrate	Merckoquant® Nitrate Tests 1.10020.0001, 1.10020.0002
Nitrate	Nitrate and Nitrite Test Strip (CZ NIT600)
Nitrate	Nitrate Test (Ammonia)
Nitrate	Nitrate Test (Gutzeit Scheme)
Nitrate	Nitron Solution
Nitrate	Safranin T Solution
Nitrate ester	Drop-Ex Plus™
Nitrate, Inorganic	Drop-Ex Plus™

(continued)

Substance	Test
Nitric acid	Nitrate and Nitrite Test Strip (CZ NIT600)
Nitrile	Drager Acrylonitrile 0.5/a
Nitrile	Hydroxylamine Hydrochloride Reagent
Nitrile	Sensidyne Tube #128SC
Nitrite	Alvarez's Reagent
Nitrite	Aniline-Phenol Reagent
Nitrite	Gas Production With Sulfuric Acid
Nitrite	Griess Test
Nitrite	Merckoquant® Nitrate Tests 1.10020.0001, 1.10020.0002
Nitrite	Merckoquant® Nitrite Tests 1.10007.0001, 1.10007.0002
Nitrite	Merckoquant® Nitrite Tests 1.10022.0001
Nitrite	Naphthylenediamine Test
Nitrite	Nitrate and Nitrite Test Strip (CZ NIT600)
Nitrite	Nitrite Test (Gutzeit Scheme)
Nitrite	Polynitroaromatic Screening Test
Nitrite	Safranin T Solution
Nitro	Ferrous Hydroxide Reagent
Nitro	Hearon-Gustavson's Reagents
Nitroaromatic	Drop-Ex Plus™
Nitrocellulose	Diphenylamine Indicator Solution
Nitrocellulose	Nitrocellulose Test
Nitrogen dioxide	Drager Chlorine 0.2/a
Nitrogen dioxide	Drager Hydrogen Sulfide 0.2/b
Nitrogen dioxide	Drager Natural Gas Odorization Tertiary Butylmercaptan
Nitrogen dioxide	Drager Nitrogen Dioxide 0.5/c
Nitrogen dioxide	Drager Nitrous Fumes 0.5/a
Nitrogen dioxide	Drager Nitrous Fumes 20/a
Nitrogen dioxide	Drager Ozone 0.05/b
Nitrogen dioxide	Drager Simultaneous Test Set I - Tube 5
Nitrogen dioxide	Drager Simultaneous Test Set II - Tube 2
Nitrogen dioxide	Sensidyne Tube #109SB
Nitrogen dioxide	Sensidyne Tube #114
Nitrogen dioxide	Sensidyne Tube #142S
Nitrogen dioxide	Sensidyne Tube #174A
Nitrogen dioxide	Sensidyne Tube #216S
Nitrogen monoxide	Drager Nitrous Fumes 0.5/a
Nitrogen monoxide	Drager Nitrous Fumes 20/a

(continued)

Substance	Test
Nitrogen monoxide	Drager Simultaneous Test Set I - Tube 5
Nitrogen monoxide	Sensidyne Tube #174A
Nitrogen mustard	Drager CDS Simultaneous Test Set I - Tube 5
Nitrogen mustard	Drager Organic Basic Nitrogen Compounds
Nitrogen mustard	M256A1 Blister Agent Test
Nitrogen oxide	Sensidyne Tube #174A
Nitrogen oxide	Sensidyne Tube #235S
Nitrogen tetroxide	Michler's Thioketone
Nitrogen trichloride	Sensidyne Tube #109SB
Nitroglycerin	Dinitrotoluene Test
Nitroso	Ferrous Hydroxide Reagent
Nitrotoluene	Nitrotoluene Test
Nitrous acid	Griess' Paper (Yellow)
Nitrous acid	Griess Test
Nitrous acid	Michler's Thioketone
Nitrous acid	Naphthylenediamine Test
Nitrous acid	Nitrate and Nitrite Test Strip (CZ NIT600)
Nitrous acid	Nitrocellulose Test
Nitrous fume	Drager Nitrous Fumes 0.5/a
Nitrous fume	Drager Nitrous Fumes 20/a
Nitrous gases	Drager Simultaneous Test Set I - Tube 5
Novichok agent	M256A1 Nerve Agent Test
Olefin	Drager Olefine 0.05%/a
Olefine	Drager Olefine 0.05%/a
Organic arsenic compound	Drager CDS Simultaneous Test Set I - Tube 4
Organic base	Nickel Dimethylglyoxime Equilibrium Solution
Organic basic nitrogen compound	Drager CDS Simultaneous Test Set I - Tube 5
Organic basic nitrogen compound	Drager Organic Basic Nitrogen Compounds
Organic compound	Sulfur Reagent
Organic gas	Sensidyne Tube #186
Organic nitro	Hearon-Gustavson's Reagents
Organic nitrogen	Ferric Thiocyanate Reagent
Organic oxygen	Ferric Thiocyanate Reagent
Organic peroxide	Merckoquant® Peroxide Test 1.10337.0001
Organic peroxide	Merckoquant® Peroxide Tests 1.10011.0001, 1.10011.0002, 1.10081.0001
Organic peroxide	Peroxide Test Strip (PER 400)
Organic sulfur	Benedict's Solution (Sulfur)
Organic sulfur	Ferric Thiocyanate Reagent
Organomercury compound	Diphenylcarbazone Reagent (Mercury)

(*continued*)

Substance	Test
Organometallic	M256A1 Lewisite Test
Organophosphate pesticide	Agri-Screen Ticket
Organophosphate pesticide	M256A1 Nerve Agent Test
Organyl tin ion	Dithizone Reagent (Tin)
Orthophosphate	Merckoquant® Phosphate Test
Osmium	Potassium Chlorate Reagent
Oxide	Nickel Dimethylglyoxime Equilibrium Solution
Oxidizer	Diphenylamine Indicator Solution
Oxidizer	Drager Hydrogen Peroxide 0.1/a
Oxidizer	Drager Nitrogen Dioxide 0.5/c
Oxidizer	Drager Nitrous Fumes 0.5/a
Oxidizer	Drager Nitrous Fumes 20/a
Oxidizer	Drager Ozone 0.05/b
Oxidizer	Drager Simultaneous Test Set I - Tube 5
Oxidizer	Drager Simultaneous Test Set II - Tube 2
Oxidizer	Hearon-Gustavson's Reagents
Oxidizer	Merckoquant® Chlorine Tests 1.17924.0001, 1.17925.0001
Oxidizer	Potassium Iodide Starch Test Strip (JD260)
Oxycodone	Nitric Acid Reagent
Oxyquinolin	Beilstein Test
Ozone	Drager Nitrogen Dioxide 0.5/c
Ozone	Drager Nitrous Fumes 0.5/a
Ozone	Drager Nitrous Fumes 20/a
Ozone	Drager Ozone 0.05/b
Ozone	Drager Simultaneous Test Set I - Tube 5
Ozone	Phenylenediamine Reagent
Ozone	Potassium Iodide Starch Test Strip (JD260)
Palladium	Aminoacetophenone Solution
Palladium	Arnold's Base
Palladium	Dimethylglyoxime Reagent
Palladium	Potassium Iodide Reagent (Gold)
Palladium	Rubeanic Acid Reagent
Palladium salt	Aminoacetophenone Reagent
Palladium salt	Rubeanic Acid (Ruthenium)
PCB	Clor-N-Oil 50
Pentobarbital	Dille-Koppanyi Reagent, Modified
Pentobarbital	Zwikker Reagent
Peptone	Ninhydrin Solution

(continued)

Substance	Test
Peracetic acid	Merckoquant® Peracetic Acid Test
Perborate	Peroxide Test Strip (PER 400)
Perchlorate	Methylene Blue Reagent
Perchlorate	Phenylanthranilic Acid Test
Perchlorate	Zwikker's Reagent
Perchloroethylene	Drager Perchloroethylene 0.1/a
Periodate	Chloric Acid Test
Permanganate	Diphenylamine Indicator Solution
Permanganate	Permanganate Test
Permanganate	Zwikker's Reagent
Peroxidase	Borinski's Reagent
Peroxide	Drop-Ex Plus™
Peroxide	Merckoquant® Peroxide Test 1.10337.0001
Peroxide	Merckoquant® Peroxide Tests 1.10011.0001, 1.10011.0002, 1.10081.0001
Peroxide	Peroxide Test Strip (PER 400)
Persulfate	Chloric Acid Test
Persulfate	Nickel Hydroxide Test
Persulfate	Zwikker's Reagent
Persulfuric acid	Nickel Hydroxide Test
Pesticide	Drager CDS Simultaneous Test Set V - Tube 5
Pesticide	Drager Phosphoric Acid Esters 0.05/a
PETN	Dichlorofluorescein Reagent
pH	pH Test For Solids
Phenobarbital	Dille-Koppanyi Reagent, Modified
Phenobarbital	Zwikker Reagent
Phenol	Aloy's Reagent
Phenol	Ceric Nitrate Reagent
Phenol	Drager Phenol 1/b
Phenol	Ferric Chloride Reagent
Phenol	Sensidyne Tube 183U
Phenol	Unsaturated Hydrocarbon Test
Phosgene	Drager Carbon Tetrachloride 5/c
Phosgene	Drager CDS Simultaneous Test Set I - Tube 2
Phosgene	Drager CDS Simultaneous Test Set V - Tube 3
Phosgene	Drager Clan Lab Simultaneous Test Set - Tube 2
Phosgene	Drager Phosgene 0.02/a
Phosgene	Drager Phosgene 0.05/a
Phosgene	Drager Simultaneous Test Set II - Tube 5
Phosgene	Sensidyne Tube #147S

(continued)

Substance	Test
Phosgene oxime	M256A1 Blister Agent Test
Phosphate	Ammonium Molybdate Solution (Deniges)
Phosphate	Merckoquant® Phosphate Test
Phosphate	Nickel Dimethylglyoxime Equilibrium Solution
Phosphine	Drager Arsine 0.05/a
Phosphine	Drager CDS Simultaneous Test Set I - Tube 4
Phosphine	Drager Clan Lab Simultaneous Test Set - Tube 1
Phosphine	Drager Cyanide 2/a
Phosphine	Drager Hydrogen Sulfide 0.2/b
Phosphine	Drager Organic Arsenic Compounds and Arsine
Phosphine	Drager Phosphine 0.1/a
Phosphine	Drager Phosphine 0.1/a
Phosphine	Drager Simultaneous Test Set I - Tube 2
Phosphine	Sensidyne Deluxe Haz Mat III Tube 131
Phosphine	Sensidyne Tube #121U
Phosphine	Sensidyne Tube #165SB
Phosphoric acid ester	Drager CDS Simultaneous Test Set V - Tube 5
Phosphoric acid ester	Drager Phosphoric Acid Esters 0.05/a
Phosphorus	Silver Nitrate Paper
Picrotoxin	Anisaldehyde Reagent
Platinum	Aqua Regia Reagent
Platinum	Rubeanic Acid Reagent
Platinum salt	Rubeanic Acid (Ruthenium)
Platinum(IV) salt	Potassium Iodide Reagent (Gold)
Polychlorinated biphenyl	Clor-N-Oil 50
Polynitroaromatic	Polynitroaromatic Screening Test
Polypeptide	Ninhydrin Solution
Polysaccharide	Lugol Starch Test
Polysaccharide	Povidone Test for Starch
Polysulfide	Sodium Sulfide Reagent
Potassium	Aurantia Solution
Potassium	Celsi's Reagent
Potassium	Merckoquant® Potassium Test
Praseodymium	Arsenazo I Reagent
Praseodymium salt	Quinalizarin Reagent (Magnesium)
Protein	BCA Protein Test
Protein	Biuret Paper
Protein	Biuret Protein Test
Protein	Bradford Reagent
Protein	Fluorescamine Test for Protein

(continued)

Substance	Test
Protein	Lowry Protein Test
Protein	Nickel Dimethylglyoxime Equilibrium Solution
Protein	Ninhydrin Solution
Protein	Nitric Acid Reagent
Pseudoephedrine	Cobalt Thiocyanate Reagent
Pyridine	Beilstein Test
Quat amine	Merckoquant® Quaternary Ammonium Compounds
Quaternary amine	Merckoquant® Quaternary Ammonium Compounds
Quaternary ammonium compound	Dragendorff's Reagent (Alkaloids)
Quaternary ammonium compound	Merckoquant® Quaternary Ammonium Compounds
Quinine	Cobalt Thiocyanate Reagent
Rare earth metal	Arsenazo I Reagent
RDX	Dichlorofluorescein Reagent
RDX	RDX Test (J-Acid)
RDX	RDX Test (Pyrolytic Oxidation)
RDX	RDX Test (Thymol)
RDX	Trinitrophenylmethylnitramine Test
Reactive organic chlorine	Silver Nitrate Reagent
Reducer	Biuret Protein Test
Reducer	Reducing Substance Test
Reducing sugar	Fehling's Solution
Reducing sugar	Luff's Reagent
Rhodium salt	Stannous Chloride Reagent (Rhodium)
Rubidium	Ball's Reagent
Ruthenium	Potassium Chlorate Reagent
Ruthenium salt	Rubeanic Acid (Ruthenium)
Salt	Marquis Reagent
Samarium	Arsenazo I Reagent
Scandium	Morin Reagent
Secobarbital	Dille-Koppanyi Reagent, Modified
Secobarbital	Zwikker Reagent
Selenite	Methylene Blue Catalytic Reduction Test
Selenite	Naphthylenediamine Test
Selenium	Guerin's Reagent
Silicate	Nickel Dimethylglyoxime Equilibrium Solution
Silver	Ammonium Dithiocarbamate Solution
Silver	Arnold's Base
Silver	Benzopurpurin 4 B Reagent

(continued)

Substance	Test
Silver	Lead(II) Test
Silver	Merckoquant® Fixing Bath Test
Silver	Merckoquant® Zinc Test
Silver	Potassium Chromate Reagent
Silver	Potassium Iodide Reagent (Thallium)
Silver	Rubeanic Acid Reagent
Silver salt	Potassium Chromate Reagent
Sodium	Bougault's Reagent
Sodium	Caley's Reagent
Sodium bicarbonate	Ferric Chloride Reagent
Stannic	Dithizone Reagent (Tin)
Stannous	Diazine Green S Solution
Stannous	Dithizone Reagent (Tin)
Stannous tin	Brucine Solution
Starch	Lugol Starch Test
Starch	Povidone Test for Starch
Strontium	Sodium Rhodizonate Solution (Barium)
Styrene	Sensidyne Tube #171SB
Sucrose	Sucrose Test
Sugar	Benedict's Solution (Carbonyl)
Sugar	Sucrose Test
Sulfate	Merckoquant® Sulfate Test
Sulfate	Sulfate Test
Sulfide	Arsine From Elemental Arsenic Test
Sulfide	Gas Production With Sulfuric Acid
Sulfide	Lead Acetate Test Strip (JD175)
Sulfide	pH Test For Solids
Sulfide	Sodium Azide - Iodine Solution
Sulfide	Sulfide Test
Sulfite	Merckoquant® Sulfite Test
Sulfite	Sulfite Test
Sulfur	Arsine From Elemental Arsenic Test
Sulfur	Sodium Nitroprusside Paper
Sulfur dioxide	Drager Carbon Dioxide 5%/A
Sulfur dioxide	Drager Natural Gas Odorization Tertiary Butylmercaptan
Sulfur dioxide	Drager Simultaneous Test Set II - Tube 1
Sulfur dioxide	Drager Simultaneous Test Set II - Tube 4
Sulfur dioxide	Sensidyne Deluxe Haz Mat III Tube 131
Sulfur dioxide	Sensidyne Tube #112SA

(continued)

Substance	Test
Sulfur dioxide	Sensidyne Tube #112SB
Sulfur dioxide	Sensidyne Tube #126SA
Sulfur dioxide	Sensidyne Tube #133A
Sulfur dioxide	Sensidyne Tube #141SB
Sulfur dioxide	Sensidyne Tube #167S
Sulfur dioxide	Sensidyne Tube #216S
Sulfur dioxide	Sensidyne Tube #232SA
Sulfur dioxide	Sensidyne Tube #239S
Sulfur dioxide	Sodium Nitroprusside Paper
Sulfur mustard	Drager CDS Simultaneous Test Set I - Tube 1
Sulfur mustard	Drager CDS Simultaneous Test Set V - Tube 2
Sulfur mustard	Drager Thioether
Sulfur mustard	M256A1 Blister Agent Test
Sulfur trioxide	Sensidyne Deluxe Haz Mat III Tube 131
Sulfur, elemental	Sodium Sulfide Reagent
Sulfur, organic	Raikow's Reagent
Sulfuric acid	Sulfuric Acid Test
Sulfurous acid	Sulfite Test
T4	RDX Test (Pyrolytic Oxidation)
Tachysterol	Antimony Trichloride Reagent
TATB	TATB Test
TDI	Drager Toluene Diisocyanate 0.02/A
TDI	Drager Toluene Diisocyanate 0.02/A
Tea	Zwikker Reagent
Technetium	Stannous Chloride Reagent (Technetium)
Terbium	Arsenazo I Reagent
Tetrachloroethylene	Sensidyne Tube #134SB
Tetrahydrocannabinol	Duquenois-Levine Reagent, Modified
Tetryl	Trinitrophenylmethylnitramine Test
Thallium	Lead(II) Test
Thallium	Potassium Bismuth Iodide Reagent
Thallium	Rhodamine B Reagent (Antimony)
Thallium(I)	Ammoniacal Silver Oxide Solution
Thallium(I)	Potassium Iodide Reagent (Thallium)
Thallium(III)	Quinizarin Reagent
Thallium(III)	Rhodamine B Reagent (Gallium)
THC	Duquenois-Levine Reagent, Modified
Thioalcohol	Isatin Reagent (Mercaptans)
Thiocarbamate pesticide	Agri-Screen Ticket
Thiocyanate	Ammonium Chloride Reagent

(continued)

Substance	Test
Thiocyanate	Ferric Solution Test
Thiocyanate	Ferrocyanide Test (Gutzeit Scheme)
Thiocyanate	ITS Cyanide ReagentStrip™
Thiocyanate	Methylene Blue Reagent
Thiocyanate	Potassium Permanganate Solution
Thiocyanate	Sodium Azide - Iodine Solution
Thiocyanate	Thiocyanate Conversion Test
Thiocyanide	Nitric Acid - Silver Nitrate Solution
Thiocyanuric acid	Sodium Azide - Iodine Solution
Thioether	Drager CDS Simultaneous Test Set I - Tube 1
Thioether	Drager CDS Simultaneous Test Set V - Tube 2
Thioether	Drager Thioether
Thiol	Gold Chloride Reagent (Thiol)
Thiol	Raikow's Reagent
Thiosulfate	Casolori's Reagent
Thiosulfate	Ferrocyanide Test (Gutzeit Scheme)
Thiosulfate	Sodium Azide - Iodine Solution
Thiosulfate	Zwikker's Reagent
Thiosulfuric acid	Sodium Azide - Iodine Solution
Thorium	Aluminon Reagent
Thorium	Morin Reagent
Thorium salt	Quinalizarin Reagent (Magnesium)
Tin	Lead(II) Test
Tin	Merckoquant® Tin Test
Tin metal	Dinitrobenzene Reagent
Tin(II)	Brucine Solution
Tin(II)	Diazine Green S Solution
Tin(II)	Dithizone Reagent (Tin)
Tin(IV)	Dithizone Reagent (Tin)
Titanium(IV)	Hydrogen Peroxide Reagent
TNT	Janovsky Test
TNT	Nitrotoluene Test
TNT	Trinitrotoluene Test
Toluene	Drager Benzene 0.5/a
Toluene	Sensidyne Tube #190U
Toluene diisocyanate	Drager Toluene Diisocyanate 0.02/A
Toluidine	Sensidyne Tube #181S
Trialkyl-lead salt	Dithizone Reagent (Lead)
Triaminotrinitrobenzene	TATB Test
Trichloroethylene	Sensidyne Tube #134SB

(continued)

Substance	Test
Trinitro-1,3,5-triazine	Trinitrophenylmethylnitramine Test
Trinitrophenylmethylnitramine	Trinitrophenylmethylnitramine Test
Trinitrotoluene	Janovsky Test
Trinitrotoluene	Nitrotoluene Test
Trinitrotoluene	Trinitrotoluene Test
Tungstate	Dipyridyl Reagent
Tungstate	Stannous Chloride Reagent (Molybdate)
Tungstate salt	Rhodamine B Reagent (Antimony)
Tungsten	Cinchonine Solution
Tungstenate	Cinchonine Solution
Unsaturated hydrocarbon	Bromine Water
Unsaturated hydrocarbon	Unsaturated Hydrocarbon Test
Uranium	Ammonium Dithiocarbamate Solution
Uranium	Fluorescein Reagent (Uranyl)
Uranium hydroxide	Alizarin Reagent (Metal)
Uranium(VI)	Pyridylazo Resorcinol (PAR) Reagent
Uranyl	Fluorescein Reagent (Uranyl)
Urea	Beilstein Test
Vanadate	Hydrogen Peroxide Reagent
Vanadium	Ammonium Molybdate Reagent
Vegetable	Melibiose Test
Vinyl chloride	Sensidyne Tube #134SB
Vitamin D	Aniline Reagent
Vitamin D	Antimony Trichloride Reagent
Vitamin D2	Antimony Trichloride Reagent
Vitamin D3	Antimony Trichloride Reagent
V-nerve agent	Drager CDS Simultaneous Test Set V - Tube 5
V-nerve agent	Drager Phosphoric Acid Esters 0.05/a
V-nerve agent	M256A1 Nerve Agent Test
Water	Water Test
Zinc	Borneolglycuronic Acid Reagent
Zinc	Dithizone Reagent (Lead)
Zinc	ITS SenSafe™ Water Metal Test
Zinc	Merckoquant® Zinc Test
Zinc	Orange IV Solution
Zinc	Resorcinol Reagent
Zinc metal	Dinitrobenzene Reagent
Zirconium	Alizarin Reagent (Zirconium)
Zirconium	Aluminon Reagent
Zirconium	Neothorin Reagent
Zirconium salt	Quinalizarin Reagent (Magnesium)

Index of Tests

Test	Substance
Agri-Screen Ticket	Acetylcholinesterase-inhibiting compound
Agri-Screen Ticket	Carbamate pesticide
Agri-Screen Ticket	Nerve agent
Agri-Screen Ticket	Organophosphate pesticide
Agri-Screen Ticket	Thiocarbamate pesticide
Alizarin Reagent (Metal)	Aluminum hydroxide
Alizarin Reagent (Metal)	Chromic hydroxide
Alizarin Reagent (Metal)	Ferric hydroxide
Alizarin Reagent (Metal)	Manganese hydroxide
Alizarin Reagent (Metal)	Uranium hydroxide
Alizarin Reagent (Zirconium)	Zirconium
Alizarin S Reagent	Aluminum
Allylthiourea Reagent	Cadmium
Aloy's Reagent	Alkaloid
Aloy's Reagent	Morphine
Aloy's Reagent	Phenol
Aluminon Reagent	Aluminum
Aluminon Reagent	Beryllium
Aluminon Reagent	Cerium
Aluminon Reagent	Iron
Aluminon Reagent	Thorium
Aluminon Reagent	Zirconium
Aluminon Reagent on Paper	Aluminum
Alvarez's Reagent	Nitrate
Alvarez's Reagent	Nitrite
Aminoacetophenone Reagent	Cerium(IV)
Aminoacetophenone Reagent	Palladium salt
Aminoacetophenone Solution	Palladium
Aminodimethylaniline Reagent	Bromine
Aminodimethylaniline Reagent	Chlorine
Aminodimethylaniline Reagent	Halogen
Aminodimethylaniline Reagent	Iodine
Aminodiphenylamine Sulfate Reagent	Benzoyl peroxide
Aminophenol Hydrochloride Solution	Copper
Aminophenol Hydrochloride Solution	Iron
Ammoniacal Silver Oxide Solution	Cerium(III)
Ammoniacal Silver Oxide Solution	Chromium(III)
Ammoniacal Silver Oxide Solution	Cobalt(II)

(*continued*)

Test	Substance
Ammoniacal Silver Oxide Solution	Iron(III)
Ammoniacal Silver Oxide Solution	Manganese(II)
Ammoniacal Silver Oxide Solution	Thallium(I)
Ammonium Chloride Reagent	Thiocyanate
Ammonium Dithiocarbamate Solution	Aluminum
Ammonium Dithiocarbamate Solution	Cobalt
Ammonium Dithiocarbamate Solution	Lead
Ammonium Dithiocarbamate Solution	Metal
Ammonium Dithiocarbamate Solution	Silver
Ammonium Dithiocarbamate Solution	Uranium
Ammonium Molybdate Reagent	Arsenic
Ammonium Molybdate Reagent	Vanadium
Ammonium Molybdate Solution (Deniges)	Arsenate
Ammonium Molybdate Solution (Deniges)	Phosphate
Ammonium Test	Ammonium
Ammonium Test (Gutzeit Scheme)	Amine
Ammonium Test (Gutzeit Scheme)	Ammonium
Aniline Reagent	Vitamin D
Aniline Sulfate Reagent	Chlorate
Aniline-Phenol Reagent	Nitrite
Anisaldehyde Reagent	Picrotoxin
Antimony Trichloride Reagent	Tachysterol
Antimony Trichloride Reagent	Vitamin D
Antimony Trichloride Reagent	Vitamin D2
Antimony Trichloride Reagent	Vitamin D3
Aqua Regia Reagent	Gold
Aqua Regia Reagent	Platinum
Arnold-Mentzel's Solution	Hydrogen peroxide
Arnold's Base	Gold
Arnold's Base	Mercury
Arnold's Base	Palladium
Arnold's Base	Silver
Arsenazo I Reagent	Dysprosium
Arsenazo I Reagent	Europium
Arsenazo I Reagent	Gadolinium
Arsenazo I Reagent	Lutetium
Arsenazo I Reagent	Neodymium
Arsenazo I Reagent	Praseodymium
Arsenazo I Reagent	Rare earth metal
Arsenazo I Reagent	Samarium

(continued)

Test	Substance
Arsenazo I Reagent	Terbium
Arsine From Elemental Arsenic Test	Alkaline earth sulfate
Arsine From Elemental Arsenic Test	Arsenic
Arsine From Elemental Arsenic Test	Sulfide
Arsine From Elemental Arsenic Test	Sulfur
Aurantia Solution	Potassium
Ball's Reagent	Cesium
Ball's Reagent	Rubidium
BCA Protein Test	Protein
Beilstein Test	Alkyl halide
Beilstein Test	Chlorinated hydrocarbon
Beilstein Test	Halogenated hydrocarbon
Beilstein Test	Oxyquinolin
Beilstein Test	Pyridine
Beilstein Test	Urea
Benedict's Solution	Hydroxy aldehyde
Benedict's Solution	Hydroxy ketone
Benedict's Solution	Keto aldehyde
Benedict's Solution (Carbonyl)	Carbonyl group
Benedict's Solution (Carbonyl)	Sugar
Benedict's Solution (Sulfur)	Mercaptan
Benedict's Solution (Sulfur)	Organic sulfur
Benzoin Reagent	Boric acid
Benzoin Reagent	Boric salt
Benzopurpurin 4 B Reagent	Mercuric
Benzopurpurin 4 B Reagent	Mercurous
Benzopurpurin 4 B Reagent	Mercury
Benzopurpurin 4 B Reagent	Silver
Bettendorf's Test	Arsenate
Bettendorf's Test	Arsenic
Bettendorf's Test	Arsenic(III)
Bettendorf's Test	Arsenic(IV)
Bettendorf's Test	Arsenite
Bisulfite Test	Bisulfite
Biuret Paper	Amino acid
Biuret Paper	Protein
Biuret Protein Test	Amine
Biuret Protein Test	Protein
Biuret Protein Test	Reducer
Borate Test (Gutzeit Scheme)	Borate

(continued)

Test	Substance
Borate Test (Gutzeit Scheme)	Boric acid
Borinski's Reagent	Peroxidase
Borneolglycuronic Acid Reagent	Zinc
Bougault's Reagent	Sodium
Bradford Reagent	Protein
Brady's Reagent	Aldehyde
Brady's Reagent	Carbonyl group
Brady's Reagent	Ketone
Bromic Acid Test	Bromate
Bromic Acid Test	Bromic acid
Bromine Water	Alkene
Bromine Water	Alkyne
Bromine Water	Aromatic
Bromine Water	Mercaptan
Bromine Water	Unsaturated hydrocarbon
Brucine Solution	Stannous tin
Brucine Solution	Tin(II)
Caley's Reagent	Sodium
Carmine Red Reagent	Boric acid
Casolori's Reagent	Thiosulfate
Catalase Test	Cheese
Catalase Test	Dairy
Catalase Test	Milk
Celsi's Reagent	Potassium
Ceric Nitrate Reagent	Alcohol
Ceric Nitrate Reagent	Dioxane
Ceric Nitrate Reagent	E85
Ceric Nitrate Reagent	Gasoline (10% ethanol)
Ceric Nitrate Reagent	Phenol
Chloric Acid Test	Chlorate
Chloric Acid Test	Chloric acid
Chloric Acid Test	Periodate
Chloric Acid Test	Persulfate
Chlorine Dioxide Test	Chlorate
Chlorine Dioxide Test	Chlorine dioxide
Chromotropic Acid Reagent	Ferric
Chromotropic Acid Reagent	Iron
Cinchonine - Potassium Iodide Solution	Bismuth
Cinchonine - Potassium Iodide Solution	Copper
Cinchonine - Potassium Iodide Solution	Lead

(*continued*)

Test	Substance
Cinchonine - Potassium Iodide Solution	Mercury
Cinchonine Solution	Tungsten
Cinchonine Solution	Tungstenate
Clor-N-Oil 50	Alkyl chloride
Clor-N-Oil 50	Chlorinated hydrocarbon
Clor-N-Oil 50	PCB
Clor-N-Oil 50	Polychlorinated biphenyl
Cobalt Thiocyanate Reagent	Alkaloid
Cobalt Thiocyanate Reagent	Cocaine
Cobalt Thiocyanate Reagent	Ephedrine
Cobalt Thiocyanate Reagent	Methadone
Cobalt Thiocyanate Reagent	Pseudoephedrine
Cobalt Thiocyanate Reagent	Quinine
Copper Metal Test	Cobalt
Copper Metal Test	Copper metal
Copper Metal Test	Nickel
Cuprion Reagent	Copper
Cuprion Reagent	Copper(I)
Cuprion Reagent	Copper(II)
Cuprous Iodide Paper	Amalgam
Cuprous Iodide Paper	Mercury
Cuprous Iodide Paper	Mercury salt
Curcumin Solution	Beryllium
Deniges' Reagent	Acetylene
Diaminobenzene Solution	Magnesium
Diazine Green S Solution	Stannous
Diazine Green S Solution	Tin(II)
Dichlorofluorescein Reagent	Explosive
Dichlorofluorescein Reagent	HMX
Dichlorofluorescein Reagent	PETN
Dichlorofluorescein Reagent	RDX
Diethylnicotinamide Reagent	Decaborane
Dihydroxyacetophenone Solution	Ferric
Dihydroxyacetophenone Solution	Iron
Dille-Koppanyi Reagent, Modified	Amobarbital
Dille-Koppanyi Reagent, Modified	Barbiturate
Dille-Koppanyi Reagent, Modified	Pentobarbital
Dille-Koppanyi Reagent, Modified	Phenobarbital
Dille-Koppanyi Reagent, Modified	Secobarbital
Dimethylaminobenzaldehyde (p-DMAB)	LSD

(continued)

Test	Substance
Dimethylaminobenzaldehyde (p-DMAB)	Lysergic acid diethylamide
Dimethylglyoxime Reagent	Bismuth
Dimethylglyoxime Reagent	Cobalt
Dimethylglyoxime Reagent	Copper
Dimethylglyoxime Reagent	Ferrous
Dimethylglyoxime Reagent	Nickel
Dimethylglyoxime Reagent	Palladium
Dinitrobenzene Reagent	Aluminum metal
Dinitrobenzene Reagent	Beryllium metal
Dinitrobenzene Reagent	Lead metal
Dinitrobenzene Reagent	Magnesium metal
Dinitrobenzene Reagent	Tin metal
Dinitrobenzene Reagent	Zinc metal
Dinitrotoluene Test	Dinitrotoluene
Dinitrotoluene Test	DNT
Dinitrotoluene Test	Explosive
Dinitrotoluene Test	Nitroglycerin
Diphenylamine Indicator Solution	Dichromate
Diphenylamine Indicator Solution	Hypochlorite
Diphenylamine Indicator Solution	Nitrate
Diphenylamine Indicator Solution	Nitrocellulose
Diphenylamine Indicator Solution	Oxidizer
Diphenylamine Indicator Solution	Permanganate
Diphenylamine Reagent	Formaldehyde
Diphenylcarbazide Reagent	Chromate
Diphenylcarbazide Reagent	Metal
Diphenylcarbazide Test	Chromate
Diphenylcarbazide Test	Chromic acid
Diphenylcarbazide Test	Chromium(III)
Diphenylcarbazide Test	Chromium(VI)
Diphenylcarbazide Test	Dichromate
Diphenylcarbazone Reagent (Germanium)	Germanium
Diphenylcarbazone Reagent (Mercury)	Mercury salt
Diphenylcarbazone Reagent (Mercury)	Mercury, organic
Diphenylcarbazone Reagent (Mercury)	Organomercury compound
Dipyridyl Reagent	Ferric
Dipyridyl Reagent	Ferrous
Dipyridyl Reagent	Iron(II)
Dipyridyl Reagent	Iron(III)
Dipyridyl Reagent	Molybdate

(continued)

Test	Substance
Dipyridyl Reagent	Tungstate
Dithizone Reagent (Lead)	Dialkyl-lead salt
Dithizone Reagent (Lead)	Lead
Dithizone Reagent (Lead)	Mercury salt
Dithizone Reagent (Lead)	Trialkyl-lead salt
Dithizone Reagent (Lead)	Zinc
Dithizone Reagent (Tin)	Organyl tin ion
Dithizone Reagent (Tin)	Stannic
Dithizone Reagent (Tin)	Stannous
Dithizone Reagent (Tin)	Tin(II)
Dithizone Reagent (Tin)	Tin(IV)
Dragendorff's Reagent (Alkaloids)	Alkaloid
Dragendorff's Reagent (Alkaloids)	Amine
Dragendorff's Reagent (Alkaloids)	Quaternary ammonium compound
Drager Acetaldehyde 100/a	Aldehyde
Drager Acetaldehyde 100/a	Ester
Drager Acetaldehyde 100/a	Ether
Drager Acetaldehyde 100/a	Hydrocarbon (petroleum)
Drager Acetaldehyde 100/a	Ketone
Drager Acetic Acid 5/a	Acid
Drager Acetone 100/b	Acetone
Drager Acetone 100/b	Aldehyde
Drager Acetone 100/b	Ammonia
Drager Acetone 100/b	Ketone
Drager Acid Test	Acid
Drager Acrylonitrile 0.5/a	Acrylonitrile
Drager Acrylonitrile 0.5/a	Nitrile
Drager Alcohol 100/a	Alcohol
Drager Amine Test	Base
Drager Ammonia 2/a	Base
Drager Aniline 0.5/a	Aniline
Drager Aniline 0.5/a	Aromatic
Drager Aniline 0.5/a	Ester
Drager Aniline 0.5/a	Ether
Drager Aniline 0.5/a	Hydrocarbon (petroleum)
Drager Aniline 0.5/a	Ketone
Drager Aniline 0.5/a	Methyl aniline
Drager Arsine 0.05/a	Antimony hydride
Drager Arsine 0.05/a	Arsine
Drager Arsine 0.05/a	Phosphine

(*continued*)

Test	Substance
Drager Benzene 0.5/a	Aromatic
Drager Benzene 0.5/a	Benzene
Drager Benzene 0.5/a	Ethyl benzene
Drager Benzene 0.5/a	Toluene
Drager Carbon Dioxide 5%/A	Carbon dioxide
Drager Carbon Dioxide 5%/A	Sulfur dioxide
Drager Carbon Monoxide 2/a	Carbon monoxide
Drager Carbon Tetrachloride 5/c	Carbon tetrachloride
Drager Carbon Tetrachloride 5/c	Chlorohydrocarbon
Drager Carbon Tetrachloride 5/c	Phosgene
Drager CDS Simultaneous Test Set I - Tube 1	Sulfur mustard
Drager CDS Simultaneous Test Set I - Tube 1	Thioether
Drager CDS Simultaneous Test Set I - Tube 2	Acetyl chloride
Drager CDS Simultaneous Test Set I - Tube 2	Carbonyl bromide
Drager CDS Simultaneous Test Set I - Tube 2	Phosgene
Drager CDS Simultaneous Test Set I - Tube 3	Hydrocyanic acid
Drager CDS Simultaneous Test Set I - Tube 4	Arsine
Drager CDS Simultaneous Test Set I - Tube 4	Lewisite
Drager CDS Simultaneous Test Set I - Tube 4	Organic arsenic compound
Drager CDS Simultaneous Test Set I - Tube 4	Phosphine
Drager CDS Simultaneous Test Set I - Tube 5	Nitrogen mustard
Drager CDS Simultaneous Test Set I - Tube 5	Organic basic nitrogen compound
Drager CDS Simultaneous Test Set V - Tube 1	Cyanogen bromide
Drager CDS Simultaneous Test Set V - Tube 1	Cyanogen chloride
Drager CDS Simultaneous Test Set V - Tube 2	Sulfur mustard
Drager CDS Simultaneous Test Set V - Tube 2	Thioether
Drager CDS Simultaneous Test Set V - Tube 3	Acetyl chloride
Drager CDS Simultaneous Test Set V - Tube 3	Carbonyl bromide
Drager CDS Simultaneous Test Set V - Tube 3	Phosgene
Drager CDS Simultaneous Test Set V - Tube 4	Chlorine
Drager CDS Simultaneous Test Set V - Tube 5	G-nerve agent
Drager CDS Simultaneous Test Set V - Tube 5	Pesticide
Drager CDS Simultaneous Test Set V - Tube 5	Phosphoric acid ester
Drager CDS Simultaneous Test Set V - Tube 5	V-nerve agent
Drager Chlorine 0.2/a	Bromine
Drager Chlorine 0.2/a	Chlorine
Drager Chlorine 0.2/a	Chlorine dioxide
Drager Chlorine 0.2/a	Nitrogen dioxide
Drager Clan Lab Simultaneous Test Set - Tube 1	Arsine
Drager Clan Lab Simultaneous Test Set - Tube 1	Phosphine

(continued)

Test	Substance
Drager Clan Lab Simultaneous Test Set - Tube 2	Phosgene
Drager Clan Lab Simultaneous Test Set - Tube 3	Amine
Drager Clan Lab Simultaneous Test Set - Tube 3	Ammonia
Drager Clan Lab Simultaneous Test Set - Tube 3	Base
Drager Clan Lab Simultaneous Test Set - Tube 4	Acid
Drager Clan Lab Simultaneous Test Set - Tube 4	Hydrochloric acid
Drager Clan Lab Simultaneous Test Set - Tube 5	Iodine
Drager Cyanide 2/a	Acid
Drager Cyanide 2/a	Cyanide
Drager Cyanide 2/a	Phosphine
Drager Cyanogen Chloride 0.25/a	Cyanogen bromide
Drager Cyanogen Chloride 0.25/a	Cyanogen chloride
Drager Ethylene 0.1/a	Alkene
Drager Ethylene 0.1/a	Carbon monoxide
Drager Ethylene 0.1/a	Ethylene
Drager Halogenated Hydrocarbon 100/a	Halogen
Drager Halogenated Hydrocarbon 100/a	Halohydrocarbon
Drager Halogenated Hydrocarbon 100/a	Hydrogen halide
Drager Hydrazine 0.2/a	Hydrazine
Drager Hydrazine 0.2/a	Hydrazine, organic
Drager Hydrazine 0.2/a	Hydrogen sulfide
Drager Hydrocyanic Acid 2/a	Hydrocyanic acid
Drager Hydrogen Fluoride 1.5/b	Hydrogen fluoride
Drager Hydrogen Fluoride 1.5/b	Hydrogen fluoride
Drager Hydrogen Peroxide 0.1/a	Hydrogen peroxide
Drager Hydrogen Peroxide 0.1/a	Oxidizer
Drager Hydrogen Sulfide 0.2/a	Hydrogen cyanide
Drager Hydrogen Sulfide 0.2/a	Hydrogen sulfide
Drager Hydrogen Sulfide 0.2/b	Arsine
Drager Hydrogen Sulfide 0.2/b	Hydrogen cyanide
Drager Hydrogen Sulfide 0.2/b	Hydrogen sulfide
Drager Hydrogen Sulfide 0.2/b	Mercaptan
Drager Hydrogen Sulfide 0.2/b	Nitrogen dioxide
Drager Hydrogen Sulfide 0.2/b	Phosphine
Drager Mercaptan 0.5/a	Mercaptan
Drager Mercury Vapour 0.1/b	Mercury
Drager Methyl Bromide 0.5/a	Halohydrocarbon
Drager Methyl Bromide 0.5/a	Methyl bromide
Drager Natural Gas Odorization Tertiary Butylmercaptan	Arsine

(continued)

Test	Substance
Drager Natural Gas Odorization Tertiary Butylmercaptan	Butylmercaptan
Drager Natural Gas Odorization Tertiary Butylmercaptan	Hydrogen sulfide
Drager Natural Gas Odorization Tertiary Butylmercaptan	Mercaptan
Drager Natural Gas Odorization Tertiary Butylmercaptan	Nitrogen dioxide
Drager Natural Gas Odorization Tertiary Butylmercaptan	Sulfur dioxide
Drager Nickel Tetracarbonyl 0.1/a	Iron tetracarbonyl
Drager Nickel Tetracarbonyl 0.1/a	Nickel tetracarbonyl
Drager Nitrogen Dioxide 0.5/c	Chlorine
Drager Nitrogen Dioxide 0.5/c	Nitrogen dioxide
Drager Nitrogen Dioxide 0.5/c	Oxidizer
Drager Nitrogen Dioxide 0.5/c	Ozone
Drager Nitrous Fumes 0.5/a	Chlorine
Drager Nitrous Fumes 0.5/a	Nitrogen dioxide
Drager Nitrous Fumes 0.5/a	Nitrogen monoxide
Drager Nitrous Fumes 0.5/a	Nitrous fume
Drager Nitrous Fumes 0.5/a	Oxidizer
Drager Nitrous Fumes 0.5/a	Ozone
Drager Nitrous Fumes 20/a	Chlorine
Drager Nitrous Fumes 20/a	Nitrogen dioxide
Drager Nitrous Fumes 20/a	Nitrogen monoxide
Drager Nitrous Fumes 20/a	Nitrous fume
Drager Nitrous Fumes 20/a	Oxidizer
Drager Nitrous Fumes 20/a	Ozone
Drager Olefine 0.05%/a	Alkene
Drager Olefine 0.05%/a	Olefin
Drager Olefine 0.05%/a	Olefine
Drager Organic Arsenic Compounds and Arsine	Arsenic, organic
Drager Organic Arsenic Compounds and Arsine	Arsine
Drager Organic Arsenic Compounds and Arsine	Phosphine
Drager Organic Basic Nitrogen Compounds	Amine
Drager Organic Basic Nitrogen Compounds	Nitrogen mustard
Drager Organic Basic Nitrogen Compounds	Organic basic nitrogen compound
Drager Ozone 0.05/b	Chlorine
Drager Ozone 0.05/b	Nitrogen dioxide
Drager Ozone 0.05/b	Oxidizer
Drager Ozone 0.05/b	Ozone

(continued)

Test	Substance
Drager Perchloroethylene 0.1/a	Bromine
Drager Perchloroethylene 0.1/a	Chlorine
Drager Perchloroethylene 0.1/a	Fluorine
Drager Perchloroethylene 0.1/a	Halogen
Drager Perchloroethylene 0.1/a	Halohydrocarbon
Drager Perchloroethylene 0.1/a	Perchloroethylene
Drager Phenol 1/b	Aromatic alcohol
Drager Phenol 1/b	Cresol
Drager Phenol 1/b	Phenol
Drager Phosgene 0.02/a	Chlorine
Drager Phosgene 0.02/a	Hydrochloric acid
Drager Phosgene 0.02/a	Phosgene
Drager Phosgene 0.05/a	Acetyl chloride
Drager Phosgene 0.05/a	Carbonyl bromide
Drager Phosgene 0.05/a	Phosgene
Drager Phosphine 0.1/a	Antimony hydride
Drager Phosphine 0.1/a	Antimony hydride
Drager Phosphine 0.1/a	Arsine
Drager Phosphine 0.1/a	Arsine
Drager Phosphine 0.1/a	Phosphine
Drager Phosphine 0.1/a	Phosphine
Drager Phosphoric Acid Esters 0.05/a	G-nerve agent
Drager Phosphoric Acid Esters 0.05/a	Pesticide
Drager Phosphoric Acid Esters 0.05/a	Phosphoric acid ester
Drager Phosphoric Acid Esters 0.05/a	V-nerve agent
Drager Simultaneous Test Set I - Tube 1	Acid
Drager Simultaneous Test Set I - Tube 2	Hydrocyanic acid
Drager Simultaneous Test Set I - Tube 2	Phosphine
Drager Simultaneous Test Set I - Tube 3	Acetylene
Drager Simultaneous Test Set I - Tube 3	Carbon monoxide
Drager Simultaneous Test Set I - Tube 3	Lewisite
Drager Simultaneous Test Set I - Tube 4	Amine
Drager Simultaneous Test Set I - Tube 4	Ammonia
Drager Simultaneous Test Set I - Tube 4	Base
Drager Simultaneous Test Set I - Tube 5	Chlorine
Drager Simultaneous Test Set I - Tube 5	Nitrogen dioxide
Drager Simultaneous Test Set I - Tube 5	Nitrogen monoxide
Drager Simultaneous Test Set I - Tube 5	Nitrous gases
Drager Simultaneous Test Set I - Tube 5	Oxidizer
Drager Simultaneous Test Set I - Tube 5	Ozone

(continued)

Test	Substance
Drager Simultaneous Test Set II - Tube 1	Hydrogen sulfide
Drager Simultaneous Test Set II - Tube 1	Sulfur dioxide
Drager Simultaneous Test Set II - Tube 2	Bromine
Drager Simultaneous Test Set II - Tube 2	Chlorine
Drager Simultaneous Test Set II - Tube 2	Chlorine dioxide
Drager Simultaneous Test Set II - Tube 2	Nitrogen dioxide
Drager Simultaneous Test Set II - Tube 2	Oxidizer
Drager Simultaneous Test Set II - Tube 3	Hydrogen sulfide
Drager Simultaneous Test Set II - Tube 4	Carbon dioxide
Drager Simultaneous Test Set II - Tube 4	Sulfur dioxide
Drager Simultaneous Test Set II - Tube 5	Phosgene
Drager Simultaneous Test Set III - Tube 1	Aldehyde
Drager Simultaneous Test Set III - Tube 1	Ketone
Drager Simultaneous Test Set III - Tube 2	Aromatic
Drager Simultaneous Test Set III - Tube 3	Alcohol
Drager Simultaneous Test Set III - Tube 3	Aldehyde
Drager Simultaneous Test Set III - Tube 3	Ester
Drager Simultaneous Test Set III - Tube 3	Ether
Drager Simultaneous Test Set III - Tube 3	Ketone
Drager Simultaneous Test Set III - Tube 4	Aliphatic
Drager Simultaneous Test Set III - Tube 4	Alkane
Drager Simultaneous Test Set III - Tube 4	Alkene
Drager Simultaneous Test Set III - Tube 4	Alkyne
Drager Simultaneous Test Set III - Tube 5	Chlorohydrocarbon
Drager Thioether	Sulfur mustard
Drager Thioether	Thioether
Drager Toluene Diisocyanate 0.02/A	TDI
Drager Toluene Diisocyanate 0.02/A	TDI
Drager Toluene Diisocyanate 0.02/A	Toluene diisocyanate
Drop-Ex Plus™	Bromate
Drop-Ex Plus™	Chlorate
Drop-Ex Plus™	Nitramine
Drop-Ex Plus™	Nitrate ester
Drop-Ex Plus™	Nitrate, Inorganic
Drop-Ex Plus™	Nitroaromatic
Drop-Ex Plus™	Peroxide
Duquenois-Levine Reagent, Modified	Cannabis
Duquenois-Levine Reagent, Modified	Hashish
Duquenois-Levine Reagent, Modified	Marijuana
Duquenois-Levine Reagent, Modified	Tetrahydrocannabinol

(*continued*)

Test	Substance
Duquenois-Levine Reagent, Modified	THC
Eriochromocyanin R	Aluminum
Ethyl-8-Hydroxytetrahydroquinoline HCl Reagent	Arsenic(III)
Ethyl-8-Hydroxytetrahydroquinoline HCl Reagent	Arsenite
Fehling's Solution	Aldehyde
Fehling's Solution	Fructose
Fehling's Solution	Glucose
Fehling's Solution	Reducing sugar
Ferric Chloride Reagent	Acetaminophen
Ferric Chloride Reagent	Chlorpromazine HCl
Ferric Chloride Reagent	Ferrocyanide
Ferric Chloride Reagent	Illegal drug
Ferric Chloride Reagent	Morphine
Ferric Chloride Reagent	Phenol
Ferric Chloride Reagent	Sodium bicarbonate
Ferric Ferricyanide Test	Alloys
Ferric Ferricyanide Test	Elemental metal
Ferric Ferricyanide Test	Metal
Ferric Solution Test	Thiocyanate
Ferric Thiocyanate Reagent	Hydrocarbon
Ferric Thiocyanate Reagent	Organic nitrogen
Ferric Thiocyanate Reagent	Organic oxygen
Ferric Thiocyanate Reagent	Organic sulfur
Ferricyanide Test	Ferricyanide
Ferrocyanide Test (Ferric Chloride)	Ferrocyanide
Ferrocyanide Test (Gutzeit Scheme)	Ferrocyanide
Ferrocyanide Test (Gutzeit Scheme)	Thiocyanate
Ferrocyanide Test (Gutzeit Scheme)	Thiosulfate
Ferrous Hydroxide Reagent	Alkyl nitrate
Ferrous Hydroxide Reagent	Alkyl nitrite
Ferrous Hydroxide Reagent	Explosive
Ferrous Hydroxide Reagent	Hydroxylamine
Ferrous Hydroxide Reagent	Nitro
Ferrous Hydroxide Reagent	Nitroso
Firearm Discharge Residue Test	Antimony
Firearm Discharge Residue Test	Barium
Firearm Discharge Residue Test	Lead
Fizz Test	Bicarbonate
Fizz Test	Carbonate
Fluorescamine Test for Protein	Protein

(continued)

Test	Substance
Fluorescein Reagent (Bromides)	Bromide
Fluorescein Reagent (Bromides)	Hydrobromic acid
Fluorescein Reagent (Bromides)	Hydroiodic acid
Fluorescein Reagent (Bromides)	Iodide
Fluorescein Reagent (Halogens)	Bromine
Fluorescein Reagent (Halogens)	Chlorine
Fluorescein Reagent (Halogens)	Iodine
Fluorescein Reagent (Uranyl)	Uranium
Fluorescein Reagent (Uranyl)	Uranyl
Fluoride Test Strip	Fluoride
Formaldoxime Reagent	Cobalt
Formaldoxime Reagent	Cupric
Formaldoxime Reagent	Ferric
Formaldoxime Reagent	Manganous
Formaldoxime Reagent	Nickel
Froede Reagent	Alkaloid
Froede Reagent	Aspirin
Froede Reagent	Codeine
Froede Reagent	LSD
Froede Reagent	Mace
Froede Reagent	Morphine
Gas Production With Sulfuric Acid	Acetate
Gas Production With Sulfuric Acid	Azide
Gas Production With Sulfuric Acid	Cyanide
Gas Production With Sulfuric Acid	Halide
Gas Production With Sulfuric Acid	Nitrite
Gas Production With Sulfuric Acid	Sulfide
Gastec Tube #180	Amine
Gold Chloride Reagent	Bromide
Gold Chloride Reagent	Iodide
Gold Chloride Reagent (Thiol)	Mercaptan
Gold Chloride Reagent (Thiol)	Mustard
Gold Chloride Reagent (Thiol)	Thiol
Griess' Paper (Yellow)	Nitrous acid
Griess Test	Nitrite
Griess Test	Nitrous acid
Guerin's Reagent	Selenium
Hearon-Gustavson's Reagents	Nitro
Hearon-Gustavson's Reagents	Organic nitro
Hearon-Gustavson's Reagents	Oxidizer

(*continued*)

Test	Substance
Hydrogen Peroxide Reagent	Chromate
Hydrogen Peroxide Reagent	Iron salt
Hydrogen Peroxide Reagent	Molybdate
Hydrogen Peroxide Reagent	Titanium(IV)
Hydrogen Peroxide Reagent	Vanadate
Hydroxylamine Hydrochloride Reagent	Amide
Hydroxylamine Hydrochloride Reagent	Cyanate
Hydroxylamine Hydrochloride Reagent	Cyanic acid
Hydroxylamine Hydrochloride Reagent	Nitrile
Hypohalogenite Differentiation Test	Hypobromite
Hypohalogenite Differentiation Test	Hypochlorite
Hypohalogenite Differentiation Test	Hypohalogenite
Hypohalogenite Differentiation Test	Hypoiodite
Hypohalogenite Test	Hypobromite
Hypohalogenite Test	Hypochlorite
Hypohalogenite Test	Hypohalogenite
Hypohalogenite Test	Hypoiodite
Indigo Carmine Reagent (Chlorate)	Chlorate
Indigo Carmine Reagent (Chromate)	Chromate
Iodic Acid Test 1	Iodate
Iodic Acid Test 1	Iodic acid
Iodic Acid Test 2	Iodate
Iodic Acid Test 2	Iodic acid
Iodine-Sulfide Reagent	Azide
Iodine-Sulfide Reagent	Hydrazoic acid
Isatin Reagent (Mercaptans)	Mercaptan
Isatin Reagent (Mercaptans)	Thioalcohol
ITS Cyanide ReagentStrip™	Cyanide
ITS Cyanide ReagentStrip™	Cyanogen chloride
ITS Cyanide ReagentStrip™	Thiocyanate
ITS SenSafe™ Water Metal Test	Arsenic
ITS SenSafe™ Water Metal Test	Cadmium
ITS SenSafe™ Water Metal Test	Chromium
ITS SenSafe™ Water Metal Test	Cobalt
ITS SenSafe™ Water Metal Test	Copper
ITS SenSafe™ Water Metal Test	Heavy metal
ITS SenSafe™ Water Metal Test	Lead
ITS SenSafe™ Water Metal Test	Mercury
ITS SenSafe™ Water Metal Test	Metal
ITS SenSafe™ Water Metal Test	Nickel

(continued)

Test	Substance
ITS SenSafe™ Water Metal Test	Zinc
Jaffe's Reagent	Antimony
Jaffe's Reagent	Bismuth
Janovsky Test	Dinitrotoluene
Janovsky Test	DNT
Janovsky Test	TNT
Janovsky Test	Trinitrotoluene
Lactase Test	Dairy
Lactase Test	Lactose
Lactase Test	Milk
Lead Acetate Test Strip (JD175)	Hydrogen sulfide
Lead Acetate Test Strip (JD175)	Sulfide
Lead Sulfide Paper	Alkali peroxide
Lead Sulfide Paper	Hydrogen peroxide
Lead(II) Test	Barium
Lead(II) Test	Cadmium
Lead(II) Test	Lead
Lead(II) Test	Silver
Lead(II) Test	Thallium
Lead(II) Test	Tin
Lowry Protein Test	Protein
Luff's Reagent	Fructose
Luff's Reagent	Glucose
Luff's Reagent	Reducing sugar
Lugol Starch Test	Polysaccharide
Lugol Starch Test	Starch
M256 Series Chemical Agent Detector Kit	Blister agent
M256 Series Chemical Agent Detector Kit	Blood agent
M256 Series Chemical Agent Detector Kit	Chemical warfare agent
M256 Series Chemical Agent Detector Kit	CX
M256 Series Chemical Agent Detector Kit	Lewisite
M256 Series Chemical Agent Detector Kit	Nerve agent
M256A1 Blister Agent Test	Blister agent
M256A1 Blister Agent Test	CX
M256A1 Blister Agent Test	H-Series blister agent
M256A1 Blister Agent Test	Nitrogen mustard
M256A1 Blister Agent Test	Phosgene oxime
M256A1 Blister Agent Test	Sulfur mustard
M256A1 Blood Agent Test	AC
M256A1 Blood Agent Test	Blood agent

(continued)

Test	Substance
M256A1 Blood Agent Test	CK
M256A1 Blood Agent Test	Cyanogen chloride
M256A1 Blood Agent Test	Hydrogen cyanide
M256A1 Lewisite Test	Lewisite
M256A1 Lewisite Test	Organometallic
M256A1 Nerve Agent Test	Carbamate pesticide
M256A1 Nerve Agent Test	G-nerve agent
M256A1 Nerve Agent Test	Nerve agent
M256A1 Nerve Agent Test	Novichok agent
M256A1 Nerve Agent Test	Organophosphate pesticide
M256A1 Nerve Agent Test	V-nerve agent
Mandelin Reagent	Acetaminophen
Mandelin Reagent	Alkaloid
Mandelin Reagent	Aspirin
Mandelin Reagent	Codeine
Mandelin Reagent	Illegal drug
Mandelin Reagent	Mace
Marquis Reagent	Amphetamine
Marquis Reagent	Aspirin
Marquis Reagent	Codeine
Marquis Reagent	Illegal drug
Marquis Reagent	Mace
Marquis Reagent	Salt
Mecke Reagent	Codeine
Mecke Reagent	Hydrocodone tartrate
Mecke Reagent	Illegal drug
Mecke Reagent	LSD
Mecke Reagent	Morphine
Melibiose Test	Bean
Melibiose Test	Melibiose
Melibiose Test	Vegetable
Merckoquant® Aluminum Test	Aluminum
Merckoquant® Ammonium Test	Ammonia
Merckoquant® Ammonium Test	Ammonium salt
Merckoquant® Arsenic Test	Arsenic
Merckoquant® Arsenic Test	Arsenic(III)
Merckoquant® Arsenic Test	Arsenic(V)
Merckoquant® Ascorbic Acid Test	Ascorbic acid
Merckoquant® Calcium Test	Cadmium
Merckoquant® Calcium Test	Calcium

(continued)

Test	Substance
Merckoquant® Calcium Test	Copper
Merckoquant® Calcium Test	Lead
Merckoquant® Chloride Test	Chloride
Merckoquant® Chlorine Tests 1.17924.0001, 1.17925.0001	Chlorine
Merckoquant® Chlorine Tests 1.17924.0001, 1.17925.0001	Oxidizer
Merckoquant® Chromate Test	Chromate
Merckoquant® Chromate Test	Chromium(III)
Merckoquant® Chromate Test	Chromium(VI)
Merckoquant® Chromate Test	Chromium, bound
Merckoquant® Chromate Test	Dichromate
Merckoquant® Cobalt Test	Cobalt
Merckoquant® Cobalt Test	Copper(II)
Merckoquant® Cobalt Test	Iron(III)
Merckoquant® Cobalt Test	Mercury(I)
Merckoquant® Copper Test	Copper
Merckoquant® Copper Test	Copper(I)
Merckoquant® Copper Test	Copper(II)
Merckoquant® Cyanide Test	Cyanide
Merckoquant® Fixing Bath Test	Silver
Merckoquant® Formaldehyde Test	Aldehyde
Merckoquant® Formaldehyde Test	Formaldehyde
Merckoquant® Glucose Test	Glucose
Merckoquant® Iron Test	Iron
Merckoquant® Iron Test	Iron(II)
Merckoquant® Iron Test	Iron(III)
Merckoquant® Lead Test	Lead
Merckoquant® Manganese Test	Manganese
Merckoquant® Molybdenum Test	Molybdenum
Merckoquant® Nickel Test	Cobalt
Merckoquant® Nickel Test	Copper
Merckoquant® Nickel Test	Mercury
Merckoquant® Nickel Test	Molybdenum
Merckoquant® Nickel Test	Nickel
Merckoquant® Nitrate Tests 1.10020.0001, 1.10020.0002	Nitrate
Merckoquant® Nitrate Tests 1.10020.0001, 1.10020.0002	Nitrite
Merckoquant® Nitrite Tests 1.10007.0001, 1.10007.0002	Nitrite

(continued)

Test	Substance
Merckoquant® Nitrite Tests 1.10022.0001	Nitrite
Merckoquant® Peracetic Acid Test	Peracetic acid
Merckoquant® Peroxide Test 1.10337.0001	Hydrogen peroxide
Merckoquant® Peroxide Test 1.10337.0001	Hydroperoxide
Merckoquant® Peroxide Test 1.10337.0001	Organic peroxide
Merckoquant® Peroxide Test 1.10337.0001	Peroxide
Merckoquant® Peroxide Tests 1.10011.0001, 1.10011.0002, 1.10081.0001	Hydrogen peroxide
Merckoquant® Peroxide Tests 1.10011.0001, 1.10011.0002, 1.10081.0001	Hydroperoxide
Merckoquant® Peroxide Tests 1.10011.0001, 1.10011.0002, 1.10081.0001	Organic peroxide
Merckoquant® Peroxide Tests 1.10011.0001, 1.10011.0002, 1.10081.0001	Peroxide
Merckoquant® Phosphate Test	Orthophosphate
Merckoquant® Phosphate Test	Phosphate
Merckoquant® Potassium Test	Potassium
Merckoquant® Quaternary Ammonium Compounds	Quat amine
Merckoquant® Quaternary Ammonium Compounds	Quaternary amine
Merckoquant® Quaternary Ammonium Compounds	Quaternary ammonium compound
Merckoquant® Sulfate Test	Sulfate
Merckoquant® Sulfite Test	Sulfite
Merckoquant® Tin Test	Tin
Merckoquant® Zinc Test	Cadmium
Merckoquant® Zinc Test	Copper
Merckoquant® Zinc Test	Lead
Merckoquant® Zinc Test	Mercury
Merckoquant® Zinc Test	Silver
Merckoquant® Zinc Test	Zinc
Methylene Blue Catalytic Reduction Test	Selenite
Methylene Blue Reagent	Chromate
Methylene Blue Reagent	Ferricyanate
Methylene Blue Reagent	Ferrocyanate
Methylene Blue Reagent	Mercurous chloride
Methylene Blue Reagent	Perchlorate
Methylene Blue Reagent	Thiocyanate
Michler's Thioketone	Bromine
Michler's Thioketone	Chlorine
Michler's Thioketone	Iodine
Michler's Thioketone	Lewisite
Michler's Thioketone	Nitrogen tetroxide

(continued)

Test	Substance
Michler's Thioketone	Nitrous acid
Molybdenum Metal Test	Molybdenum metal
Morin Reagent	Aluminum
Morin Reagent	Beryllium
Morin Reagent	Gallium
Morin Reagent	Indium
Morin Reagent	Scandium
Morin Reagent	Thorium
Naphthylenediamine Test	Nitrite
Naphthylenediamine Test	Nitrous acid
Naphthylenediamine Test	Selenite
Neothorin Reagent	Hafnium
Neothorin Reagent	Zirconium
Nickel Dimethylglyoxime Equilibrium Solution	Carbonates
Nickel Dimethylglyoxime Equilibrium Solution	Nickel
Nickel Dimethylglyoxime Equilibrium Solution	Organic base
Nickel Dimethylglyoxime Equilibrium Solution	Oxide
Nickel Dimethylglyoxime Equilibrium Solution	Phosphate
Nickel Dimethylglyoxime Equilibrium Solution	Protein
Nickel Dimethylglyoxime Equilibrium Solution	Silicate
Nickel Hydroxide Test	Bromite
Nickel Hydroxide Test	Chlorite
Nickel Hydroxide Test	Iodite
Nickel Hydroxide Test	Persulfate
Nickel Hydroxide Test	Persulfuric acid
Ninhydrin Solution	Adrenalin
Ninhydrin Solution	Amino acid
Ninhydrin Solution	Peptone
Ninhydrin Solution	Polypeptide
Ninhydrin Solution	Protein
Nitrate and Nitrite Test Strip (CZ NIT600)	Nitrate
Nitrate and Nitrite Test Strip (CZ NIT600)	Nitric acid
Nitrate and Nitrite Test Strip (CZ NIT600)	Nitrite
Nitrate and Nitrite Test Strip (CZ NIT600)	Nitrous acid
Nitrate Test (Ammonia)	Nitrate
Nitrate Test (Gutzeit Scheme)	Nitrate
Nitric Acid - Silver Nitrate Solution	Azide
Nitric Acid - Silver Nitrate Solution	Bromide
Nitric Acid - Silver Nitrate Solution	Chloride
Nitric Acid - Silver Nitrate Solution	Cyanide

(continued)

Test	Substance
Nitric Acid - Silver Nitrate Solution	Iodide
Nitric Acid - Silver Nitrate Solution	Thiocyanide
Nitric Acid Reagent	Heroin
Nitric Acid Reagent	Illegal drug
Nitric Acid Reagent	Mescaline
Nitric Acid Reagent	Morphine
Nitric Acid Reagent	Oxycodone
Nitric Acid Reagent	Protein
Nitrite Test (Gutzeit Scheme)	Nitrite
Nitrocellulose Test	Nitrocellulose
Nitrocellulose Test	Nitrous acid
Nitron Solution	Nitrate
Nitrophenylhydrazine Solution	Aldehyde
Nitrophenylhydrazine Solution	Carbonyl
Nitrophenylhydrazine Solution	Ketone
Nitroso-R-Salt Solution	Cobalt
Nitrotoluene Test	Dinitrotoluene
Nitrotoluene Test	DNT
Nitrotoluene Test	Explosive
Nitrotoluene Test	Nitrotoluene
Nitrotoluene Test	TNT
Nitrotoluene Test	Trinitrotoluene
Offord's Reagent	Chlorate
Offord's Reagent	Cobalt
Offord's Reagent	Cupric
Offord's Reagent	Ferric
Offord's Reagent	Nickel
Orange IV Solution	Zinc
Palladium Chloride Paper	Amalgam
Palladium Chloride Paper	Mercury vapor
Palladous Chloride Reagent	Hydroiodic acid
Palladous Chloride Reagent	Iodide
Permanganate Test	Permanganate
Peroxide (10%) Test Strip	Hydrogen peroxide
Peroxide Test Strip (PER 400)	Hydrogen peroxide
Peroxide Test Strip (PER 400)	Hydroperoxide
Peroxide Test Strip (PER 400)	Organic peroxide
Peroxide Test Strip (PER 400)	Perborate
Peroxide Test Strip (PER 400)	Peroxide
pH Test For Solids	Ammonium

(*continued*)

Test	Substance
pH Test For Solids	Carbonate
pH Test For Solids	Hydroxide
pH Test For Solids	pH
pH Test For Solids	Sulfide
Phenylanthranilic Acid Test	Chlorate
Phenylanthranilic Acid Test	Chloride
Phenylanthranilic Acid Test	Perchlorate
Phenylenediamine Reagent	Aldehyde
Phenylenediamine Reagent	Carbonyl group
Phenylenediamine Reagent	Ketone
Phenylenediamine Reagent	Ozone
Polynitroaromatic Screening Test	Explosive
Polynitroaromatic Screening Test	Nitrite
Polynitroaromatic Screening Test	Polynitroaromatic
Potassium Bismuth Iodide Reagent	Cesium
Potassium Bismuth Iodide Reagent	Thallium
Potassium Chlorate Reagent	Osmium
Potassium Chlorate Reagent	Ruthenium
Potassium Chromate Reagent	Silver
Potassium Chromate Reagent	Silver salt
Potassium Ethyl Xanthate Reagent	Molybdate
Potassium Iodide Reagent (Gold)	Gold(III)
Potassium Iodide Reagent (Gold)	Palladium
Potassium Iodide Reagent (Gold)	Platinum(IV) salt
Potassium Iodide Reagent (Thallium)	Lead
Potassium Iodide Reagent (Thallium)	Mercury
Potassium Iodide Reagent (Thallium)	Silver
Potassium Iodide Reagent (Thallium)	Thallium(I)
Potassium Iodide Starch Test Strip (JD260)	Chlorine
Potassium Iodide Starch Test Strip (JD260)	Hydrogen peroxide
Potassium Iodide Starch Test Strip (JD260)	Iodine
Potassium Iodide Starch Test Strip (JD260)	Oxidizer
Potassium Iodide Starch Test Strip (JD260)	Ozone
Potassium Permanganate Solution	Alkene
Potassium Permanganate Solution	Alkyne
Potassium Permanganate Solution	Thiocyanate
Povidone Test for Starch	Polysaccharide
Povidone Test for Starch	Starch
Procke-Uzel's Reagent	Lithium
Prussian Blue Test	Cyanide

(continued)

Test	Substance
Prussian Blue Test	Ferrocyanide
Pyridine Reagent	Auric
Pyridine Reagent	Gold
Pyridylazo Resorcinol (PAR) Reagent	Niobium
Pyridylazo Resorcinol (PAR) Reagent	Uranium(VI)
Quinalizarin Reagent (Aluminum)	Aluminum
Quinalizarin Reagent (Aluminum)	Aluminum salt
Quinalizarin Reagent (Aluminum)	Beryllium
Quinalizarin Reagent (Magnesium)	Beryllium salt
Quinalizarin Reagent (Magnesium)	Cerium salt
Quinalizarin Reagent (Magnesium)	Lanthanum salt
Quinalizarin Reagent (Magnesium)	Magnesium
Quinalizarin Reagent (Magnesium)	Magnesium salt
Quinalizarin Reagent (Magnesium)	Neodymium salt
Quinalizarin Reagent (Magnesium)	Praseodymium salt
Quinalizarin Reagent (Magnesium)	Thorium salt
Quinalizarin Reagent (Magnesium)	Zirconium salt
Quinizarin Reagent	Gallium(III)
Quinizarin Reagent	Indium(III)
Quinizarin Reagent	Thallium(III)
Raikow's Reagent	Mercaptan
Raikow's Reagent	Sulfur, organic
Raikow's Reagent	Thiol
RDX Test (J-Acid)	C-4
RDX Test (J-Acid)	Cyclonite
RDX Test (J-Acid)	Cyclotrimethylenetrinitramine
RDX Test (J-Acid)	Cyclotrimethylenetrinitramine
RDX Test (J-Acid)	Formaldehyde
RDX Test (J-Acid)	Hexogen
RDX Test (J-Acid)	RDX
RDX Test (Pyrolytic Oxidation)	C-4
RDX Test (Pyrolytic Oxidation)	Cyclonite
RDX Test (Pyrolytic Oxidation)	Cyclotrimethylenetrinitramine
RDX Test (Pyrolytic Oxidation)	Hexogen
RDX Test (Pyrolytic Oxidation)	RDX
RDX Test (Pyrolytic Oxidation)	T4
RDX Test (Thymol)	C-4
RDX Test (Thymol)	Cyclonite
RDX Test (Thymol)	Cyclotrimethylenetrinitramine
RDX Test (Thymol)	Hexogen

(continued)

Test	Substance
RDX Test (Thymol)	HMX
RDX Test (Thymol)	RDX
Reducing Substance Test	Alcohol
Reducing Substance Test	Aldehyde
Reducing Substance Test	Carbohydrate
Reducing Substance Test	Dimethylsulfoxide
Reducing Substance Test	DMSO
Reducing Substance Test	Reducer
Resorcinol Reagent	Zinc
Rhodamine B Reagent (Antimony)	Antimony
Rhodamine B Reagent (Antimony)	Bismuth chloride
Rhodamine B Reagent (Antimony)	Gold
Rhodamine B Reagent (Antimony)	Mercury chlorides
Rhodamine B Reagent (Antimony)	Molybdate salt
Rhodamine B Reagent (Antimony)	Thallium
Rhodamine B Reagent (Antimony)	Tungstate salt
Rhodamine B Reagent (Gallium)	Antimony
Rhodamine B Reagent (Gallium)	Gallium
Rhodamine B Reagent (Gallium)	Gold
Rhodamine B Reagent (Gallium)	Thallium(III)
Rubeanic Acid (Cobalt)	Cobalt
Rubeanic Acid (Cobalt)	Copper
Rubeanic Acid (Cobalt)	Nickel
Rubeanic Acid (Ruthenium)	Palladium salt
Rubeanic Acid (Ruthenium)	Platinum salt
Rubeanic Acid (Ruthenium)	Ruthenium salt
Rubeanic Acid Reagent	Copper
Rubeanic Acid Reagent	Ferric
Rubeanic Acid Reagent	Mercurous
Rubeanic Acid Reagent	Palladium
Rubeanic Acid Reagent	Platinum
Rubeanic Acid Reagent	Silver
Safranin O Solution	Hypobromite
Safranin O Solution	Hypobromous acid
Safranin O Solution	Hypochloric acid
Safranin O Solution	Hypochlorite
Safranin O Solution	Hypohalogenous acid
Safranin O Solution	Hypoiodite
Safranin O Solution	Hypoiodous acid
Safranin T Solution	Nitrate

(continued)

Test	Substance
Safranin T Solution	Nitrite
Schiff's Reagent	Aldehyde
Schiff's Reagent	Ketone
Sensidyne Deluxe Haz Mat III Tube 131	Acetic acid
Sensidyne Deluxe Haz Mat III Tube 131	Amine
Sensidyne Deluxe Haz Mat III Tube 131	Ammonia
Sensidyne Deluxe Haz Mat III Tube 131	Arsine
Sensidyne Deluxe Haz Mat III Tube 131	Bromine
Sensidyne Deluxe Haz Mat III Tube 131	Chlorine
Sensidyne Deluxe Haz Mat III Tube 131	Hydrazine
Sensidyne Deluxe Haz Mat III Tube 131	Hydrogen bromide
Sensidyne Deluxe Haz Mat III Tube 131	Hydrogen chloride
Sensidyne Deluxe Haz Mat III Tube 131	Hydrogen cyanide
Sensidyne Deluxe Haz Mat III Tube 131	Hydrogen sulfide
Sensidyne Deluxe Haz Mat III Tube 131	Phosphine
Sensidyne Deluxe Haz Mat III Tube 131	Sulfur dioxide
Sensidyne Deluxe Haz Mat III Tube 131	Sulfur trioxide
Sensidyne Test #165SA	Acetylene
Sensidyne Test #165SA	Ethyl mercaptan
Sensidyne Test #165SA	Hydrogen sulfide
Sensidyne Test #165SA	Methyl mercaptan
Sensidyne Tube #101S	Acetylene
Sensidyne Tube #101S	Ammonia
Sensidyne Tube #101S	Butadiene
Sensidyne Tube #101S	Carbon disulfide
Sensidyne Tube #101S	Chlorine
Sensidyne Tube #101S	Hydrogen
Sensidyne Tube #102SC	Acetone
Sensidyne Tube #102SC	Acrolein
Sensidyne Tube #102SC	Aldehyde
Sensidyne Tube #102SC	Ketone
Sensidyne Tube #102SD	Acetone
Sensidyne Tube #102SD	Alcohol
Sensidyne Tube #102SD	Alkane
Sensidyne Tube #102SD	Aromatic
Sensidyne Tube #102SD	Ester
Sensidyne Tube #102SD	Ketone
Sensidyne Tube #105SB	Amine
Sensidyne Tube #105SB	Ammonia

(*continued*)

Test	Substance
Sensidyne Tube #105SB	Hydrazine
Sensidyne Tube #105SH	Ammonia
Sensidyne Tube #109SB	Bromine
Sensidyne Tube #109SB	Chlorine
Sensidyne Tube #109SB	Chlorine dioxide
Sensidyne Tube #109SB	Nitrogen dioxide
Sensidyne Tube #109SB	Nitrogen trichloride
Sensidyne Tube #111U	Acetate
Sensidyne Tube #111U	Alcohol
Sensidyne Tube #111U	Ester
Sensidyne Tube #111U	Ethyl acetate
Sensidyne Tube #111U	Hydrocarbon (larger aliphatic)
Sensidyne Tube #111U	Ketone
Sensidyne Tube #112SA	Acetone
Sensidyne Tube #112SA	Carbon disulfide
Sensidyne Tube #112SA	Dicyanide
Sensidyne Tube #112SA	Hydrogen cyanide
Sensidyne Tube #112SA	Hydrogen sulfide
Sensidyne Tube #112SA	Sulfur dioxide
Sensidyne Tube #112SB	Hydrogen cyanide
Sensidyne Tube #112SB	Hydrogen sulfide
Sensidyne Tube #112SB	Sulfur dioxide
Sensidyne Tube #114	Bromine
Sensidyne Tube #114	Chlorine
Sensidyne Tube #114	Chlorine dioxide
Sensidyne Tube #114	Nitrogen dioxide
Sensidyne Tube #118SB	Benzene
Sensidyne Tube #120SB	Hydrogen sulfide
Sensidyne Tube #121U	Arsine
Sensidyne Tube #121U	Diborane
Sensidyne Tube #121U	Hydrogen cyanide
Sensidyne Tube #121U	Hydrogen selenide
Sensidyne Tube #121U	Mercaptan
Sensidyne Tube #121U	Phosphine
Sensidyne Tube #126SA	Carbon dioxide
Sensidyne Tube #126SA	Hydrogen cyanide
Sensidyne Tube #126SA	Hydrogen sulfide
Sensidyne Tube #126SA	Sulfur dioxide
Sensidyne Tube #126UH	Carbon dioxide

(continued)

Test	Substance
Sensidyne Tube #128SC	Acrylonitrile
Sensidyne Tube #128SC	Hydrogen cyanide
Sensidyne Tube #128SC	Nitrile
Sensidyne Tube #133A	Acetaldehyde
Sensidyne Tube #133A	Acetone
Sensidyne Tube #133A	Acrolein
Sensidyne Tube #133A	Methyl ethyl ketone
Sensidyne Tube #133A	Methyl isobutyl ketone
Sensidyne Tube #133A	Sulfur dioxide
Sensidyne Tube #134SB	Dichloroethylene
Sensidyne Tube #134SB	Halohydrocarbon
Sensidyne Tube #134SB	Tetrachloroethylene
Sensidyne Tube #134SB	Trichloroethylene
Sensidyne Tube #134SB	Vinyl chloride
Sensidyne Tube #137U	Alcohol
Sensidyne Tube #137U	Hydrogen
Sensidyne Tube #141SB	Acid
Sensidyne Tube #141SB	Carbon disulfide
Sensidyne Tube #141SB	Hydrogen sulfide
Sensidyne Tube #141SB	Sulfur dioxide
Sensidyne Tube #142S	Hydrogen sulfide
Sensidyne Tube #142S	Mercury
Sensidyne Tube #142S	Nitrogen dioxide
Sensidyne Tube #147S	Carbon tetrachloride
Sensidyne Tube #147S	Chlorohydrocarbons
Sensidyne Tube #147S	Chloropicrin
Sensidyne Tube #147S	Phosgene
Sensidyne Tube #156S	Acid
Sensidyne Tube #156S	Chlorine
Sensidyne Tube #156S	Hydrogen chloride
Sensidyne Tube #156S	Hydrogen fluoride
Sensidyne Tube #157SB	Halogen
Sensidyne Tube #157SB	Halohydrocarbon
Sensidyne Tube #157SB	Methyl bromide
Sensidyne Tube #165SB	Arsine
Sensidyne Tube #165SB	Ethyl mercaptan
Sensidyne Tube #165SB	Hydrogen cyanide
Sensidyne Tube #165SB	Hydrogen selenide
Sensidyne Tube #165SB	Mercaptan

(continued)

Test	Substance
Sensidyne Tube #165SB	Phosphine
Sensidyne Tube #167S	Arsine
Sensidyne Tube #167S	Hydrogen selenide
Sensidyne Tube #167S	Hydrogen sulfide
Sensidyne Tube #167S	Mercury
Sensidyne Tube #167S	Nickel carbonyl
Sensidyne Tube #167S	Sulfur dioxide
Sensidyne Tube #171SB	Acetaldehyde
Sensidyne Tube #171SB	Formaldehyde
Sensidyne Tube #171SB	Monomer
Sensidyne Tube #171SB	Styrene
Sensidyne Tube #174A	Chlorine
Sensidyne Tube #174A	Nitrogen dioxide
Sensidyne Tube #174A	Nitrogen monoxide
Sensidyne Tube #174A	Nitrogen oxide
Sensidyne Tube #181S	Aniline
Sensidyne Tube #181S	Toluidine
Sensidyne Tube #186	Alcohol
Sensidyne Tube #186	Aromatic
Sensidyne Tube #186	Hydrocarbon
Sensidyne Tube #186	Hydrogen sulfide
Sensidyne Tube #186	Ketone
Sensidyne Tube #186	Organic gas
Sensidyne Tube #190U	Alcohol
Sensidyne Tube #190U	Butanol
Sensidyne Tube #190U	Toluene
Sensidyne Tube #214S	Alcohol
Sensidyne Tube #214S	Alkane
Sensidyne Tube #214S	Aromatic
Sensidyne Tube #214S	Dichlorobenzene
Sensidyne Tube #214S	Ester
Sensidyne Tube #214S	Halohydrocarbon
Sensidyne Tube #215S	Dichlorobenzene
Sensidyne Tube #216S	Acetic acid
Sensidyne Tube #216S	Chlorine
Sensidyne Tube #216S	Hydrogen chloride
Sensidyne Tube #216S	Nitrogen dioxide
Sensidyne Tube #216S	Sulfur dioxide
Sensidyne Tube #232SA	Aldehyde

(continued)

Test	Substance
Sensidyne Tube #232SA	Ethylene glycol
Sensidyne Tube #232SA	Ethylene oxide
Sensidyne Tube #232SA	Hydrogen sulfide
Sensidyne Tube #232SA	Sulfur dioxide
Sensidyne Tube #235S	Dichloroethane
Sensidyne Tube #235S	Halogen
Sensidyne Tube #235S	Halohydrocarbon
Sensidyne Tube #235S	Nitrogen oxide
Sensidyne Tube #239S	Carbon disulfide
Sensidyne Tube #239S	Carbonyl sulfide
Sensidyne Tube #239S	Hydrogen sulfide
Sensidyne Tube #239S	Sulfur dioxide
Sensidyne Tube #247S	Hydrogen peroxide
Sensidyne Tube 107SA	Acetylene
Sensidyne Tube 107SA	Diethyl ether
Sensidyne Tube 183U	Amine
Sensidyne Tube 183U	Ammonia
Sensidyne Tube 183U	Cresol
Sensidyne Tube 183U	Phenol
Sensidyne Tube 282S	Hydrogen sulfide
Sensidyne Tube 282S	Hydrogen sulfide
Sensidyne Tube 282S	Mercaptan
Sensidyne Tube 282S	Mercaptan
Silver Ferrocyanide Paper	Bromide
Silver Ferrocyanide Paper	Chloride
Silver Ferrocyanide Paper	Iodide
Silver Nitrate Paper	Arsenic
Silver Nitrate Paper	Arsenite
Silver Nitrate Paper	Chromate
Silver Nitrate Paper	Phosphorus
Silver Nitrate Reagent	Chloride
Silver Nitrate Reagent	Reactive organic chlorine
Simon's Reagent	Amphetamine
Simon's Reagent	Ecstasy
Simon's Reagent	MDMA HCl
Simon's Reagent	Methamphetamine HCl
Simon's Reagent	Methylenedioxy-N-methamphetamine
Simon's Reagent	Methylphenidate HCl
Sodium Arsenite Reagent	Lithium

(continued)

Test	Substance
Sodium Azide - Iodine Solution	Sulfide
Sodium Azide - Iodine Solution	Thiocyanate
Sodium Azide - Iodine Solution	Thiocyanuric acid
Sodium Azide - Iodine Solution	Thiosulfate
Sodium Azide - Iodine Solution	Thiosulfuric acid
Sodium Nitroprusside Paper	Hydrogen sulfide
Sodium Nitroprusside Paper	Mercaptan
Sodium Nitroprusside Paper	Sulfur
Sodium Nitroprusside Paper	Sulfur dioxide
Sodium Rhodizonate Solution (Barium)	Barium
Sodium Rhodizonate Solution (Barium)	Strontium
Sodium Rhodizonate Solution (Calcium)	Calcium salt
Sodium Sulfide Reagent	Polysulfide
Sodium Sulfide Reagent	Sulfur, elemental
Sodium Thiosulfate Reagent	Cobalt
Stannous Chloride Reagent (Molybdate)	Molybdate
Stannous Chloride Reagent (Molybdate)	Tungstate
Stannous Chloride Reagent (Rhodium)	Rhodium salt
Stannous Chloride Reagent (Technetium)	Technetium
Sucrose Test	Sucrose
Sucrose Test	Sugar
Sulfanilic Acid Reagent	Bromate
Sulfanilic Acid Reagent	Bromic acid
Sulfate Test	Sulfate
Sulfide Test	Sulfide
Sulfite Test	Sulfite
Sulfite Test	Sulfurous acid
Sulfur Reagent	Hydrocarbon
Sulfur Reagent	Hydrogen (organic)
Sulfur Reagent	Organic compound
Sulfuric Acid Test	Sulfuric acid
TATB Test	Explosive
TATB Test	TATB
TATB Test	Triaminotrinitrobenzene
Thiocyanate Conversion Test	Thiocyanate
Thiocyanate Solution Test (Cobalt)	Cobalt
Thiocyanate Solution Test (Cobalt)	Copper(II)
Thiocyanate Solution Test (Cobalt)	Iron(III)
Trinitrophenylmethylnitramine Test	Cyclonite

(continued)

Test	Substance
Trinitrophenylmethylnitramine Test	RDX
Trinitrophenylmethylnitramine Test	Tetryl
Trinitrophenylmethylnitramine Test	Trinitro-1,3,5-triazine
Trinitrophenylmethylnitramine Test	Trinitrophenylmethylnitramine
Trinitrotoluene Test	Dinitrotoluene
Trinitrotoluene Test	DNT
Trinitrotoluene Test	Explosive
Trinitrotoluene Test	TNT
Trinitrotoluene Test	Trinitrotoluene
Triphenylmethylarsonium Reagent	Antimony
Triphenylmethylarsonium Reagent	Barium
Triphenylmethylarsonium Reagent	Lead
Universal pH Test Strip	Acid
Universal pH Test Strip	Base
Universal pH Test Strip	Caustic
Universal pH Test Strip	Corrosive
Unsaturated Hydrocarbon Test	Aliphatic amine
Unsaturated Hydrocarbon Test	Alkene
Unsaturated Hydrocarbon Test	Alkyne
Unsaturated Hydrocarbon Test	Methyl ethyl ketone
Unsaturated Hydrocarbon Test	Phenol
Unsaturated Hydrocarbon Test	Unsaturated hydrocarbon
Van Eck's Reagent	Chromic acid
Van Eck's Reagent	Chromium(VI)
Villiers-Fayolle's Reagent	Chlorine
Water Test	Water
Watersafe® Lead Test	Lead
Zirconium-Alizarin Reagent	Fluoride
Zirconium-Alizarin Reagent	Hydrofluoric acid
Zwikker Reagent	Barbiturate
Zwikker Reagent	Pentobarbital
Zwikker Reagent	Phenobarbital
Zwikker Reagent	Secobarbital
Zwikker Reagent	Tea
Zwikker's Reagent	Chromate
Zwikker's Reagent	Perchlorate
Zwikker's Reagent	Permanganate
Zwikker's Reagent	Persulfate
Zwikker's Reagent	Thiosulfate

Agri-Screen Ticket	
Detects	**Test Principle**
• Acetylcholinesterase-inhibiting compound • Carbamate pesticide • Nerve agent • Organophosphate pesticide • Thiocarbamate pesticide	Acetylcholinesterase-inhibiting materials deactivate acetylcholinesterase, blocking the conversion of indoxyl acetate to a blue-colored reaction product.

	Test Phase	
	✓	Solid
	✓	Liquid
	✓	Gas/Vapor

Sensitivity
7 ppm carbaryl; 3 ppm DDVP; 2 ppm diazinon in water

Directions and Comments

A capsule of 2% bromine solution ("bromine water," CAS 7726-95-6) that is included with the test is added to the sample only to create a detectable derivative if testing for thiocarbamates; this step is not necessary for other pesticides and nerve agents. An absorbent surface coated with stabilized eel acetylcholinesterase (CAS 9000-81-1) is soaked with an aqueous sample for 1 minute. The test ticket is folded so that another pad, impregnated with indoxyl acetate (CAS 608-08-2), touches the acetylcholinesterase pad. If active acetylcholinesterase remains, the indoxyl acetate will be converted to a blue reaction product. If the acetylcholinesterase on the pad has been poisoned by a nerve agent or acetylcholinesterase-inhibiting pesticide, the blue reaction product will not form and the test surface color will remain white. This test is less expensive and less rugged than the M256 kit. The manufacturer places a 2-year expiration date on the test when stored at room temperature. It is estimated the test would degrade in a week if stored in a vehicle without air conditioning in the summer. The test is adaptable to vapor testing (moisten the pad, collect vapors for 10 minutes and perform the test) and wipe testing (wipe a surface with a moistened pad and perform the test). See color images 2.1, 2.2, and 2.3.

References	**Manufacturer**
Goodson, L., et al., United States Patent 4324858, April 13, 1982. Lupo, T., Notes from correspondence, Neogen Corporation, Lansing, Michigan, December 8, 2006.	Neogen Corporation, Lansing, Michigan, USA (www.neogen.com)

Alizarin Reagent (Metal)	
Detects	**Test Principle**
• Aluminum hydroxide • Chromic hydroxide • Ferric hydroxide • Manganese hydroxide • Uranium hydroxide	Precipitated aluminum hydroxide absorbs alizarin to form a red lake.

	Test Phase	
		Solid
	✓	Liquid
		Gas/Vapor

Sensitivity
4 µg/ml aluminum in the presence of 800 µg/ml iron; 0.5 µg/ml aluminum in the presence of 100 µg/ml uranium.

(*continued*)

Alizarin Reagent (Metal) (*Continued*)

Directions and Comments

Dissolve 10 mg alizarin (CAS 72-48-0) in 100 ml of 95% alcohol. The test will not detect aluminum directly; it detects aluminum hydroxide. Aluminum metal must be oxidized to Al_2O_3, which is then dissolved in sodium hydroxide solution to form aluminum hydroxide. Aluminum hydroxide is precipitated from solution by adding acid until the solution is neutral or slightly acidic. A small piece of precipitate is isolated. A drop of the alizarin solution is added to the precipitate; a red color indicates aluminum. Ferric hydroxide produces a violet lake; chromic hydroxide produces a maroon lake; uranium and manganese hydroxides may also form colored lakes. Interfering metals, such as ferric and chromium ions, are separated on filter paper with 5% potassium ferrocyanide (CAS 14459-95-1), which traps the metals on paper and allows the aluminum to migrate with water to the outside of the ring where it is detected. Alizarin is also used as a pH indicator; yellow 5.5–6.8 red. Compare to the Alizarin S Test. Zirconium may interfere; compare to Alizarin Reagent (Zirconium).

References	Manufacturer
Welcher, F., *Chemical Solutions: Reagents Useful to the Chemist, Biologist, and Bacteriologist,* D. Van Nostrand Company, Inc., New York, 1942. *Chem. Abstracts* 15, 2598 (1921). Feigl, F. and Anger, V., *Spot Tests in Inorganic Analysis,* Sixth Edition, Elsevier Publishing Company, Amsterdam, 1972.	

Alizarin Reagent (Zirconium)

Detects	Test Principle	
• Zirconium	Zirconium and several metals react with alizarin to form a red to violet color. Only the zirconium retains color after treatment with hydrochloric acid.	
	Test Phase	
		Solid
	✓	Liquid
		Gas/Vapor
	Sensitivity	
	0.5 µg/ml zirconium	

Directions and Comments

Dissolve 1 mg alizarin (CAS 72-48-0) in 10 ml of 95% alcohol. Dissolve the sample in water and assure the pH is very close to neutral. Add 0.5 ml of test solution and 0.5 ml of sample solution to a test tube. Heat to boiling and then cool. Zirconium and other metals produce a red to violet color; low concentrations may form red flakes. Add 1 ml of 1N hydrochloric acid. Only the zirconium compound will remain unaffected.

References	Manufacturer
Feigl, F. and Anger, V., *Spot Tests in Inorganic Analysis,* Sixth Edition, Elsevier Publishing Company, Amsterdam, 1972.	

Alizarin S Reagent		
Detects	**Test Principle**	
• Aluminum	Aluminum reacts with an aqueous solution of alizarin S to form a red lake in ammonia fume.	
	Test Phase	
		Solid
	✓	Liquid
		Gas/Vapor
	Sensitivity	
	0.67 μg/ml aluminum	

Directions and Comments	
Mix a 0.5% (w/v) sodium alizarinsulfonate (CAS 130-22-3) in water. Moisten filter paper with the solution and then add a drop of sample and hold the paper in ammonia fume. A bright red lake indicates aluminum, but the result is dependent on pH and concentration. It is preferable to add the test solution directly to a slightly alkaline sample suspected of containing aluminum followed by addition of acetic acid; serial dilutions may be necessary since this alternate test does not indicate high concentration of aluminum. Manganese, chromium, cobalt, and iron interfere due to the color of their solutions. Alizarin S is also used as a pH indicator; yellow 3.7–5.2 violet. Compare to Alizarin Reagent (Metals), which uses a different reagent.	

References	Manufacturer
Welcher, F., *Chemical Solutions: Reagents Useful to the Chemist, Biologist, and Bacteriologist,* D. Van Nostrand Company, Inc., New York, 1942. *Chem. Abstracts* 9, 3186 (1915); 24, 1818 (1930). *J. Soc. Chem. Ind.,* 34, 936 (1915). *Mikrochemie,* 7, 221 (1929).	

Allylthiourea Reagent		
Detects	**Test Principle**	
• Cadmium	Cadmium forms a yellow precipitate with allylthiourea in basic solution.	
	Test Phase	
	✓	Solid
	✓	Liquid
		Gas/Vapor
	Sensitivity	
	Not available	

Directions and Comments
The test solution is a 5% (w/v) solution of allylthiourea (CAS 109-57-9) in water. On a spot plate add one drop of sample, a drop of test solution or a few milligrams of solid, and then a drop of 30% (w/v) sodium hydroxide. A yellow precipitate forms more so in the presence of cadmium. Simultaneously perform the test in a second spot plate but replace the sample with a drop of distilled water. Compare the test to the blank since the blank may form a yellow color, but more slowly than the cadmium solution. Copper does not interfere.

(continued)

Allylthiourea Reagent (*Continued*)	
References	**Manufacturer**
Welcher, F., *Chemical Solutions: Reagents Useful to the Chemist, Biologist, and Bacteriologist*, D. Van Nostrand Company, Inc., New York, 1942. *Chemical Abstracts* 23, 4644 (1929).	

Aloy's Reagent		
Detects	**Test Principle**	
• Alkaloid • Morphine • Phenols	Uranium precipitates alkaloids and other compounds. Morphine additionally reduces the uranium to produce a unique red color.	
	Test Phase	
	✓	**Solid**
	✓	**Liquid**
		Gas/Vapor
	Sensitivity	
	0.1% phenol	

Directions and Comments

Dissolve "a little" uranium nitrate (CAS 10102-06-4) or uranium acetate (CAS 541-09-3) in water and add directly to a drop of liquid sample or a few milligrams of solid sample. Alkaloids will precipitate; morphine will produce a bright red color. The test is an early presumptive test for morphine. Phenol compounds can produce a red color in a basic environment. If the test solution is formed from 1% aqueous uranium acetate and 0.1% formaldehyde and then mixed with concentrated sulfuric acid, codeine base or salt as well as ethyl morphine will form a blue color; morphine yields a violet color.

References	**Manufacturer**
Welcher, F., *Chemical Solutions: Reagents Useful to the Chemist, Biologist, and Bacteriologist*, D. Van Nostrand Company, Inc., New York, 1942. *Chemical Abstracts* 23, 4644 (1929). *Bull. Soc. Chim.* 29, 610 (1903). *Chemical Abstracts* 21, 2358 (1921).	

Aluminon Reagent		
Detects	**Test Principle**	
• Aluminum • Beryllium • Cerium • Iron • Thorium • Zirconium	Aluminum salts react with aluminon to form a red lake.	
	Test Phase	
		Solid
	✓	Liquid
		Gas/Vapor
	Sensitivity	
	Less than 10 µg Al^{3+}/ml	

(*continued*)

Aluminon Reagent (*Continued*)

Directions and Comments

Mix a 0.1% (w/v) aqueous solution of aluminon (ammonium salt of aurintricarboxylic acid, CAS 569-58-4). The presence of aluminum and several other metals are indicated by the formation of a red lake. The test can be made more specific by buffering the sample to pH 4.5–5.5 with ammonium acetate. Heat the sample gently (do not boil) and raise the pH to 7.1–7.3 with a solution of ammonium hydroxide and ammonium carbonate. If these extra steps are performed, only iron will commonly interfere. Uncommon metals that can form similar lakes are the hydroxides or basic acetates of beryllium, yttrium, lanthanum, cerium, neodymium, europium, zirconium, and thorium. Scandium can possibly interfere.

References	Manufacturer
Welcher, F., *Chemical Solutions: Reagents Useful to the Chemist, Biologist, and Bacteriologist*, D. Van Nostrand Company, Inc., New York, 1942. *J. Am. Chem. Soc.* 47, 142 (1925). *Official and Tentative Methods of Analysis of the Association of Agricultural Chemists,* Fourth Edition, Association of Agricultural Chemists, Washington, DC, 1935. Middleton, A., Journal of the American Chemical Society 48, 2125–2126 (1925).	

Aluminon Reagent on Paper

Detects	Test Principle		
• Aluminum	Aluminum(III) ions react with aluminon to form a red color.		
	Test Phase		
			Solid
	✓		Liquid
			Gas/Vapor
	Sensitivity		
	0.16 µg/ml aluminum(III)		

Directions and Comments

Mix an aqueous solution (w/v) of 0.1% aluminon (aurintricarboxylic acid ammonium salt, CAS 569-58-4) and 1% ammonium acetate (CAS 631-61-8). Moisten filter paper with the test solution and set aside. The sample is prepared by dissolving aluminum(III) salts in acid; aluminum metal must first be dissolved in sodium hydroxide or ammonium hydroxide and then acidified. Add a drop of test solution to the filter paper. Red or other colors may indicate aluminum or other metals. Hold the paper over ammonia fume; the red color of aluminum will remain while color caused by interfering metals will fade.

References	Manufacturer
Feigl, F. and Anger, V., *Spot Tests in Inorganic Analysis,* Sixth Edition, Elsevier Publishing Company, Amsterdam, 1972. *J. Chem. Education,* 14, 281 (1937).	

Alvarez's Reagent

Detects	Test Principle
• Nitrate • Nitrite	Nitrate or nitrite reacts with diphenylamine and resorcinol in concentrated sulfuric acid to form various colors upon manipulation.

(continued)

Alvarez's Reagent (*Continued*)		
	Test Phase	
	✓	Solid
	✓	Liquid
		Gas/Vapor
	Sensitivity	
	Not available, but the test was originally designed for use on a visible crystal.	

Directions and Comments
Prepare the test solution in proportions of 1 g diphenylamine (CAS 122-39-4), 1 g resorcinol (CAS 108-46-3), and 100 ml concentrated sulfuric acid. Add a few drops of the test solution to a crystal of the sample. Nitrate will produce a yellow-green color with a blue margin; adding alcohol forms a yellow to orange solution. Nitrite will produce a dark blue color with a red margin; adding alcohol forms a red solution. Liquid samples may be tested, but the colors of the margins will not be apparent and individual colors less discernable.

References	Manufacturer
Welcher, F., *Chemical Solutions: Reagents Useful to the Chemist, Biologist, and Bacteriologist,* D. Van Nostrand Company, Inc., New York, 1942. *Chem. News* 91, 155 (1905).	

Aminoacetophenone Reagent		
Detects	**Test Principle**	
• Cerium(IV) • Palladium salt	4′-Aminoacetophenone reacts with cerium(IV) in acid solution to form a purple color that slowly changes to a red-brown precipitate.	
	Test Phase	
		Solid
	✓	Liquid
		Gas/Vapor
	Sensitivity	
	1 μg/ml cerium(IV)	

Directions and Comments
Mix a 0.01M solution of 4′-aminoacetophenone (CAS 99-92-3) in 0.01M hydrochloric acid. The sample is dissolved in 1–5M hydrochloric or sulfuric acid. The test solution is added to a small amount of sample solution. A purple color that develops immediately and then slowly fades to red-brown indicates cerium(IV). Palladium salts produce similar results.

References	Manufacturer
Feigl, F. and Anger, V., *Spot Tests in Inorganic Analysis,* Sixth Edition, Elsevier Publishing Company, Amsterdam, 1972. Somidevamma, G. and Sarma, M., *Chemist-Analyst*, 56, 92 (1967).	

Aminoacetophenone Solution	
Detects	**Test Principle**
• Palladium	p-Aminoacetophenone solution reacts with palladium to form a yellow turbidity.

	Test Phase	
		Solid
✓		Liquid
		Gas/Vapor

Sensitivity
0.005 mg palladium chloride in 5 ml water

Directions and Comments
Dissolve 1 g p-aminoacetophenone (CAS 99-92-3) in 40 ml of warm 0.6N hydrochloric acid. Heat gently if needed and after cooling dilute to 100 ml with water. The sample must be in aqueous solution for this test. Add equal amounts of test solution and sample solution. A yellow precipitate or turbidity is indicative of palladium. The test is specific to palladium.

References	**Manufacturer**
Welcher, F., *Chemical Solutions: Reagents Useful to the Chemist, Biologist, and Bacteriologist*, D. Van Nostrand Company, Inc., New York, 1942. *Mikrochemie* 24, 20 (1938).	

Aminodimethylaniline Reagent	
Detects	**Test Principle**
• Bromine • Chlorine • Halogen • Iodine	p-Aminodimethylaniline reacts with free halogen in water to form a pink color at varying sensitivities.

	Test Phase	
		Solid
✓		Liquid
		Gas/Vapor

Sensitivity
Chlorine 1:65,000. Bromine 1:1,300,000. Iodine 1:400,000.

Directions and Comments
Dissolve 5 g p-aminodimethylaniline (CAS 99-98-9) in 100 ml of alcohol. Place three drops of the solution in 5 ml of water. Slowly add the diluted test solution dropwise to the sample. A clear pink color indicates the threshold of detection for free chlorine, bromine, or iodine.

References	**Manufacturer**
Welcher, F., *Chemical Solutions: Reagents Useful to the Chemist, Biologist, and Bacteriologist*, D. Van Nostrand Company, Inc., New York, 1942. *Ind. Eng. Chem., Anal. Ed.* 225 (1931).	

Aminodiphenylamine Sulfate Reagent

Detects	Test Principle	
• Benzoyl peroxide	Di-p-aminodiphenylamine sulfate reacts with benzoyl peroxide to form a green color.	
	Test Phase	
	✓	Solid
	✓	Liquid
		Gas/Vapor
	Sensitivity	
	100 ppm benzoyl peroxide in flour	

Directions and Comments

Mix 1 g of di-p-aminodiphenylamine sulfate (CAS 4477-28-5) in 100 ml of alcohol. It may help to crush the powder into a small amount of alcohol before diluting to 100 ml. The mixture must be refluxed for 30 minutes. Storage time is not provided, but the mixture must be shaken before use. The test was originally developed to detect benzoyl peroxide in flour, so the instructions call for 0.7 g of flour to be shaken in 2.5 ml petroleum ether and then mixed with 1 ml of the test solution followed by shaking and a few minutes of standing. A green color indicates the presence of benzoyl peroxide. For general peroxide detection refer to Merckoquant® Peroxide Test 1.10337.0001.

References	Manufacturer
Welcher, F., *Chemical Solutions: Reagents Useful to the Chemist, Biologist, and Bacteriologist*, D. Van Nostrand Company, Inc., New York, 1942. *Chemical Abstracts* 19, 740 (1925).	

Aminophenol Hydrochloride Solution

Detects	Test Principle	
• Copper • Iron	p-Aminophenol hydrochloride solution reacts with copper and iron to form a blue-violet precipitate.	
	Test Phase	
		Solid
	✓	Liquid
		Gas/Vapor
	Sensitivity	
	0.0002 mg copper; 0.000069 mg iron.	

Directions and Comments

Dissolve 3 g of p-aminophenol hydrochloride (CAS 51-78-5) in 120 ml of alcohol. Add the test solution dropwise to a liquid sample containing copper or iron ions. A blue-violet precipitate indicates the presence of copper or iron. The test is specific; other ions do not interfere.

References	Manufacturer
Welcher, F., *Chemical Solutions: Reagents Useful to the Chemist, Biologist, and Bacteriologist*, D. Van Nostrand Company, Inc., New York, 1942. *Mikrochemie* 17, 118 (1935).	

Ammoniacal Silver Oxide Solution		
Detects	**Test Principle**	
• Cerium(III) • Chromium(III) • Cobalt(II) • Iron(III) • Manganese(II) • Thallium(I)	Ammoniacal silver oxide solution reacts with thallium(I), manganese(II), iron(III), cobalt(II), cerium(III), chromium(III) salts in the absence of sulfide ion to produce a black precipitate.	
	Test Phase	
	✓	Solid
	✓	Liquid
		Gas/Vapor
	Sensitivity	
	1 µg/ml cerium	

Directions and Comments

The test forms black precipitates with ammoniacal silver oxide solution and thallium(I), manganese(II), iron(III), cobalt(II), cerium(III), chromium(III) salts in the absence of sulfide ion. The absence of sulfide ion can be confirmed by adding acid and testing for hydrogen sulfide gas with a lead acetate test strip. The ammoniacal silver oxide solution is prepared by carefully adding strong sodium hydroxide solution to a 2% aqueous solution of silver nitrate (CAS 7761-88-8) which will form a brown precipitate. Ammonium hydroxide solution is added dropwise until the solution clears. A small amount of this solution is added to the solid or liquid sample. A black precipitate indicates a positive result.

References	Manufacturer
Feigl, F., *Spot Tests in Inorganic Analysis,* Fifth Edition, Elsevier Publishing Company, New York, 1958.	

Ammonium Chloride Reagent		
Detects	**Test Principle**	
• Thiocyanate	Ammonium chloride heated to 200–300°C with thiocyanate will produce hydrogen sulfide.	
	Test Phase	
	✓	Solid
	✓	Liquid
		Gas/Vapor
	Sensitivity	
	10 µg/ml potassium thiocyanate	

Directions and Comments

Place 0.1 g of solid sample or two drops of liquid sample in a test tube with 0.1 g of ammonium chloride (CAS 12125-02-9). Suspend a strip of lead acetate paper over or in the opening of the test tube. Heat the base of the tube gently to 200–300°C. The appearance of a brown to black spot on the lead acetate paper from hydrogen sulfide production indicates thiocyanate in the sample. See color images 2.4 and 2.5.

References	Manufacturer
Feigl, F. and Anger, V., *Spot Tests in Inorganic Analysis,* Sixth Edition, Elsevier Publishing Company, Amsterdam, 1972. *Mikrochim. Acta* 85 (1954).	

Ammonium Dithiocarbamate Solution	
Detects	**Test Principle**
• Aluminum • Cobalt • Lead • Metal • Silver • Uranium	Ammonium dithiocarbamate is added to aqueous solutions containing metal ions resulting in the formation of various colors.

Test Phase	
	Solid
✓	Liquid
	Gas/Vapor

Sensitivity
Various

Directions and Comments

Ammonium dithiocarbamate (CAS 513-74-6) is available commercially and its use is recommended; however, there is no guidance on preparing the reagent for use. (Ammonium dithiocarbamate is prepared, according to the reference, by adding 35 ml of concentrated ammonium hydroxide to 10 ml of carbon disulfide and warming slightly. After standing, the solution is decanted from the unchanged carbon disulfide.) When a few drops of test solution are added to solutions containing metal salts, the following colors or precipitates form: cobalt (green); lead (red); silver (black); zinc (white); aluminum (white); copper (brown); uranium (light yellow); iron (red to red-brown); tin (orange-red); nickel (bright red); bismuth (orange-yellow); manganese (yellow slowly changing to brown). Although the test uses hazardous chemicals, it is a powerful tool to rapidly sort metal ions in certain applications.

References	**Manufacturer**
Welcher, F., *Chemical Solutions: Reagents Useful to the Chemist, Biologist, and Bacteriologist*, D. Van Nostrand Company, Inc., New York, 1942. *Chemical Abstracts* 18, 3569 (1924).	

Ammonium Molybdate Reagent	
Detects	**Test Principle**
• Arsenic • Vanadium	Ammonium molybdate reacts with arsenic to form arsenomolybdate, which is then reduced to molybdenum blue.

Test Phase	
✓	**Solid**
✓	**Liquid**
	Gas/Vapor

Sensitivity
Not available, but used as a colorimetric titration test.

Directions and Comments

Dissolve 2.5 g ammonium molybdate tetrahydrate (CAS 12054-85-2) in 100 ml water and mix with the sample. Add stannous chloride pentahydrate (CAS 10026-06-9) or other suitable reducing agent. The appearance of a blue color indicates the presence of arsenic or vanadium. Molybdenum blue is from several complex cluster formations and is obtained by reduction of acidified molybdate(VI) solutions. Basic sample solutions should be made slightly acidic. Compare to Ammonium Molybdate Solution (Deniges).

(continued)

Ammonium Molybdate Reagent (*Continued*)	
References	Manufacturer
Welcher, F., *Chemical Solutions: Reagents Useful to the Chemist, Biologist, and Bacteriologist*, D. Van Nostrand Company, Inc., New York, 1942. *Ind. Eng. Chem., Anal. Ed.* 1 136–139 (1929). *J. Am. Chem. Soc.* 23, 105 (1901).	

Ammonium Molybdate Solution (Deniges)		
Detects	Test Principle	
• Phosphate • Arsenate	Ammonium molybdate reacts with phosphate or arsenate to form an intermediate product, which is then reduced to molybdenum blue.	
	Test Phase	
	✓	Solid
	✓	Liquid
		Gas/Vapor
	Sensitivity	
	1 mg/l phosphate; 2 mg/l arsenate	

Directions and Comments

Add 50 ml concentrated sulfuric acid to a 10% (w/v) solution of ammonium molybdate tetrahydrate (CAS 12054-85-2). Add a few drops of the test solution to 5 ml of the sample solution or 0.5 g of solid sample and then add a few drops of aqueous stannous chloride pentahydrate (CAS 10026-06-9) solution. A blue color indicates phosphate or arsenate. Molybdenum blue is from several complex cluster formations and is obtained by reduction of acidified molybdate(VI) solutions. Basic sample solutions not made slightly acidic by the sulfuric acid should be adjusted to a slightly acidic pH. Compare to Ammonium Molybdate Reagent.

References	Manufacturer
Welcher, F., *Chemical Solutions: Reagents Useful to the Chemist, Biologist, and Bacteriologist*, D. Van Nostrand Company, Inc., New York, 1942. *Chemical Abstracts* 15, 218 (1921).	

Ammonium Test		
Detects	Test Principle	
• Ammonium	Ammonium salts in aqueous solution react with the addition of strong base to form ammonia gas.	
	Test Phase	
	✓	Solid
	✓	Liquid
		Gas/Vapor
	Sensitivity	
	2 ppm ammonia with Drager Ammonia 2/a	

(*continued*)

Ammonium Test (*Continued*)
Directions and Comments
Ammonium is soluble in water. The addition of sodium hydroxide causes a shift in equilibrium and ammonia gas forms. Gentle heating of the liquid will decrease ammonia solubility and increase ammonia production. Ammonia gas is lighter than air and is detected in air above the liquid with a pH test strip wetted with deionized water. A basic pH will be indicated. Low concentration of ammonia can be detected with a colorimetric tube, such as Drager Ammonia 2/a. Volatile amines will cause the pH strip to indicate base with increasing production of vapor during heating. Heating of the solution caused by addition of a sodium hydroxide pellet can be reduced if a 6M solution of sodium hydroxide is used instead. Cyanides will also produce ammonia when warmed with alkalis.

References	Manufacturer
Petrucci, R. and Wismer, R., *General Chemistry with Qualitative Analysis,* Second Edition, Macmillan Publishing Company, New York, USA, 1987.	

Ammonium Test (Gutzeit Scheme)		
Detects	**Test Principle**	
• Amine • Ammonium	Ammonium reacts with Nessler's reagent to form yellow-brown $Hg_2I_3NH_2$.	
	Test Phase	
	✓	Solid
	✓	Liquid
		Gas/Vapor
	Sensitivity	
	Not available	
Directions and Comments		
In this spot test a small amount of sample is heated in a test tube with a small piece of solid sodium hydroxide. Black filter paper moistened with Nessler's reagent (dipotassium tetraiodomercurate(II) in dilute sodium hydroxide, CAS 7783-33-7) is held over the test tube and ammonia is detected by a yellow color. Alternatively, a wetted pH strip can be used to detect basic ammonia vapor at reduced sensitivity. Amines and other basic vapors produce a basic result on a pH strip.		

References	Manufacturer
Gutzeit, G., *Helv. Chim. Acta*, 12, 829 (1929).	

Aniline Reagent (Shear)		
Detects	**Test Principle**	
• Vitamin D	Vitamin D oil reacts with aniline and hydrochloric acid to form a red color.	
	Test Phase	
	✓	Solid
	✓	Liquid
		Gas/Vapor
	Sensitivity	
	Not available	

(*continued*)

Aniline Reagent (Shear) (*Continued*)

Directions and Comments

Add 1 ml of hydrochloric acid to 5 ml of aniline (CAS 62-53-3). Mix the test reagent and sample in equal volumes as liquid. If the sample is solid, add the test solution to a small amount of sample. The production of a red color upon standing indicates vitamin D. One reference directs the test to be boiled for 30 seconds and allowed to stand.

References	Manufacturer
Welcher, F., *Chemical Solutions: Reagents Useful to the Chemist, Biologist, and Bacteriologist*, D. Van Nostrand Company, Inc., New York, 1942. *Proc. Soc. Exptl. Biol. Med.* 23, 546 (1926). Shear, M. J., *Proceedings of the Society for Experimental Biology and Medicine* 23, 546-549 (1926).	

Aniline Sulfate Reagent

Detects	Test Principle	
• Chlorate	Aniline sulfate reacts with chlorates in the presence of sulfuric acid to form a blue color.	
	Test Phase	
	✓	Solid
	✓	Liquid
		Gas/Vapor
	Sensitivity	
	Not available	

Directions and Comments

Dissolve 0.67 g aniline sulfate (CAS 542-16-5) in 10 ml water. Alternately, 4.5 ml aniline (CAS 62-53-3) may be mixed with 8.5 ml 6N sulfuric acid and then diluted with 100 ml water. Place the liquid sample or a small amount of solid sample in a spot plate. Add a drop of test solution followed by a drop of concentrated sulfuric acid. A blue color indicates the presence of chlorate.

References	Manufacturer
Welcher, F., *Chemical Solutions: Reagents Useful to the Chemist, Biologist, and Bacteriologist*, D. Van Nostrand Company, Inc., New York, 1942. Treadwell, E. and Hall, W., *Analytical Chemistry*, Seventh Edition, Wiley, New York, 1930.	

Aniline-Phenol Reagent

Detects	Test Principle	
• Nitrite	Aniline-phenol mixture forms an intense yellow color when mixed with alkaline nitrite solution.	
	Test Phase	
	✓	Solid
	✓	Liquid
		Gas/Vapor
	Sensitivity	
	Not available	

(*continued*)

Aniline-Phenol Reagent (*Continued*)
Directions and Comments
Add 1 ml of aniline (CAS 62-53-3) and 1 g of phenol (CAS 108-95-2) to 15 ml of concentrated hydrochloric acid. Dilute with 150 ml water. Add a few drops of test solution to the sample. An intense yellow color indicates nitrite.

References	Manufacturer
Welcher, F., *Chemical Solutions: Reagents Useful to the Chemist, Biologist, and Bacteriologist*, D. Van Nostrand Company, Inc., New York, 1942. *Chemist Analyst*, J.T. Baker, January 1933.	

Anisaldehyde Reagent (Minovici)	
Detects	**Test Principle**
• Picrotoxin	Picrotoxin ($C_{30}H_{34}O_{13}$) consists of picrotoxinin ($C_{15}H_{16}O_6$) and picrotin ($C_{15}H_{18}O_7$) and when treated with sulfuric acid reacts with anisaldehyde to form a blue color.

Test Phase	
✓	Solid
✓	Liquid
	Gas/Vapor

Sensitivity
1:5000 picrotoxin

Directions and Comments
Dissolve 1 g p-anisaldehyde (CAS 123-11-5) in 5 ml absolute alcohol (CAS 64-17-5). Treat the sample with two drops of sulfuric acid and allow the solution to react for 1 minute. Add one drop of the test solution. An indigo-blue color that gradually changes to blue indicates picrotoxin. The test is made more sensitive if performed at 80°C, but very low concentrations of picrotoxin may appear as a pale red color. Picrotoxin is a toxin found in *Anamirta cocculus* seeds. It is used as a central nervous system stimulant, antidote, convulsant, and gamma aminobutyric acid (GABA) antagonist.

References	Manufacturer
Welcher, F., *Chemical Solutions: Reagents Useful to the Chemist, Biologist, and Bacteriologist*, D. Van Nostrand Company, Inc., New York, 1942. *Zeitschr. Untersuch. Nahr.-u. Genussm.* 687 (1900).	

Antimony Trichloride Reagent	
Detects	**Test Principle**
• Tachysterol • Vitamin D • Vitamin D2 • Vitamin D3	Vitamin D reacts with antimony trichloride in chloroform to produce an orange-yellow color.

(*continued*)

Antimony Trichloride Reagent (*Continued*)

	Test Phase	
	✓	Solid
	✓	Liquid
		Gas/Vapor
	Sensitivity	
	Not available	

Directions and Comments

Mix a saturated solution of about 25 g of antimony trichloride (CAS 10025-91-9) in anhydrous chloroform (CAS 67-66-3). Add the test solution to the sample. A yellow-orange color indicates vitamin D2, vitamin D3, or tachysterol. The test was an approved method for vitamin D detection in food. It is not clear if the test works in nonfood environment.

References	Manufacturer
Welcher, F., *Chemical Solutions: Reagents Useful to the Chemist, Biologist, and Bacteriologist*, D. Van Nostrand Company, Inc., New York, 1942. Jacobs, M., *The Chemical Analysis of Foods and Food Products,* D. Van Nostrand, New York, 1938.	

Aqua Regia Reagent

Detects	Test Principle	
• Gold	Gold and platinum are soluble in aqua regia.	
• Platinum	**Test Phase**	
	✓	Solid
		Liquid
		Gas/Vapor
	Sensitivity	
	Not available	

Directions and Comments

Aqua regia (CAS 8007-56-5) consists of one volume of concentrated nitric acid (CAS 7697-37-2) and three volumes of concentrated hydrochloric acid (CAS 7647-01-0). Aqua regia will dissolve gold and platinum while other acids will not.

References	Manufacturer
Welcher, F., *Chemical Solutions: Reagents Useful to the Chemist, Biologist, and Bacteriologist*, D. Van Nostrand Company, Inc., New York, 1942. *Handbook of Chemistry and Physics,* 24th Edition, The Chemical Rubber Publishing Company, Cleveland, 1941.	

Arnold-Mentzel's Solution

Detects	Test Principle
• Hydrogen peroxide	Vanadic acid and sulfuric acid react with hydrogen peroxide to form a red color.

(*continued*)

Arnold-Mentzel's Solution (*Continued*)

	Test Phase	
		Solid
	✓	Liquid
		Gas/Vapor
	Sensitivity	
	6 ppm hydrogen peroxide	

Directions and Comments

Dissolve 1 g vanadic acid (sodium metavanadate, CAS 13718-26-8) in 100 g dilute sulfuric acid. Mix a drop of test solution with a drop of sample solution. The appearance of a red color indicates hydrogen peroxide.

References	Manufacturer
Welcher, F., *Chemical Solutions: Reagents Useful to the Chemist, Biologist, and Bacteriologist*, D. Van Nostrand Company, Inc., New York, 1942. Leach and Winton, *Food Inspection and Analysis,* Fourth Edition, Wiley, New York, 1920.	

Arnold's Base

Detects	Test Principle	
• Gold • Mercury • Palladium • Silver	Arnold's base reacts with gold, silver, mercury, or palladium to form a blue to purple color in mildly acidic solution.	
	Test Phase	
	✓	Solid
	✓	Liquid
		Gas/Vapor
	Sensitivity	
	Not available	

Directions and Comments

Dissolve 0.5 g Arnold's base (4,4′-methylenebis[N,N-dimethylaniline], CAS 101-61-1) and 1 g of citric acid (CAS 77-92-9) in 100 ml water. Add a few drops of test solution to the sample. A blue to violet color indicates gold, silver, mercury, and/or palladium.

References	Manufacturer
Welcher, F., *Chemical Solutions: Reagents Useful to the Chemist, Biologist, and Bacteriologist*, D. Van Nostrand Company, Inc., New York, 1942. *J. Am. Chem. Soc.* 34, 32 (1912). *Chemical Abstracts* 25, 2380 (1931).	

Arsenazo I Reagent

Detects	Test Principle
• Dysprosium • Europium	Arsenazo I dyestuff reacts with all rare earth metals to produce a red color.

(continued)

Arsenazo I Reagent (*Continued*)

Detects	Test Phase	
• Gadolinium • Lutetium • Neodymium • Praseodymium • Rare earth metal • Samarium • Terbium		Solid
	✓	Liquid
		Gas/Vapor

Test Phase

	Solid
✓	Liquid
	Gas/Vapor

Sensitivity

0.2 µg/ml rare earth salt

Directions and Comments

Mix a 0.05% aqueous solution of arsenazo I (2-[1,8-Dihydroxy-3,6-disulfo-2-naphthylazo]benzenearsonic acid trisodium salt, CAS 66019-20-3). Mix the sample in water and adjust pH to nearly neutral. Add a drop of sample solution to a spot plate and then add a drop of test solution. A red color indicates a rare earth salt containing dysprosium, europium, gadolinium, lutetium, neodymium, praseodymium, samarium, or terbium. Thorium (violet at pH 8), titanium, uranium (blue at pH 8), and zirconium interfere. Aluminum interferes but can be avoided by first treating the sample with sulfosalicylic acid (CAS 5965-83-3).

References	Manufacturer
Feigl, F. and Anger, V., *Spot Tests in Inorganic Analysis,* Sixth Edition, Elsevier Publishing Company, Amsterdam, 1972.	

Arsine Generation Test

Detects	Test Principle
• Alkaline earth sulfate • Arsenic • Sulfide • Sulfur	Elemental arsenic reacts with hydrogen to form arsine, which is detected with a test paper. Hydrogen is supplied through the thermal decomposition of sodium formate in the presence of sodium hydroxide.

Test Phase

✓	Solid
✓	Liquid
	Gas/Vapor

Sensitivity

Not available, but lead acetate test strips detect 1–2 ppm hydrogen sulfide.

Directions and Comments

The test uses about 0.25 g each of powdered sodium hydroxide and sodium formate (CAS 141-53-7) with a similar amount of sample in a test tube. The opening of the tube is covered with a paper strip moistened with a 5% solution of silver nitrate (CAS 7761-88-8) and the base of the tube is heated. Hydrogen will become available to react with arsenic above 205°C. A black color on the strip indicates arsenic, free sulfur, sulfide, or alkaline earth sulfate. To determine the sulfur compounds, repeat the test but use a moistened lead acetate test strip; a black color indicates one or more of the sulfur compounds. The absence of a black color on the lead acetate strip and the presence of a black color on the silver nitrate strip indicate elemental arsenic. The test is only reliable for free arsenic, not necessarily other bound forms of arsenic such as arsenic(III), arsenic(V), arsenate, or arsenite.

References	Manufacturer
Vournasos, A.C., *Ber.*, 43, 2264 (1910). Feigl, F., *Spot Tests in Inorganic Analysis,* Fifth Edition, Elsevier Publishing Company, New York, 1958.	

Aurantia Solution	
Detects	**Test Principle**
• Potassium	Potassium will form an orange-red precipitate with a solution of aurantia dye and sodium carbonate.

Test Phase	
✓	Solid
✓	Liquid
	Gas/Vapor

Sensitivity
0.003 mg potassium

Directions and Comments

The reference states to mix 1 g aurantia dye (dipicrylamine, CAS 131-73-7) and 10 ml of 1N sodium carbonate solution (CAS 497-19-8) to 100 ml water and boil until dissolved. Caution: dipicrylamine is a shock-sensitive explosive material. The aqueous solution is stable but the pure material should not be used as a field reagent. Mix the test solution with the sample. An orange-red precipitate forms in the presence of potassium. Ammonium ion interferes, but alkali and alkaline earth metal ions do not.

References	**Manufacturer**
Welcher, F., *Chemical Solutions: Reagents Useful to the Chemist, Biologist, and Bacteriologist*, D. Van Nostrand Company, Inc., New York, 1942. *Mikrochemie* 14, 265 (1934). *Chemical Abstracts* 28, 2642 (1934).	

Ball's Reagent	
Detects	**Test Principle**
• Cesium • Rubidium	Cesium or rubidium forms a yellow precipitate with sodium nitrite and bismuth nitrate.

Test Phase	
✓	Solid
✓	Liquid
	Gas/Vapor

Sensitivity
Not available

Directions and Comments

Dissolve 50 g of sodium nitrite (CAS 7632-00-0) in 100 ml water and make slightly acidic if necessary with nitric acid. Add 10 g bismuth nitrate (CAS 10035-06-0), shake, and filter. Acidify again with nitric acid prior to use. Add the test solution to the sample. A yellow precipitate indicates the presence of cesium or rubidium.

References	**Manufacturer**
Welcher, F., *Chemical Solutions: Reagents Useful to the Chemist, Biologist, and Bacteriologist*, D. Van Nostrand Company, Inc., New York, 1942. *J. Chem. Soc.* 95, 2126 (1909).	

Beilstein Test	
Detects	**Test Principle**
<ul><li>Alkyl halide</li><li>Chlorinated hydrocarbon</li><li>Halogenated hydrocarbon</li><li>Urea</li><li>Pyridine</li><li>Oxyquinoline</li></ul>	Organic molecules containing chlorine, bromine, or iodine react with copper oxide in a torch flame to produce volatile copper halides that will display a characteristic green or blue flame.

Test Phase	
✓	Solid
✓	Liquid
✓	Gas/Vapor

Sensitivity
Various. Halogen is detected in 2 mg of sample using the Dijkstra method.

Directions and Comments
Heat a thin copper wire (or a thicker wire hammered flat) in a torch flame until contaminants burn away. Continue to heat the wire in the hottest part of the flame (the tip of the inner, light blue flame) until the wire does not discolor the flame. A coating of copper oxide will form. Remove the copper and allow it to cool. Apply the sample to the cooled copper and place it in the hottest part of the flame. A green, blue-green, or dark blue flame is caused by the presence of a copper halide, thus indicating organo chlorine, bromine, or iodine. The test does not work on organofluorine material. Interference is caused by some nitrogen-containing material such as urea, pyridine, and oxyquinolines, acid amides and cyano compounds, which will form volatile copper cyanide and produce a blue to green flame color. The presence of some acids, especially halogen acids, produces a green flame color. Acids and some nitrogen compounds (e.g., urea), can be screened by pH. If the test material is a polychloroarsene, highly toxic chloro-dioxins may be formed. A modification of this test by Dijkstra calls for about 2 mg of sample to be absorbed into activated charcoal. The charcoal is placed in a small glass tube and air is blown through the tube and into the torch flame ahead of the copper wire. Nonvolatile materials may be heated in the glass tube and forced into the airstream. Alternately, soak a cotton swab with liquid sample and hold it near the torch air intake while holding the copper wire in the flame. See color images 2.6 and 2.7.

References	**Manufacturer**
Beilstein, F., "Ueber den Nachweis von Chlor, Brom und Jod in organischen Substanzen". *Ber. Dtsch. Chem. Ges.* 5, 620–621 (1872). Dijkstra, D.W., *Chem. Weekblad*, 34, 351 (1937). Feigl, F., *Spot Tests in Inorganic Analysis,* Fifth Edition, Elsevier Publishing Company, New York, 1958. Scholz-Böttcher, B.M., *Müfit Bahadir,* Henning Hopf, 1992. "The Beilstein Test: An Unintentional Dioxin Source in Analytical and Research Laboratories". Angewandte Chemie International Edition in English 31, 443–444.	

Benedict's Solution (Carbonyl)	
Detects	**Test Principle**
<ul><li>α-Hydroxy aldehyde</li><li>α-Hydroxy ketone</li><li>α-Keto aldehyde</li></ul>	Cupric ions in a citrate complex are reduced by certain carbonyl-containing substances.

(continued)

Benedict's Solution (Carbonyl) (*Continued*)

• Carbonyl group	**Test Phase**	
• Sugar	✓	Solid
	✓	Liquid
		Gas/Vapor
	Sensitivity	
	0.01% glucose	

Directions and Comments

Benedict's solution is prepared by mixing 1.73 g of sodium citrate (CAS 6132-04-3) and 1 g of anhydrous sodium carbonate (CAS 497-19-8, 2.7 g of sodium carbonate decahydrate [CAS 6132-02-1] may be substituted) in 8 ml of water. Heat and stir until dissolved. Mix 0.173 g of copper sulfate (CAS 7758-99-8), in 1 ml water until dissolved. Slowly mix the two solutions while stirring and then add water to dilute the solution to 10 ml. Dissolve 0.2 g of sample in 5 ml of water and then add 5 ml of the previously mixed Benedict's solution. A red, yellow, or yellow-green color indicates a positive result. Heat to boiling if no color forms spontaneously. The colors produced depend on the nature and amount of reducing agent present. The copper in Benedict's solution selectively oxidizes α-hydroxy aldehydes, α-hydroxy ketones, and α-keto aldehydes. It is not a general reagent for oxidizing simple aliphatic or aromatic aldehydes. Sulfur compounds interfere. Some sugars require treatment with hydrochloric acid and neutralization with sodium hydroxide before testing with Benedict's solution.

References	Manufacturer
Shriner, R., Fuson, R. and Curtin, D., *The Systematic Identification of Organic Compounds, A Laboratory Manual,* Fifth Edition, John Wiley and Sons, Inc., New York, 1964. *J. Chem. Educ.,* 37, 205 (1960).	

Benedict's Solution (Sulfur)

Detects	**Test Principle**	
• Mercaptans • Organic sulfur	Cupric ions in a citrate complex are reduced by certain sulfur-containing substances to produce various reaction products.	
	Test Phase	
		Solid
	✓	Liquid
		Gas/Vapor
	Sensitivity	

Directions and Comments

Benedict's solution is prepared by mixing 1.73 g of sodium citrate (CAS 6132-04-3) and 1 g of anhydrous sodium carbonate (CAS 497-19-8, 2.7 g of sodium carbonate decahydrate [CAS 6132-02-1] may be substituted) in 8 ml of water. Heat and stir until dissolved. Mix 0.173 g of copper sulfate (CAS 7758-99-8), in 1 ml water until dissolved. Slowly mix the two solutions while stirring and then add water to dilute the solution to 10 ml. Dissolve 0.2 g of sample in 5 ml of water and then add 5 ml of Benedict's solution. Heat to boiling for 1 minute if no color forms spontaneously. Aliphatic and aromatic mercaptans produce a greasy yellow ball upon boiling. Tertiary alkyl mercaptans and compounds containing sulfides or disulfide groups produce a blue color. Thioglycolic acid, mercaptoethanol, thiosalicylic acid, dialkyl sulfones, diaryl sulfones, and sulfonates do not react. Complex biological compounds may undergo extensive degradation and produce black cupric sulfide.

(*continued*)

Benedict's Solution (Sulfur) (*Continued*)

References	Manufacturer
Shriner, R., Fuson, R. and Curtin, D., *The Systematic Identification of Organic Compounds, A Laboratory Manual,* Fifth Edition, John Wiley and Sons, Inc., New York, 1964. *J. Chem. Educ., 37,* 205 (1960).	

Benzoin Reagent

Detects	Test Principle		
• Boric acid • Boric salts	Benzoin in alkali solution reacts with boric acid or boric salts to form a green-yellow fluorescent color in UV light.		
	Test Phase		
			Solid
		✓	Liquid
			Gas/Vapor
	Sensitivity		
	0.04 µg/ml boric acid		

Directions and Comments

Mix a 0.5% alcoholic solution of benzoin (CAS 119-53-9). Dissolve the sample in 0.5N sodium hydroxide and place a drop in a test tube. Evaporate gently to dryness with a torch. Allow to cool and add a drop of water followed by two drops of benzoin solution. A green-yellow fluorescence under UV light indicates boric acid or boric salts. Bromides, chlorides, iodides, carbonates, cyanides, ferrocyanides, thiocyanides, nitrates, nitrites, phosphates, sulfides, sulfates, and sulfites do not interfere. Oxidizers interfere and may be eliminated by adding excess sulfite before testing.

References	Manufacturer
Feigl, F. and Anger, V., *Spot Tests in Inorganic Analysis,* Sixth Edition, Elsevier Publishing Company, Amsterdam, 1972.	

Benzopurpurin 4 B Reagent

Detects	Test Principle		
• Mercuric • Mercurous • Mercury • Silver	Benzopurpurin 4 B reacts with silver, mercurous, or mercuric ion to form a colored ring around a red spot.		
	Test Phase		
			Solid
		✓	Liquid
			Gas/Vapor
	Sensitivity		
	0.03 mg silver; 0.04 mercurous; 0.015 mercuric ions.		

(continued)

Benzopurpurin 4 B Reagent (*Continued*)

Directions and Comments

Dissolve 0.2 g benzopurpurin 4 B (CAS 992-59-6) in 100 ml water. Place a drop of sample solution in a spot plate and add a drop of test solution. A brown ring surrounding a red spot indicates silver ion. A blue-gray ring indicates mercuric ion; a red-violet ring indicates mercurous ion. Benzopurpurin 4 B is also a pH indicator; blue-violet 1.3–4.0 red.

References	Manufacturer
Welcher, F., *Chemical Solutions: Reagents Useful to the Chemist, Biologist, and Bacteriologist*, D. Van Nostrand Company, Inc., New York, 1942. *Chemical Abstracts* 34, 51 (1940).	

Bettendorf's Test

Detects	Test Principle	
• Arsenic • Arsenate • Arsenic(III) • Arsenic(IV) • Arsenite	Arsenic reacts with stannous chloride in hydrochloric acid to form a brown precipitate.	
	Test Phase	
		Solid
	✓	Liquid
		Gas/Vapor
	Sensitivity	
	1 µg/ml arsenic; 1 mg/l arsenic oxide.	

Directions and Comments

Mix 1 g stannous chloride pentahydrate (CAS 10026-06-9) in 10 ml of concentrated hydrochloric acid and the clear portion is decanted and used as the test solution. Add the test solution to the sample solution in a proportion of 1:5 and warm gently. Arsenic will cause a brown discoloration or precipitate. Addition of 1 ml of sulfuric acid may accelerate the reaction. If necessary to remove interfering mercury compounds, prepare the sample as follows: Mix in a test tube one drop of sample, two drops of ammonium hydroxide, one drop of 10% hydrogen peroxide (CAS 7722-84-1), and one drop of 10% magnesium chloride (CAS 7786-30-3). Evaporate slowly to residue and then strongly pyrolize. This will also reduce mercury and noble metal compounds, but not antimony compounds. Test the ash with Bettendorf's solution.

References	Manufacturer
Welcher, F., *Chemical Solutions: Reagents Useful to the Chemist, Biologist, and Bacteriologist*, D. Van Nostrand Company, Inc., New York, 1942. *Chemical Abstracts* 7, 2026–2027 (1913). Winkler, L. W., *Angewandte Chemie*, 26(Aufsatz) 143–144 (1913). Feigl, F. and Anger, V., *Spot Tests in Inorganic Analysis*, Sixth Edition, Elsevier Publishing Company, Amsterdam, 1972.	

Bicinchoninic Acid (BCA) Protein Test

Detects	Test Principle
• Protein • Amino acids • Peptides	Bicinchoninic acid (BCA) combined with copper sulfate produces a clear apple green solution. Protein will reduce copper(II) to copper(I), which then complexes with BCA to form a purple color.

(*continued*)

Bicinchoninic Acid (BCA) Protein Test (*Continued*)

	Test Phase	
	✓	Solid
	✓	Liquid
		Gas/Vapor
	Sensitivity	
	1% protein (visual result), much lower if an optical reader is used	

Directions and Comments

Copper sulfate (CAS 7758-99-8) mixed with BCA (bicinchoninic acid, no CAS listing) solution produces a clear apple green solution. Protein reduces copper(II) and the resulting copper(I) ion complexes with a pair of BCA molecules to form an intense purple color. The color change from green to purple is easily discernable without the use of a reader with protein concentration as low as 1%. This solution is the simplest of the general protein tests to use, is stable and may be sprayed directly on a suspected protein or applied to a white absorbent test bed of paper, cotton, or polypropylene. The most effective detection is obtained by drying a 10% solution of copper sulfate on a polypropylene substrate. The sample is applied directly to the polypropylene surface or is collected with tape and then pressed against the polypropylene surface. Next, BCA solution is applied to one portion of the polypropylene and allowed to wick through the substrate and sample. See color images 2.8 and 2.9.

References	Manufacturer
Smith, P.K. et al., Measurement of protein using bicinchoninic acid, *Anal. Biochem.* 150, 76–85 (1985).	BCA is available from Thermo Scientific as BCA Protein Assay Reagent A, part no. 23221 (www.piercenet.com) or from Sigma-Aldrich as Bicinchoninic Acid solution, part no. B9643 (www.sigmaaldrich.com).

Bisulfite Test

Detects	Test Principle	
• Bisulfite	Bisulfite reacts with barium sulfate to form an acidic solution; sulfite forms a neutral solution. Bisulfite and sulfite are discerned by pH upon the same treatment.	
	Test Phase	
	✓	Solid
	✓	Liquid
		Gas/Vapor
	Sensitivity	
	Not available	

Directions and Comments

This test should only be used after sulfites have been shown to exist in the sample from another test, such as the Sulfite Test. A 1% barium chloride (CAS 10361-37-2) solution will react with sulfites to form a neutral solution; check the pH before and after addition of the barium. No change or a slight change indicates bisulfite is not present and only sulfite is present. An acidic change indicates bisulfite is present. Fresh 3% hydrogen peroxide (medical grade, CAS 7722-84-1) may be used in place of barium chloride.

References	Manufacturer
Villiers, A., *Bull. Soc. Chim.* 47, 546 (1887). Feigl, F., *Spot Tests in Inorganic Analysis,* Fifth Edition, Elsevier Publishing Company, New York, 1958.	

Biuret Paper	
Detects	**Test Principle**
• Amino acid • Protein	Protein reduces cupric ion and forms a complex that is pink-violet to purple-violet in color.

	Test Phase	
	✓	Solid
	✓	Liquid
		Gas/Vapor
	Sensitivity	
	Not available, but used as a spot test	

Directions and Comments

Biuret reagent (mixture, CAS not available) is made by mixing 25 ml of 3% (w/v) copper sulfate pentasulfate (CAS 7758-99-8) solution with 1L of 10% (w/v) solution of potassium hydroxide (CAS 1310-58-3). Biuret reagent is also available commercially (e.g., Carolina Biological Supply Company). Biuret reagent is dried on paper strips. Liquid sample that is neutral or basic is applied directly; solid sample is placed on the paper and a drop of water is added. Moist powders that are neutral to basic may be tested directly. The Biuret reagent may be used in a test tube with less sensitivity. A pink-violet to purple-violet color from the original blue indicates protein. The complex will also form with amino acids. Do not confuse with another compound called Biuret reagent (CAS 108-19-0). See color image 2.10.

References	Manufacturer
Welcher, F., *Chemical Solutions: Reagents Useful to the Chemist, Biologist, and Bacteriologist*, D. Van Nostrand Company, Inc., New York, 1942. *J. Biol. Chem.*, 7, 60 (1910). *J. Biol. Chem.*, 7, 11 (1910). Material Safety Data Sheet Biuret Reagent, Carolina Biological Supply Company, Burlington, North Carolina, November 27, 2007.	

Biuret Protein Test	
Detects	**Test Principle**
• Amine • Protein • Reducing compound	Protein reduces copper(II) sulfate to copper(I). The original copper sulfate blue changes to a purple color.

	Test Phase	
	✓	Solid
	✓	Liquid
		Gas/Vapor
	Sensitivity	
	3–4% protein visually; much lower if an optical reader is used	

Directions and Comments

The test uses an aqueous solution of 1.7% copper sulfate (CAS# 7758-99-8), 9.3% ethylene glycol (CAS# 107-21-1), and 16.7% sodium hydroxide (CAS# 1310-73-2). This deep blue solution can be diluted with water in a test tube. A small sample may be added to the test tube or spot plate, or the dilute solution may be added directly to a powder on a surface. If protein is present, a purple color (540 nm) will develop from blue. The color change is difficult to see in the case of low concentrations of protein. The result is sensitive to the presence of nitrogen-containing compounds. This solution is tradenamed "Biuret Reagent" and has no CAS listing. Do not confuse it with Biuret Reagent (CAS 108-19-0).

(continued)

Biuret Protein Test (*Continued*)	
References	**Manufacturer**
Biuret Reagent Material Safety Data Sheet, Carolina Biological Supply Company, Burlington, North Carolina, November 27, 2007.	Carolina Biological Supply Company, 2700 York Road, Burlington, North Carolina, USA (www.carolina.com).

Borate Test (Gutzeit Scheme)		
Detects	**Test Principle**	
• Borate • Boric acid	Turmeric solution reacts with borate in acid to form a red rosocyanine complex. Sodium carbonate will react with the rosocyanine complex to form a blue-black color.	
	Test Phase	
		Solid
	✓	Liquid
		Gas/Vapor
	Sensitivity	
	6 ppm boric acid on test strips	
Directions and Comments		
The sample is first reacted with a soluble carbonate compound such as sodium carbonate (CAS 497-19-8) and then acetic acid (CAS 64-19-7) is added until no more carbon dioxide is produced. Turmeric contains curcumin ([1E,6E]-1,7-bis[4-hydroxy-3-methoxyphenyl]-1,6-heptadiene-3,5-dione, CAS 458-37-7), which forms a red complex in the presence of borate. Addition of sodium carbonate will shift the pH to base and a blue-black color is formed from the rosocyanine complex. Gutzeit specifies sodium carbonate, but other dilute basic solutions may be used. Paper strips may be soaked in the solution and then dried and stored. The paper turns red in contact with moist boric acid materials and when treated with sodium hydroxide a blue-green to black color forms.		
References	**Manufacturer**	
Gutzeit, G., *Helv. Chim. Acta*, 12, 829 (1929). *U.S. Pharmacopia* XI, 449; *Official and Tentative Methods of Analysis of the Association of Agricultural Chemists*, Fourth Edition, 436, Association of Agricultural Chemists, Washington, DC, 1935. Feigl, F. and Anger, V., *Spot Tests in Inorganic Analysis,* Sixth Edition, Elsevier Publishing Company, Amsterdam, 1972.		

Borinski's Reagent	
Detects	**Test Principle**
• Peroxidase	Peroxidase enzyme will cleave hydrogen peroxide to form water and oxygen, which in turn reacts with guaiac resin to form a blue reaction product.

(*continued*)

Borinski's Reagent (*Continued*)

	Test Phase	
		Solid
✓		Liquid
		Gas/Vapor
Sensitivity		
Detects natural peroxidase from 1 part raw milk in 10 parts pasteurized milk. The reaction and sensitivity are similar to a medical test for occult blood in stool.		

Directions and Comments

Mix 0.85 g of guaiac resin (CAS 9000-29-7) in 70% alcohol and shake for 30 minutes. Add 10 ml dilute phenol (CAS 108-95-2) and 5 ml 3% hydrogen peroxide (medical grade, CAS 7722-84-1). Add 10 drops of test solution to 5 ml of sample. A blue color indicates peroxidase. Peroxidase is a family of complex enzymes; most will cleave hydrogen peroxide and some will cleave specific organic peroxides.

References	Manufacturer
Welcher, F., *Chemical Solutions: Reagents Useful to the Chemist, Biologist, and Bacteriologist*, D. Van Nostrand Company, Inc., New York, 1942. *Chemical Abstracts* 20, 3752 (1926).	

Bougault's Reagent

Detects	Test Principle	
• Sodium	Antimony chloride and potassium carbonate in hydrogen peroxide solution will react with sodium to form a precipitate.	
	Test Phase	
		Solid
	✓	Liquid
		Gas/Vapor
	Sensitivity	
	Not available	

Directions and Comments

Mix 1 g antimony chloride (CAS 10025-91-9), 10 ml of 33% (w/v) potassium carbonate (CAS 584-08-7) solution and 45 ml 3% hydrogen peroxide (medical grade, CAS 7722-84-1). Heat while stirring, cool, and filter. Add the test solution to the sample solution. Sodium will cause a precipitate to form.

References	Manufacturer
Welcher, F., *Chemical Solutions: Reagents Useful to the Chemist, Biologist, and Bacteriologist*, D. Van Nostrand Company, Inc., New York, 1942. *J. Pharm. Chim.*, 437 (1905).	

Bradford Reagent	
Detects	**Test Principle**
• Protein	A complex is formed by protein and an aqueous solution of a dye, Brilliant Blue G, in the presence of phosphoric acid and methanol.

	Test Phase	
	✓	Solid
	✓	Liquid
		Gas/Vapor

Sensitivity
1–2% protein visually; much lower if an optical reader is used

Directions and Comments

Bradford reagent is 0.004% Brilliant Blue G (no CAS listing), 10% phosphoric acid (CAS 7664-38-2) and 4% methanol (CAS 67-56-1). Protein is mixed with the Bradford solution and then read at 595 nm for quantitative result. The amount of absorption is proportional to the protein present. A color change from red to blue is sufficient for qualitative results. A short incubation at room temperature may be required for lower concentrations of protein. Use of a blank helps visualize low concentration if not using a reader.

References	**Manufacturer**
Bradford, M.M., *Anal. Biochem.*, 72, 248 (1976). Material Safety Data Sheet Bradford Reagent, Sigma-Aldrich, St. Louis, Missouri, January 4, 2008.	Sigma-Aldrich Corp., St. Louis, Missouri, USA (www.sigmaaldrich.com)

Brady's Reagent	
Detects	**Test Principle**
• Carbonyl group • Ketone • Aldehyde	2,4-Dinitrophenylhydrazine reacts with carbonyl group (C=O) in aldehydes and ketones to form a yellow, orange, or red precipitate.

	Test Phase	
	✓	Solid
	✓	Liquid
		Gas/Vapor

Sensitivity
Not available, but used as a laboratory method of derivatizing aldehydes and ketones

Directions and Comments

Mix a 1% solution (w/v) of 2,4-dinitrophenylhydrazine (CAS 119-26-6) in water. Add the test mixture to the liquid or solid sample; a yellow, orange, or red oil or precipitate indicates aldehyde or ketone. Carboxylic acids, amides, and esters do not interfere. Some allyl alcohol derivatives are oxidized by Brady's reagent to a ketone or aldehyde, which then give a positive result. Some alcohols may form ketones or aldehydes after prolonged exposure to air; additional testing for alcohol is suggested (see Ceric Nitrate Reagent). 2,4-Dinitrophenylhydrazine becomes a shock-sensitive explosive if allowed to dry to less than 30% water content. A 0.2 M (4%) nonexplosive solution is available commercially as a substitute for the explosive solid (Sigma Aldrich 42215, CAS 125038-14-4).

(continued)

Brady's Reagent (*Continued*)	
References	Manufacturer
Welcher, F., *Chemical Solutions: Reagents Useful to the Chemist, Biologist, and Bacteriologist*, D. Van Nostrand Company, Inc., New York, 1942. Johanson, G.D., *J. Am. Chem. Soc.* 73, 5888 (1951). Shriner, R., Fuson, R. and Curtin, D., *The Systematic Identification of Organic Compounds, A Laboratory Manual*, Fifth Edition, John Wiley and Sons, Inc., New York, 1964. *J. Chem. Educ.*, 37, 205 (1960).	

Bromic Acid Test		
Detects	**Test Principle**	
• Bromic acid • Bromate	Bromates react with manganous sulfate in the presence of sulfuric acid to form a red color. Further reaction with alkali acetate forms a brown precipitate.	
	Test Phase	
	✓	Solid
	✓	Liquid
		Gas/Vapor
	Sensitivity	
	20 µg/ml potassium bromate when benzidine is used	
Directions and Comments		
Mix a 2% solution of manganous sulfate monohydrate (CAS 10034-96-5) and sulfuric acid. Add a drop of this solution to a few milligrams of solid sample or a drop of liquid sample in a test tube. A red color should form indicating bromate. Confirm bromate by adding sodium acetate (CAS 127-09-3), which should produce a brown precipitate of hydrated manganese oxide. The original test used benzidine to improve sensitivity by producing a deep blue color from faintly visible results. Benzidine (4,4'-diaminobiphenyl) is a carcinogen. Chlorates and iodates do not interfere.		
References	**Manufacturer**	
Feigl, F., *Spot Tests in Inorganic Analysis*, Fifth Edition, Elsevier Publishing Company, New York, 1958.		

Bromine Water Solution	
Detects	**Test Principle**
• Alkene • Alkyne • Aromatic • Mercaptan • Unsaturated hydrocarbon	Bromine undergoes addition reaction with unsaturated hydrocarbons, thus clearing the light brown color of the solution.

(continued)

Bromine Water Solution (*Continued*)

	Test Phase
✓	Solid
✓	Liquid
	Gas/Vapor
Sensitivity	
Not available	

Directions and Comments

Bromine water (CAS 7726-95-6) forms a saturated solution at about 3% in water. Dissolve 0.1 g of solid sample or two drops of liquid sample in water. Add test solution by the drop (with gentle agitation) to the sample solution until assured the bromine color will not clear. If the brown color clears, unsaturated hydrocarbon compounds are indicated. Mercaptans are converted to disulfides and may be detected by the appropriate test. This test is more dependable for causing an addition reaction with aromatics than is the Unsaturated Hydrocarbon Test and further, this test does not evolve hydrogen bromide gas due to the presence of water. Bromine water is available commercially as a reagent. It is also packaged in a glass capsule in the Agri-Screen Ticket. See color images 2.11 and 2.12.

References	Manufacturer
Shriner, R., Fuson, R. and Curtin, D., *The Systematic Identification of Organic Compounds, A Laboratory Manual,* Fifth Edition, John Wiley and Sons, Inc., New York, 1964. *J. Chem. Educ.,* 37, 205 (1960).	Neogen Corporation, Lansing, Michigan, USA (www.neogen.com).

Brucine Solution (Tin)

Detects	Test Principle
• Stannous • Tin(II)	Brucine and nitric acid will react with stannous ions to form red-violet color.

	Test Phase
✓	Solid
✓	Liquid
	Gas/Vapor
Sensitivity	
Not available	

Directions and Comments

Dissolve 0.1 g brucine (10,11-dimethoxystrychnine, CAS 357-57-3) in 1 ml cold nitric acid and 50 ml water, and then boil for 15 minutes. The test solution will be yellow. Add the test solution to the sample. A red-violet color indicates stannous ion.

References	Manufacturer
Welcher, F., *Chemical Solutions: Reagents Useful to the Chemist, Biologist, and Bacteriologist*, D. Van Nostrand Company, Inc., New York, 1942. *Rev. Intern. Falsific.,* 98 (1895).	

Caley's Reagent	
Detects	**Test Principle**
• Sodium	Uranyl acetate and cobaltous acetate in acetic acid react with sodium ions to form a yellow precipitate.

	Test Phase	
		Solid
	✓	Liquid
		Gas/Vapor

Sensitivity
Not available

Directions and Comments
Dissolve 4 g uranyl acetate (CAS 541-09-3) and 3 g acetic acid (CAS 64-19-7) in water, dilute to 50 ml and heat to 75°C until solution is complete. Dissolve 20 g cobaltous acetate (CAS 6147-53-1) and 3 g acetic acid in water, dilute to 50 ml and heat to 75°C until the solution is complete. Mix the warm solutions, cool, let stand for 2 hours and filter. The reagent can be stored in an amber glass container. The test is performed by adding 1 part sample solution to 20 parts test solution and shaking. A yellow precipitate indicates sodium ion.

References	**Manufacturer**
Welcher, F., *Chemical Solutions: Reagents Useful to the Chemist, Biologist, and Bacteriologist*, D. Van Nostrand Company, Inc., New York, 1942. *J. Am. Chem. Soc.* 51, 1965 (1929).	

Carmine Red Reagent	
Detects	**Test Principle**
• Boric acid	Boric acid reacts with carmine red to produce a blue color from the original red.

	Test Phase	
	✓	Solid
	✓	Liquid
		Gas/Vapor

Sensitivity
0.018 mg boric acid per ml

Directions and Comments
Dissolve 0.05 g carmine red (CAS 1390-65-4) in 100 g concentrated sulfuric acid. Place a drop of red test solution on the sample. A blue color indicates boric acid.

References	**Manufacturer**
Welcher, F., *Chemical Solutions: Reagents Useful to the Chemist, Biologist, and Bacteriologist*, D. Van Nostrand Company, Inc., New York, 1942. *Chemical Abstracts* 31, 2124 (1937).	

Casolori's Reagent		
Detects	**Test Principle**	
• Thiosulfate	Nitroprusside reacts with thiosulfate to form a blue color.	
	Test Phase	
	✓	Solid
	✓	Liquid
		Gas/Vapor
	Sensitivity	
	0.01N thiosulfate	
Directions and Comments		
Dissolve 5 g sodium nitroprusside dihydrate (CAS 13755-38-9) in 95 ml water. Leave the solution exposed to air and light until it turns brown. To eliminate the waiting time, add 2 drops of a solution of potassium ferricyanide (CAS 13746-66-2) and 1 drop of sodium hydroxide. Filter if necessary. Add a few drops of the test solution to the sample. A blue color forms with thiosulfate. Dilute solutions (<0.01N thiosulfate) will turn green in 30 minutes and will not attain the blue color. Sulfites and tetrathionates are not detected.		
References	**Manufacturer**	
Welcher, F., *Chemical Solutions: Reagents Useful to the Chemist, Biologist, and Bacteriologist*, D. Van Nostrand Company, Inc., New York, 1942. *Chemical Abstracts* 6, 2632 (1911).		

Catalase Test		
Detects	**Test Principle**	
• Dairy product • Milk • Cheese	Catalase cleaves hydrogen peroxide to form water and oxygen, which will bubble from the solution in higher concentration.	
	Test Phase	
	✓	Solid
	✓	Liquid
		Gas/Vapor
	Sensitivity	
Directions and Comments		
Most organisms contain catalase (CAS 9001-05-2), an enzyme that decomposes hydrogen peroxide to water and oxygen. Dairy bacteria do not contain catalase and do not effervesce when in contact with hydrogen peroxide solution. Catalase can be used to differentiate dairy products from other food and biological products.		
References	**Manufacturer**	
Boon, E., Downs, A., Marcey, D., *Proposed Mechanism of Catalase in Catalase: H_2O_2: H_2O_2 Oxidoreductase*. Catalase Structural Tutorial Text. April 15, 2008.		

Celsi's Reagent	
Detects	**Test Principle**
• Potassium	Potassium ions react with proportioned amounts of cobalt nitrate and sodium thiosulfate to produce a sky blue color.

	Test Phase
✓	Solid
✓	Liquid
	Gas/Vapor

Sensitivity
Not available

Directions and Comments

The test uses two solutions. Solution A is 7 g of cobalt nitrate hexahydrate (CAS 10026-22-9) in 50 ml of 80% methanol. Solution B is 19 g of sodium thiosulfate pentahydrate (CAS 10102-17-7) in 50 ml of water. The test is performed by adding one drop of each solution to 10 ml of methanol and waiting until a violet color develops. Next, a few drops of liquid sample or a few milligrams of solid sample are added. Production of a sky-blue precipitate indicates potassium.

References	**Manufacturer**
Welcher, F., *Chemical Solutions: Reagents Useful to the Chemist, Biologist, and Bacteriologist*, D. Van Nostrand Company, Inc., New York, 1942. *Chemical Abstracts* 28, 3026 (1934).	

Ceric Nitrate Reagent	
Detects	**Test Principle**
• Alcohol • Phenol • Dioxane • E85 fuel • Gasoline (10% ethanol)	Ceric nitrate reacts with alcoholic hydroxyl to change the yellow solution to red.

	Test Phase
✓	Solid
✓	Liquid
	Gas/Vapor

Sensitivity
Not available

Directions and Comments

Mix 2 g of ceric nitrate (ammonium ceric nitrate, CAS 16774-21-3) in 5 ml of 2N nitric acid (CAS 7697-37-2) and heat as necessary to dissolve. Dilute 0.5 ml of the ceric nitrate solution with 3 ml of water in a test tube and add 0.1 g of solid sample or 5 drops of liquid sample. A color change from yellow to red indicates alcohol. Brown to green-brown indicates phenols. Deep red to brown indicates dioxane. The test is limited to compounds with a carbon to hydroxyl ratio of about 10:1. The test also detects hydroxy acids, hydroxy aldehydes and other compounds containing alcoholic hydroxyl within the 10:1 limit. Amino alcohols are not detected due to precipitation caused by high pH. Organic compounds containing carbon, hydrogen, oxygen and halogen do not interfere. Other compounds may cause precipitates or colors to form; look for the immediate color change of yellow to red, even if the solution then clears, as an indication of alcohol. Gasoline containing 10% ethanol or E85 fuel can be determined by this reagent by placing 1 ml of reagent in a test tube followed by 1 ml of sample. Ethanol will mix into the lower, reagent layer and cause a red color; gasoline will remain in an upper, insoluble layer according to its proportion (about 90% will remain for 10% ethanol fuel and only a thin upper gasoline layer, if any, will remain for E85 fuel).

(continued)

Ceric Nitrate Reagent (*Continued*)	
References	**Manufacturer**
Shriner, R., Fuson, R. and Curtin, D., The Systematic Identification of Organic Compounds, A Laboratory Manual, Fifth Edition, John Wiley and Sons, Inc., New York, 1964. J. Chem. Educ., 37, 205 (1960). R. Houghton, unpublished studies, St. Johns, Michigan, 2008.	

Chloric Acid Test		
Detects	**Test Principle**	
• Chloric acid • Chlorate • Persulfate • Periodate	Chlorates react with manganese sulfate in strong phosphoric acid to form purple complex manganese(III) phosphate ions.	
	Test Phase	
		Solid
	✓	Liquid
		Gas/Vapor
	Sensitivity	
	0.05 µg/ml chlorate with diphenylcarbazide	

Directions and Comments

A drop of saturated manganous sulfate monohydrate (CAS 10034-96-5) solution is added to a drop of concentrated phosphoric acid (CAS 7664-38-2) and then mixed with a drop of the sample. The mixture is then warmed briefly and allowed to cool. A purple color indicates chlorate. If necessary, the addition of a drop of 1% alcoholic solution of diphenylcarbazide (CAS 140-22-7) will intensify the color of very dilute samples. Persulfates and periodates react similarly. Nitrates may react similarly, but can be determined with a nitrate test strip.

References	**Manufacturer**
Feigl, F., *Spot Tests in Inorganic Analysis,* Fifth Edition, Elsevier Publishing Company, New York, 1958. Feigl, F. and Anger, V., *Spot Tests in Inorganic Analysis,* Sixth Edition, Elsevier Publishing Company, Amsterdam, 1972.	

Chlorine Dioxide Test		
Detects	**Test Principle**	
• Chlorine dioxide • Chlorate	Diphenylamine with ethyl acetate and trichloroacetic acid will react with chlorine dioxide to produce a green color.	
	Test Phase	
	✓	Solid
	✓	Liquid
		Gas/Vapor
	Sensitivity	
	3 µg/ml chlorate	

(*continued*)

Chlorine Dioxide Test (*Continued*)

Directions and Comments

This test paper may be used to detect chlorine dioxide gas directly or to determine chlorate from solution. The paper is prepared by mixing a saturated solution of diphenylamine (CAS 122-39-4) in ethyl acetate (CAS 141-78-6) and then adding 0.5 g of trichloroacetic acid (CAS 76-03-9). (Diphenylamine is oxidized by many compounds to its blue form but the presence of ethyl acetate and trichloroacetic acid make the color change to blue resistant to oxidizers containing bromine or iodine.) Place a few milligrams of solid sample or one drop of liquid sample in a test tube with a drop of concentrated sulfuric acid. Moisten a filter paper with the solution (the solution must be prepared just before use) and place it over the opening of the test tube. Place the tube in a boiling water bath. The appearance of a green color on the filter paper within a few minutes indicates chlorine dioxide formed from chlorate. Bromates and iodates do not interfere.

References	Manufacturer
Feigl, F. and Anger, V., *Spot Tests in Inorganic Analysis*, Sixth Edition, Elsevier Publishing Company, Amsterdam, 1972. Jungeis, E. and Ben-Dor, L., *Talanta*, 11, 718 (1964).	

Chromotropic Acid Reagent (Iron)

Detects	Test Principle	
• Ferric • Iron(III)	Chromotropic acid reacts with ferric ions to produce a green color that is then discharged with stannous chloride.	
	Test Phase	
	✓	Solid
	✓	Liquid
		Gas/Vapor
	Sensitivity	
	Not available	

Directions and Comments

The test solution is a 2% (w/v) aqueous solution of chromotropic acid (4,5-dihydroxy-2,7-naphthalenedisulfonic acid, disodium salt, dihydrate, CAS 5808-22-0). The test solution is added to the sample; a green color indicates ferric and other ions. Addition of stannous chloride pentahydrate solution (CAS 10026-06-9) will clear the green color if the color was caused by ferric ions. No other ions are known to react similarly.

References	Manufacturer
Welcher, F., *Chemical Solutions: Reagents Useful to the Chemist, Biologist, and Bacteriologist*, D. Van Nostrand Company, Inc., New York, 1942. Belcher, J. and Williams, G., *A Course in Qualitative Analysis*, Houghton-Mifflin, Boston, 1938.	

Cinchonine - Potassium Iodide Solution

Detects	Test Principle	
• Bismuth • Copper • Lead • Mercury	Cinchonine, potassium iodide, and bismuth react to produce a double salt visible as a red precipitate.	
	Test Phase	
		Solid
	✓	Liquid
		Gas/Vapor
	Sensitivity	
	0.14 µg/ml bismuth	

(continued)

Cinchonine - Potassium Iodide Solution (*Continued*)

Directions and Comments

Mix 1 g cinchonine (CAS 118-10-5) with just enough concentrated nitric acid to form a thick paste. Dissolve the paste into 10 ml water. Separately, dissolve 5 g potassium iodide (CAS 7681-11-0) in 50 ml of water. Combine the two solutions and dilute to 100 ml with water. Optimally, allow the solution to stand for 48 hours, then filter and store in amber glass; the solution will keep indefinitely. The test solution should be slightly acidic; increased acidity will decrease the sensitivity of the test. The test is performed by adding a drop of test solution to a few milligrams of solid sample or a drop of liquid sample in a spot plate. Bismuth will cause an orange-red precipitate. Cadmium does not interfere. Interfering metals are cupric (brown), mercury (white), and lead (yellow). These interfering colors can be separated if the test is performed on filter paper; an orange-red area indicates bismuth. Cadmium does not interfere. Cinchonine is an alkaloid and it has been suggested that other alkaloids, such as caprolactam (CAS 105-60-2) could be substituted for cinchonine in this test for bismuth. Likewise, bismuth is often used as a test for alkaloids.

References	Manufacturer
Welcher, F., *Chemical Solutions: Reagents Useful to the Chemist, Biologist, and Bacteriologist*, D. Van Nostrand Company, Inc., New York, 1942. *Chemical Abstracts* 17, 2687 (1923). Engelder, C., Dunkelberger, T. and Schiller, W., *Semi-Micro Qualitative Analysis*, Wiley, New York, 1936.	

Cinchonine Solution (Tungsten)

Detects	Test Principle		
• Tungsten • Tungstenate	Cinchonine reacts with tungstenate to rapidly form a precipitate.		
	Test Phase		
			Solid
	✓		Liquid
			Gas/Vapor
	Sensitivity		
	Not available		

Directions and Comments

Mix 50 ml of water with 50 ml of concentrated hydrochloric acid and then dissolve 12.5 g cinchonine (CAS 118-10-5) in the solution. Add the test solution to the sample solution; a precipitate indicates tungstenate.

References	Manufacturer
Welcher, F., *Chemical Solutions: Reagents Useful to the Chemist, Biologist, and Bacteriologist*, D. Van Nostrand Company, Inc., New York, 1942. Hillebrand and Lundell, *Applied Inorganic Analysis*, Wiley, New York, 1929.	

Clor-N-Oil 50

Detects	Test Principle
• Alkyl chloride • Chlorinated hydrocarbon	Chlorine is removed from polychlorinated biphenyls (PCBs) by an organo-sodium reagent and measured with a colorimetric indicator that changes from blue to yellow.

(*continued*)

Clor-N-Oil 50 (*Continued*)

• PCB • Polychlorinated Biphenyl	**Test Phase**	
	✓	Solid
	✓	Liquid
		Gas/Vapor
	Sensitivity	
	50 ppm PCB in oil	

Directions and Comments

A sample of the oil to be tested is reacted with a mixture of metallic sodium catalyzed with naphthalene and diglyme at ambient temperature. This process converts all organic halogens to their respective sodium halides. All halides in the treated mixture, including those present prior to the reaction, are then extracted into an aqueous buffer, a premeasured amount of mercuric nitrate is added, followed by a solution of diphenylcarbazone as the indicator. The normally blue indicator becomes yellow in the presence of more than 50 ppm PCB in oil. The test is intended as a screening test for PCB in transformer oil. Quantitative results may vary for PCB in other substances and for halogenated hydrocarbons other than PCB. Since the test uses metallic sodium, water and sulfur interfere. Consult the test insert for detailed information.

References	Manufacturer
Method 9079 - Screening Test Method for Polychlorinated Biphenyls in Transformer Oil, http://www.epa.gov/sw-846/pdfs/9079.pdf, United States Environmental Protection Agency, Washington, DC, December 1996. Dexsil® Corporation, www.dexsil.com, Hamden, Connecticut, March 4, 2008.	Dexsil® Corporation, Hamden, CT, www.dexil.com.

Cobalt Thiocyanate Reagent

Detects	Test Principle	
• Alkaloid • Ephedrine • Methadone • Pseudoephedrine • Quinine	Cobalt thiocyanate reacts with nearly all alkaloids to form water-insoluble precipitates of varying shades of blue.	
	Test Phase	
	✓	Solid
	✓	Liquid
		Gas/Vapor
	Sensitivity	
	60 µg cocaine HCl; 250 µg methadone HCl.	

Directions and Comments

This test is a presumptive field test for illegal drugs. The test reagent is a 2% aqueous solution of cobalt(II) thiocyanate (CAS 3017-60-5), which is mixed with a small amount of sample. A brilliant green-blue color is produced by the hydrochloride salts of benzphetamine, chlordiazepoxide, chlorpromazine, doxepin, methadone, and methylphenidate as well as brompheniramine maleate and hydrocodone tartrate. A strong, but less intense green-blue color is produced by the hydrochloride salts of cocaine, diacetylmorphine, ephedrine, meperidine, phencyclidine, procaine, propoxyphene, and pseudoephedrine. Quinine HCl produces a strong blue color. Alkaloids present in legal drugs and other materials are also indicated. Refer to the appendix for drug screen results. See color images 2.13 and 2.14.

(continued)

Cobalt Thiocyanate Reagent (*Continued*)	
References	Manufacturer
Scott, L., *Specific Field Test for Cocaine*, μg, VI, 179(1973). U.S. Department of Justice. Color Test Reagents/Kits for Preliminary Identification of Drugs of Abuse, NIJ Standard–0604.01, National Institute of Justice, National Law Enforcement and Corrections Technology Center, Rockville, MD, July 2000.	Several commercial test kits are available.

Copper Metal Test		
Detects	**Test Principle**	
• Copper metal • Cobalt metal • Nickel metal	Copper metal is dissolved into solution with ammonium hydroxide and then mixed with rubeanate acid to form an olive-green precipitate of copper rubeanate.	
	Test Phase	
	✓	Solid
	✓	Liquid
		Gas/Vapor
	Sensitivity	
	0.006 μg/ml copper. Detects trace amounts of copper metal deposited by copper-coated bullets.	

Directions and Comments	
The copper sample may be collected by pressing a piece of paper wetted with ammonium hydroxide against the sample for a minute. Do not rub if using the pattern to determine bullet trajectory. The test solution is formed by mixing 10 mg rubeanic acid (dithiooxamide, CAS 79-40-3) per 1 ml ethanol. Wet the sample on the paper with a few drops of test solution; an olive-green to black color indicates the presence of copper. If used in conjunction with a rhodizonate test for lead, perform the lead test first to avoid interference from ammonia. Nickel and cobalt react as copper, but with less sensitivity. Nickel produces a blue-red color. Cobalt produces a brown color. Large amounts of ammonium salts interfere by reducing sensitivity.	
References	Manufacturer
Jungreis, E., *Spot Test Analysis: Clinical, Environmental, Forensic, and Geochemical Applications*, Second Edition, John Wiley & Sons, Inc., New York, 1997. Lekstrom, J. and Koons, R., *J. Forensic Sci.*, 4, 1283 (1986).	

Cuprion Reagent	
Detects	**Test Principle**
• Copper • Copper(I) • Copper(II)	Copper(I) ions react with cuprion to form a violet color. Copper(II) is reduced to copper(I) by a reducing agent, such as hydroxylamine, and detected by cuprion.

(*continued*)

Cuprion Reagent (*Continued*)

	Test Phase	
	✓	Solid
	✓	Liquid
		Gas/Vapor
	Sensitivity	
	0.05 µg/ml copper(I)	

Directions and Comments

Mix a saturated solution of cuprion (2,2'-biquinoline, CAS 119-91-5) in alcohol; cuprion is insoluble in water. Acid may be used to dissolve metallic samples but must be adjusted to pH >3 before testing. To determine copper(I) ions, test directly with cuprion. A red-purple color indicates copper(I). To determine copper(II), place a drop of aqueous sample solution in a spot plate or test tube and add a few grains of hydroxylamine hydrochloride (CAS 5470-11-1) followed by a drop of the cuprion solution. A red-purple color indicates copper(I) from reduced copper(II). This test is specific to copper(I) ions. If large quantities of potentially interfering or strongly colored ions are present, perform the test in a test tube with a larger volume and then add a water insoluble solvent, such as hexane, and mix. The water insoluble copper(I) - cuprion complex will color the hexane layer while interfering colored ions collect in the water layer. Compare to Merckoquant® Copper Test.

References	Manufacturer
Feigl, F. and Anger, V., *Spot Tests in Inorganic Analysis,* Sixth Edition, Elsevier Publishing Company, Amsterdam, 1972. *Anal. Abstracts*, 16 (1969).	

Cuprous Iodide Paper

Detects	Test Principle	
• Amalgam • Mercury • Mercury salts	Mercury reacts with colorless cuprous iodide to form red cupro tetraiodo mercuriate.	
	Test Phase	
	✓	Solid
	✓	Liquid
	✓	Gas/Vapor
	Sensitivity	
	Not available but described as sensitive	

Directions and Comments

This test is similar to palladium chloride paper but also detects mercury salts. 5 g of copper sulfate (CAS 7758-99-8) is dissolved in 75 ml of water and then mixed with 5 g sodium sulfite (CAS 7757-83-7) and 11 g potassium iodide (CAS 7681-11-0) dissolved in 75 ml of water. A precipitate will form, which is collected by centrifuge or filter, washed well with water and stored moist. Just before the test, form a thin slurry from the paste and place on the paper strip. The paper is held over a sample that is slowly heated to red hot. A red color forms in the paste to indicate mercury.

References	Manufacturer
Feigl, F., *Spot Tests in Inorganic Analysis,* Fifth Edition, Elsevier Publishing Company, New York, 1958.	

Curcumin Solution

Detects	Test Principle
• Beryllium	Curcumin forms an orange-red lake with beryllium in slightly alkaline solution.

	Test Phase	
		Solid
✓		Liquid
		Gas/Vapor

Sensitivity
Not available

Directions and Comments

Dissolve 0.1 g curcumin ([1E,6E]-1,7-bis[4-hydroxy-3-methoxyphenyl]-1,6-heptadiene-3,5-dione, CAS 458-37-7) in 100 ml of alcohol. Beryllium is toxic, especially by inhalation of dust. Solid beryllium may be dissolved into solution by use of hydrochloric or sulfuric acid; beryllium resists oxidation and nitric acid. Adjust the pH of the sample solution to slightly alkaline with sodium carbonate or other suitable compound. Iron and aluminum interfere; both are removed by the addition of sodium fluoride to the sample solution. Add a few drops of test solution to the slightly alkaline sample solution. An orange-red lake indicates beryllium. Compare to Borate Test (Gutzeit Scheme).

References	Manufacturer
Welcher, F., *Chemical Solutions: Reagents Useful to the Chemist, Biologist, and Bacteriologist,* D. Van Nostrand Company, Inc., New York, 1942. *J. Am. Chem. Soc.* 50, 393–395 (1928).	

Deniges' Reagent

Detects	Test Principle
• Acetylene	Acetylene reacts with cupric sulfate in a solution of ammonium chloride and hydrochloric acid on paper to form red copper acetylide.

	Test Phase	
		Solid
		Liquid
✓		Gas/Vapor

Sensitivity
Not available, but the paper quickly turns red in a weak stream of acetylene in air.

Directions and Comments

Dissolve 5 g of ammonium chloride (CAS 12125-02-9), 2.5 g cupric sulfate (CAS 7758-99-8), and a drop of hydrochloric acid in a little water and then dilute with water to 25 ml. Just before the test boil 5 ml of the solution with a few copper turnings until the solution is nearly colorless. Add 1 ml water and cool in ice. The cooled solution is used to wet paper strips, which will turn red in the presence of acetylene gas as long as the strips are still wet. Dry copper acetylide is a shock-sensitive explosive but is not overtly dangerous when present in small quantities on the paper strip.

References	Manufacturer
Welcher, F., *Chemical Solutions: Reagents Useful to the Chemist, Biologist, and Bacteriologist*, D. Van Nostrand Company, Inc., New York, 1942. *Chemical Abstracts* 16, 395 (1922).	

Diaminobenzene Solution	
Detects	**Test Principle**
• Magnesium	p-Diaminobenzene forms a red-violet precipitate with magnesium ion in basic solution.

		Test Phase
		Solid
✓		Liquid
		Gas/Vapor

Sensitivity
Not available

Directions and Comments

Mix a 0.5% (w/v) aqueous solution of p-diaminobenzene (CAS 624-18-0). Add a few drops of the test solution to the sample and then add concentrated potassium hydroxide solution by the drop. A red-violet precipitate indicates magnesium salt. The test may be used to detect magnesium metal if the metal is first dissolved with an acid.

References	**Manufacturer**
Welcher, F., *Chemical Solutions: Reagents Useful to the Chemist, Biologist, and Bacteriologist*, D. Van Nostrand Company, Inc., New York, 1942. *Chemical Abstracts* 27, 1840 (1933).	

Diazine Green S Solution	
Detects	**Test Principle**
• Stannous • Tin(II)	Diazine Green S reacts with stannous salts to form a red color from blue.

		Test Phase
✓		Solid
✓		Liquid
		Gas/Vapor

Sensitivity
Not available

Directions and Comments

Mix 10 mg of Diazine Green S (3-Diethylamino-7-[4-dimethylaminophenylazo]-5-phenylphenazinium chloride, CAS 2869-83-20) in 100 ml of water. Add hydrochloric acid by the drop until the solution turns a pure blue color. Add the test solution to the sample solution or a few milligrams of solid sample in a spot plate or test tube. A violet color that eventually becomes red indicates stannous ion. Antimony does not interfere.

References	**Manufacturer**
Welcher, F., *Chemical Solutions: Reagents Useful to the Chemist, Biologist, and Bacteriologist*, D. Van Nostrand Company, Inc., New York, 1942. *Chemical Abstracts* 3859 (1928).	

Dichlorofluorescein Reagent	
Detects	**Test Principle**
• Explosive • HMX • PETN • RDX	Dichlorofluorescein causes a yellow fluorescence while tetra-n-butylammonium hydroxide in DMSO reacts with certain explosive compounds to form a blue to black color. An ultraviolet lamp is used to increase contrast between the yellow fluorescence and the blackened positive result.

	Test Phase	
	✓	Solid
	✓	Liquid
		Gas/Vapor

Sensitivity
Not available. Qualitative result only.

Directions and Comments

The test solution is made by mixing a 2 ml solution (w/v) of 10% tetra-n-butylammonium hydroxide (CAS 2052-49-5) in methanol and then adding 50 mg of dichlorofluorescein (2′,7′- dichlorofluorescein, CAS 76-54-0). Finally, mix the solution with 98 ml of DMSO (dimethylsulfoxide, CAS 67-68-5) to form 100 ml of test solution. The test is most sensitive when performed on filter paper. Place a drop of liquid sample or a few milligrams of solid sample on the paper. Add two drops of test solution. A blue to black color is presumptive of RDX, HMX, or PETN. If the yellow color of the test solution remains, shine a UV lamp (optimal wavelength is 254 nm) to make the yellow color fluoresce and cause a sharper contrast with any slight, positive blue to black color. This presumptive test for compounds similar to RDX and was developed for use at Los Alamos National Laboratory. The paper filter may be used as a wipe to collect the sample.

References	**Manufacturer**
Baytos, J., *Field Spot-Test Kit for Explosives,* Los Alamos National Laboratory, Los Alamos, New Mexico, July 1991.	

Diethylnicotinamide Reagent	
Detects	**Test Principle**
• Decaborane	N,N-diethylnicotinamide reacts selectively with decaborane to produce an orange-red complex.

	Test Phase	
	✓	Solid
	✓	Liquid
		Gas/Vapor

Sensitivity
1 μg/ml decaborane

Directions and Comments

This is a selective test for decaborane, such as that found in laundry detergent, among other boron containing compounds. Mix a saturated aqueous solution of N,N-diethylnicotinamide (CAS 59-26-7). Add a drop of test solution to a drop of sample solution. An orange-red color indicates decaborane. Boric acid, boron salts, diborane, pentaborane and proteins do not interfere.

References	**Manufacturer**
Anal. Chem., 28, 133 (1956).	

Dihydroxyacetophenone Solution

Detects	Test Principle
• Ferric • Iron	2,4-Dihydroxyacetophenone reacts with ferric salts to produce a red color.

	Test Phase
✓	Solid
✓	Liquid
	Gas/Vapor

Sensitivity

2 ppm iron if no other metals are present

Directions and Comments

Mix a 10% (w/v) solution of 2,4-dihydroxyacetophenone (2′,4′-dihydroxyacetophenone, CAS 89-84-9) in 95% alcohol. pH should be mildly acidic for maximum sensitivity. Add a drop of test solution to a drop of sample solution or a few milligrams of solid sample; a red color indicated ferric ions. Colored solutions may be diluted and the test performed with a control to facilitate the immediate formation of the red color. Some metals will form a precipitate with the test solution, but the red color indicating iron becomes visible before the precipitate forms.

References	Manufacturer
Welcher, F., *Chemical Solutions: Reagents Useful to the Chemist, Biologist, and Bacteriologist*, D. Van Nostrand Company, Inc., New York, 1942. *Ind. Eng. Chem., Anal. Ed.* 9, 334 (1937).	

Dille-Koppanyi Reagent, Modified

Detects	Test Principle
• Amobarbital • Barbiturate • Pentobarbital • Phenobarbital • Secobarbital	Barbiturates react with cobalt(II) to form a light purple colored complex.

	Test Phase
✓	Solid
✓	Liquid
	Gas/Vapor

Sensitivity

15 µg Phenobarbital

Directions and Comments

This test is a presumptive field test for illegal drugs. The test uses two solutions. Solution A is made by mixing 0.1 g of cobalt(II) acetate dihydrate (CAS 6147-53-1 is acceptable) in 100 ml of methanol followed by 0.2 ml of glacial acetic acid (CAS 64-19-7). Solution B is a 5% solution of isopropylamine (CAS 75-31-0) in methanol (v/v). The sample is placed in a spot plate and two drops of Solution A are added followed by one drop of Solution B. A light purple color is produced by barbiturates such as amobarbital, pentobarbital, phenobarbital, and secobarbital. Refer to the appendix for drug screen results.

References	Manufacturer
U.S. Department of Justice. Color Test Reagents/Kits for Preliminary Identification of Drugs of Abuse, NIJ Standard–0604.01, National Institute of Justice, National Law Enforcement and Corrections Technology Center, Rockville, MD, July 2000.	Several commercial test kits are available.

Dimethylaminobenzaldehyde (p-DMAB)

Detects	Test Principle
• LSD • Lysergic acid diethylamide	p-Dimethylaminobenzaldehyde (p-DMAB) reacts with LSD to form a deep purple color.

Test Phase	
✓	Solid
✓	Liquid
	Gas/Vapor

Sensitivity
6 µg LSD

Directions and Comments

The p-DMAB test solution consists of 2 g of p-DMAB (CAS 2929-84-2) in 50 ml of 95% ethanol and 50 ml of concentrated hydrochloric acid. The test solution is added to the sample. A deep purple color indicates LSD. Refer to the appendix for drug screen results.

References	Manufacturer
U.S. Department of Justice. Color Test Reagents/Kits for Preliminary Identification of Drugs of Abuse, NIJ Standard–0604.01, National Institute of Justice, National Law Enforcement and Corrections Technology Center, Rockville, MD, July 2000.	Several commercial test kits are available.

Dimethylglyoxime Reagent

Detects	Test Principle
• Bismuth • Cobalt • Copper • Ferrous • Iron(I) • Nickel • Palladium	Dimethylglyoxime reacts with nickel to form a red precipitate in basic solution.

Test Phase	
	Solid
✓	Liquid
	Gas/Vapor

Sensitivity
0.4 µg/ml ferrous ions

Directions and Comments

Mix 1 g of dimethylglyoxime (CAS 95-45-4) in 80 ml of 95% alcohol. Add a drop of test solution to a drop of sample solution in a microplate or test tube. Metal containing nickel may be dissolved with acid and then pH adjusted to slightly acid before addition of the test solution. Use ammonium hydroxide as necessary. A red color develops if nickel or ferrous ions are present. To differentiate nickel from ferrous, add a little potassium cyanide; the red caused by nickel will disappear but red caused by ferrous remains. Copper, cobalt, and bismuth produce a brown, sometimes violet, color. Palladium produces a yellow complex that is acid insoluble and ammonia soluble. Lead may interfere. Gold salts are reduced to gold metal. Compare to Merckoquant® Nickel Test. Ferric ion may be detected if first reduced with hydroxylamine hydrochloride (CAS 5470-11-1).

References	Manufacturer
Welcher, F., *Chemical Solutions: Reagents Useful to the Chemist, Biologist, and Bacteriologist*, D. Van Nostrand Company, Inc., New York, 1942. Hillebrand and Lundell, *Applied Inorganic Analysis,* Wiley, New York, 1929. Feigl, F. and Anger, V., *Spot Tests in Inorganic Analysis,* Sixth Edition, Elsevier Publishing Company, Amsterdam, 1972. *Chemist-Analyst,* 50, 77 (1961).	

Dinitrobenzene Reagent

Detects	Test Principle
<ul><li>Aluminum metal</li><li>Beryllium metal</li><li>Lead metal</li><li>Magnesium metal</li><li>Tin metal</li><li>Zinc metal</li></ul>	Certain metals react with alkali hydroxide to produce hydrogen gas. Hydrogen reduces colorless o-dinitrobenzene to a violet alkali salt.

Test Phase

✓	Solid
	Liquid
	Gas/Vapor

Sensitivity

Not available

Directions and Comments

Mix a 1% alcoholic solution of o-dinitrobenzene (1,2 dinitrobenzene, CAS 528-29-0). Caution: Dry dinitrobenzene is heat and shock sensitive and toxic. Carry the solution for field use. The test is performed by placing a few milligrams of sample in a test tube and adding a drop of test solution and a drop of 0.5N sodium hydroxide solution. The test tube is placed in a boiling water bath. A violet color indicates the presence of metallic aluminum, zinc, lead, tin, or beryllium. Magnesium forms an interfering thin coat of magnesium oxide; this can be removed by washing with acid, rinsing with water, and immediately performing the test. The test can also be performed on a large metallic surface by mixing the test solution and alkali solution in equal volumes and then placing a drop on the metal.

References	Manufacturer
Feigl, F. and Anger, V., *Spot Tests in Inorganic Analysis,* Sixth Edition, Elsevier Publishing Company, Amsterdam, 1972.	

Dinitrotoluene Test

Detects	Test Principle
<ul><li>Dinitrotoluene</li><li>DNT</li><li>Explosive</li><li>Nitroglycerin</li></ul>	Dinitrotoluene reacts with tetramethylammonium hydroxide in acetone and alcohol to form a blue color.

Test Phase

✓	Solid
✓	Liquid
	Gas/Vapor

Sensitivity

Not available

Directions and Comments

Place a few milligrams of solid sample or a drop of liquid sample in a spot plate and add one drop of a solution of equal parts acetone and ethanol followed by a drop of 20% tetramethylammonium hydroxide (CAS 10424-65-4). Formation of a blue color indicates DNT; the color will become dark green if nitroglycerin is present.

References	Manufacturer
General Information Bulletin, 74-8, 4, Federal Bureau of Investigation, Washington, DC, 1974. Jungreis, E., *Spot Test Analysis: Clinical, Environmental, Forensic, and Geochemical Applications,* Second Edition, John Wiley & Sons, Inc., New York, 1997.	

Diphenylamine Indicator Solution		
Detects	**Test Principle**	
• Dichromate • Hypochlorite • Nitrate • Nitrocellulose • Oxidizers • Permanganate	Acidified diphenylamine solution is clear in the reduced form while the oxidized form is violet.	
	Test Phase	
	✓	Solid
	✓	Liquid
		Gas/Vapor
	Sensitivity	
	5 µg/ml antimony; 0.5 µg/ml nitrate.	

Directions and Comments	
Dissolve 1 g of diphenylamine (CAS 122-39-4) in 100 g of concentrated sulfuric acid. Carefully add a drop of the test solution to a drop of the sample solution or a few milligrams of the solid sample. A violet color indicates the diphenylamine has been oxidized. This solution was developed for the titration of oxidizing agents such as nitrate, permanganate, dichromate, ceric salts, ferrous sulfate, and vanadate. Tungstate interferes with the test. This test is also used to detect nitrocellulose. Organic antimony compounds may be ignited and the residue tested with the diphenylamine solution directly; a blue color appears within a few minutes. Use this test for organic antimony only if the Rhodamine B reagent fails.	
References	**Manufacturer**
Welcher, F., *Chemical Solutions: Reagents Useful to the Chemist, Biologist, and Bacteriologist*, D. Van Nostrand Company, Inc., New York, 1942. *J. Am. Chem. Soc.* 46, 263 (1924). Feigl, F. and Anger, V., *Spot Tests In Inorganic Analysis*, Sixth Edition, Elsevier Publishing Company, Amsterdam, 1972. *Ann. de Chinzie Analyt.*, ii, 12l.	

Diphenylamine Reagent		
Detects	**Test Principle**	
• Formaldehyde	Strongly acidified diphenylamine reacts with formaldehyde to form a green color.	
	Test Phase	
		Solid
	✓	Liquid
		Gas/Vapor
	Sensitivity	
	Not available	

Directions and Comments	
Mix a 1% solution (w/w) of diphenylamine (CAS 122-39-4) in concentrated sulfuric acid. Add a drop of the test solution to a microplate or test tube and then carefully add one drop of the test solution. A green zone formed when the two solutions meet indicates formaldehyde. Compare to Diphenylamine Indicator Solution.	
References	**Manufacturer**
Welcher, F., *Chemical Solutions: Reagents Useful to the Chemist, Biologist, and Bacteriologist*, D. Van Nostrand Company, Inc., New York, 1942.	

Diphenylcarbazide Reagent (Chromate)

Detects	Test Principle		
• Chromate • Metals	Diphenylcarbazide reacts with various metal salts to produce various colors. Chromate produces a violet color.		
	Test Phase		
		Solid	
	✓	Liquid	
		Gas/Vapor	
	Sensitivity		
	0.01 ppm chromate		

Directions and Comments

Mix a saturated solution of diphenylcarbazide (CAS 1 40-22-7) in alcohol. Add the test solution to the sample solution. This test is applicable to acidic samples; acetic acid is preferred. Chromate causes a violet color; iron does not interfere. Other metals produce various colors in various basic and neutral pH. Compare to Diphenylcarbazide Test.

References	Manufacturer
Welcher, F., *Chemical Solutions: Reagents Useful to the Chemist, Biologist, and Bacteriologist,* D. Van Nostrand Company, Inc., New York, 1942. *J. Am. Chem. Soc.* 50, 2363 (1928). *Chemical Abstracts* 1818 (1930). *Chemist Analyst,* J.T. Baker, January 1936.	

Diphenylcarbazide Test

Detects	Test Principle		
• Chromate • Chromic acid • Chromium(III) • Chromium(VI) • Dichromate	Chromate in strong acid solution reacts with diphenylcarbazide to produce a blue-violet to red color.		
	Test Phase		
		Solid	
	✓	Liquid	
		Gas/Vapor	
	Sensitivity		
	0.5 µg/ml potassium chromate (Feigl)		

Directions and Comments

Chromate in acidic aqueous solution is mixed with a 1% alcoholic solution of diphenylcarbazide (CAS 140-22-7). A blue-violet to red color indicates chromate depending on the amount present. Jungreis advises use of a saturated solution of diphenylcarbazide for post-blast detection of chromium(VI). Chromium(III) is also detected if first oxidized to chromate by adding an equal volume of oxidizing solution prepared by mixing 8 ml bromine water (CAS 7726-95-6 and packaged in the Agri-Screen Ticket) with 2 ml 5N sodium hydroxide and 2 g potassium bromide (CAS 7758-02-3). The oxidizing solution is allowed to react for one minute and then a drop of concentrated sulfuric acid and then two drops of 20% 5-sulfosalicylic acid (CAS 5965-83-3) are added. Finally, the diphenylcarbazide solution is added, with a violet color indicating any form of chromium. Mercury, vanadium, and molybdenum salts interfere. Colored compounds such as some copper, cobalt, nickel, and iron salts obscure the violet result; a blank test solution may be helpful in determining color shift to violet. Mercury interference can be avoided by addition of hydrochloric acid to the sample before testing. Molybdate interference can be avoided by addition of oxalic acid to the sample before testing.

(continued)

Diphenylcarbazide Test (*Continued*)	
References	**Manufacturer**
Feigl, F., *Spot Tests in Inorganic Analysis*, Fifth Edition, Elsevier Publishing Company, New York, 1958. Jungreis, E., *Spot Test Analysis: Clinical, Environmental, Forensic, and Geochemical Applications*, Second Edition, John Wiley & Sons, Inc., New York, 1997. Feigl, F. and Anger, V., *Spot Tests in Inorganic Analysis*, Sixth Edition, Elsevier Publishing Company, Amsterdam, 1972.	

Diphenylcarbazone Reagent (Germanium)		
Detects	**Test Principle**	
• Germanium	Germanium-molybdate mixture reacts with diphenylcarbazone to produce a violet color.	
	Test Phase	
		Solid
	✓	Liquid
		Gas/Vapor
	Sensitivity	
	0.01 µg/ml germanium	

Directions and Comments

This test is presumptive for germanium. Dissolve the sample in water and add a few grains of ammonium molybdate (CAS 12054-85-2) or other molybdate salt. Add a drop of saturated alcoholic diphenylcarbazone (CAS 538-62-5, also present in the Clor-N-Oil Test). A purple color indicates germanium. The purple color will intensify upon addition of a drop of concentrated hydrochloric or sulfuric acid.

References	**Manufacturer**
Feigl, F. and Anger, V., *Spot Tests in Inorganic Analysis*, Sixth Edition, Elsevier Publishing Company, Amsterdam, 1972. *Anal. Abstracts*, 16 (1969). *Anal. Abstracts*, 2, 2684 (1955).	

Diphenylcarbazone Reagent (Mercury)		
Detects	**Test Principle**	
• Mercury salt • Mercury, organic • Organomercury compounds	Diphenylcarbazone reacts with mercurous ions to form a blue to violet color.	
	Test Phase	
		Solid
	✓	Liquid
		Gas/Vapor
	Sensitivity	
	1 µg/ml mercury	

(*continued*)

Diphenylcarbazone Reagent (Mercury) (*Continued*)
Directions and Comments
This test is specific for mercurous ions in 0.2N nitric acid solution. Other metals give colors with this test at other pH. The test loses sensitivity with decreasing pH. Mix a 1% alcoholic solution of diphenylcarbazone (CAS 538-62-5, also present in the Clor-N-Oil Test) and place a drop on a filter paper. Add a drop of test solution to the filter paper. A blue to violet color indicates mercurous ions. Molybdate may be masked by adding oxalic acid (CAS 6153-56-6); chromate is eliminated by reduction via hydrogen peroxide (CAS 7722-84-1). Cobalt, copper, and iron also produced colors with diphenylcarbazone, but at lower acidity. Chloride decreases sensitivity.

References	Manufacturer
Feigl, F. and Anger, V., *Spot Tests in Inorganic Analysis,* Sixth Edition, Elsevier Publishing Company, Amsterdam, 1972. *Anal. Abstracts,* 16 (1969).	

Dipyridyl Reagent		
Detects	**Test Principle**	
• Ferric • Ferrous • Iron(II) • Iron(III) • Molybdate • Tungstate	Dipyridyl reacts with ferrous ions to form a red precipitate in acidic solution.	
	Test Phase	
	✓	Solid
	✓	Liquid
		Gas/Vapor
	Sensitivity	
	0.03 µg/ml iron(II)	

Directions and Comments
Mix a 2% solution of a,a′-dipyridyl (2,2′-dipyridyl, CAS 366-18-7) in alcohol. Dissolve the sample in 3N hydrochloric acid. Place a drop of liquid sample or a few milligrams of solid sample in a spot plate and add a drop of test solution. A pink to red color indicates ferrous ion. Ferric ion does not interfere. Ferric ion can be detected by reduction with hydroxylamine hydrochloride (CAS 5470-11-1) to ferrous ion. The test may be performed on filter paper impregnated with the test solution, which has been dried. Molybdate is detected with a,a′-dipyridyl (1,10-phenanthroline, CAS 66-71-7, may be substituted) if combined with stannous chloride (Tin[II] chloride dihydrate, CAS 10025-69-1) by the formation of violet molybdenum chloride - dipyridyl complex. Tungstates interfere by forming a blue color, but can be removed by complex formation with tartaric acid (CAS 87-69-4).

References	Manufacturer
Welcher, F., *Chemical Solutions: Reagents Useful to the Chemist, Biologist, and Bacteriologist*, D. Van Nostrand Company, Inc., New York, 1942. Feigl, F. and Anger, V., *Spot Tests in Inorganic Analysis,* Sixth Edition, Elsevier Publishing Company, Amsterdam, 1972. *Mikrochimica Acta,* 1, 264–266 (1937).	

Dithizone Reagent (Lead)	
Detects	**Test Principle**
• Dialkyl-lead salts • Lead	Lead forms a red complex with dithizone in neutral solution.

(continued)

Dithizone Reagent (Lead) (*Continued*)

	Test Phase	
• Mercury salts		
• Trialkyl-lead salts		Solid
• Zinc	✓	Liquid
		Gas/Vapor

Sensitivity
0.04 µg/ml lead or 0.1 µg/ml dialkyl-lead or trialkyl-lead salt in neutral solution with carbon tetrachloride as solvent

Directions and Comments

Lead and other cations form water-insoluble salts with dithizone. Carbon tetrachloride (carcinogenic) was originally used as the dithizone solvent; hexane or methanol is a safer alternative. The sample must be nearly neutral; if acidic, add sodium carbonate (CAS 497-19-8). Mix the test solution in a ratio of 1 mg of dithizone (CAS 60-10-6) in 50 ml of hexane or methanol. The test uses a drop of liquid sample or a small piece of precipitate in a drop or two of dithizone solution, which is then shaken. A red color indicates lead; other colors are possible. Dialkyl-lead salts produce a yellow-orange color; trialkyl-lead salts produce a red-orange color. Mercury salts form an orange color. Silver, nickel, zinc (purple-red), cadmium, and antimony interfere but can be treated with cyanide. Tin, bismuth, and antimony also interfere but can be separated with a qualitative analysis scheme. Oxidizers can be treated with hydroxylamine (CAS 5470-11-1). See color images 2.15 and 2.16.

References	Manufacturer
H. Fischer, *Z. Angew. Chem.*, 42, 1025 (1929); *Mikrochemie*, 8, 319 (1930). Feigl, F. and Anger, V., *Spot Tests in Inorganic Analysis*, Sixth Edition, Elsevier Publishing Company, Amsterdam, 1972. *Anal. Abstracts*, 16 (1969). Welcher. *Ric. Sci.*, 30, 1988 (1960). *Anal. Abstracts*, 8, 5072 (1961).	

Dithizone Reagent (Tin)

Detects	Test Principle
• Organyl tin ions	Monoalkyl, dialkyl, and trialkyl or aryl tin(IV) compounds react with dithizone to form various colored compounds.
• Stannic	
• Stannous	

Detects	Test Phase	
• Tin(II)	✓	Solid
• Tin(IV)	✓	Liquid
		Gas/Vapor

Sensitivity
"µg amounts" of organyl tin ions.

Directions and Comments

This test was intended for laboratory use as a paper chromatograph. Mix the test solution in a ratio of 1 mg of dithizone (CAS 60-10-6) in 50 ml of hexane or methanol. Dissolve 0.1 g of sample into 1 ml of a suitable solvent and place a drop on the center of a filter paper. Add a drop or two more of the solvent to force the sample to migrate through the paper. Allow a few minutes for the sample to separate and then spray the dithizone solution onto the paper.

(*continued*)

Dithizone Reagent (Tin)

A violet color on a green background indicates inorganic tin(IV). A blue-red color indicates mono-substituted tin compounds. A yellow color indicates di- and tri-substituted tin compounds. Inorganic tin(IV) and monoalkylated tin(IV) compounds can be masked with EDTA (CAS 60-00-4).

References	Manufacturer
Feigl, F. and Anger, V., *Spot Tests in Inorganic Analysis,* Sixth Edition, Elsevier Publishing Company, Amsterdam, 1972.	

Dragendorff's Reagent		

Detects	Test Principle	
• Alkaloid • Amine • Amine, Quat • Amine, Secondary • Amine, Tertiary • Quaternary ammonium compound	Bismuth subnitrate reacts with several types of amines in acetic acid and potassium iodide to produce a bright orange color.	
	Test Phase	
	✓	Solid
	✓	Liquid
		Gas/Vapor
	Sensitivity	
	Not available. Qualitative result only.	

Directions and Comments

Dragendorff's Reagent is a two-part solution that will keep for only a few days once mixed. Solution A is 1.7 g of bismuth subnitrate (CAS 1304-85-4) in a 20% solution (v/v) of acetic acid. Solution B is a 40% solution (w/v) of potassium iodide (CAS 7681-11-0) in water. Separately these two solutions will store indefinitely. Add a drop of the mixed Dragendorff's Reagent to a drop of liquid sample or a few milligrams of solid sample. The mixed Dragendorff's Reagent is a clear orange color. The appearance of an brilliant, opaque orange color indicates a secondary, tertiary, or quaternary amine. Primary amines and heterocycles such as pyridine reportedly produce a more brown orange color. Phosphine compounds and some ethers may also be detected. See color images 2.17, 2.18, and 2.19.

References	Manufacturer
Welcher, F., *Chemical Solutions: Reagents Useful to the Chemist, Biologist, and Bacteriologist,* D. Van Nostrand Company, Inc., New York, 1942. *Chemical Abstracts* 19, 223 (1925).	

Drager Acetaldehyde 100/a		

Detects	Test Principle	
• Aldehyde • Ester • Ether • Hydrocarbon (petroleum) • Ketone	Aldehydes convert yellow-orange chromium(VI) to green chromium(III).	
	Test Phase	
		Solid
		Liquid
	✓	Gas/Vapor
	Sensitivity	
	Not available. Qualitative result only.	

(*continued*)

Drager Acetaldehyde 100/a (*Continued*)
Directions and Comments
This is a broad screening test. The most practical detection of these compounds at higher concentration will be made with a combustible gas indicator. More complex molecules give a reduced sensitivity. Some aldehydes, ethers, ketones, petroleum hydrocarbons, and esters are indicated. The package insert may contain additional information.

References	Manufacturer
Drager-Tube/CMS Handbook, 13th Edition, Drager Safety AG & Co. KGaA, Lubeck, Germany, 2004.	Drager Safety AG & Co. KGaA, Lubeck, Germany

Drager Acetic Acid 5/a		
Detects	**Test Principle**	
• Acid	This test is a pH indicator.	
	Test Phase	
		Solid
		Liquid
	✓	Gas/Vapor
	Sensitivity	
	5 ppm acetic acid	
Directions and Comments		
This is a sensitive pH test in air. Other organic acid gases produce a yellow color from blue. Mineral acids produce a red discoloration. The package insert may contain additional information.		

References	Manufacturer
Drager-Tube/CMS Handbook, 13th Edition. Drager Safety AG & Co. KGaA, Lubeck, Germany, 2004.	Drager Safety AG & Co. KGaA, Lubeck, Germany

Drager Acetone 100/b		
Detects	**Test Principle**	
• Acetone • Aldehyde • Ammonia • Ketone	Carbonyl group reacts with 2,4-dinitrophenylhydrazine to produce yellow hydrazone.	
	Test Phase	
		Solid
		Liquid
	✓	Gas/Vapor
	Sensitivity	
	Not available. Qualitative result only.	
Directions and Comments		
This test is designed as a screen for ketones and aldehydes, which produce a yellow color. Ammonia causes a yellow-brown discoloration. The package insert may contain additional information.		

References	Manufacturer
Drager-Tube/CMS Handbook, 13th Edition. Drager Safety AG & Co. KGaA, Lubeck, Germany, 2004.	Drager Safety AG & Co. KGaA, Lubeck, Germany

Drager Acid Test	
Detects	Test Principle
• Acid	This test is a pH indicator.

Test Phase		
		Solid
		Liquid
	✓	Gas/Vapor

Sensitivity
Qualitative only. No specific sensitivity listed.

Directions and Comments
This test cannot discern individual acids. Any discoloration from yellow to pink indicates an acid gas or vapor, but it is impossible to differentiate them with this test. The package insert may contain additional information.

References	Manufacturer
Drager-Tube/CMS Handbook, 13th Edition. Drager Safety AG & Co. KGaA, Lubeck, Germany, 2004.	Drager Safety AG & Co. KGaA, Lubeck, Germany

Drager Acrylonitrile 0.5/a	
Detects	Test Principle
• Acrylonitrile • Nitrile	Chromium(VI) generates hydrogen cyanide from nitriles, which then reacts with mercuric chloride to form hydrochloric acid. The newly formed acid reacts with methyl red to produce a red color.

Test Phase		
		Solid
		Liquid
	✓	Gas/Vapor

Sensitivity
0.5 ppm acrylonitrile

Directions and Comments
This test is designed for quantitative measurement of acrylonitrile. The method should be adaptable to qualitative results for other nitriles. The package insert may contain additional information.

References	Manufacturer
Drager-Tube/CMS Handbook, 13th Edition. Drager Safety AG & Co. KGaA, Lubeck, Germany, 2004.	Drager Safety AG & Co. KGaA, Lubeck, Germany

Drager Alcohol 100/a	
Detects	Test Principle
• Alcohol	Alcohols convert yellow-orange chromium(VI) to green chromium(III).

Test Phase		
		Solid
		Liquid
	✓	Gas/Vapor

(continued)

Drager Alcohol 100/a (Continued)	
	Sensitivity
	Qualitative only. No specific sensitivity listed.
Directions and Comments	
This is a broad screening test. The most practical detection of these compounds at higher concentration will be with a combustible gas sensor. More complex alcohols ($>C_4$) give a markedly reduced sensitivity. Some aldehydes, ethers, ketones, and esters are indicated. The package insert may contain additional information.	
References	**Manufacturer**
Drager-Tube/CMS Handbook, 13th Edition. Drager Safety AG & Co. KGaA, Lubeck, Germany, 2004.	Drager Safety AG & Co. KGaA, Lubeck, Germany

Drager Amine Test		
Detects	**Test Principle**	
• Amine	This test is a pH indicator.	
• Ammonia	**Test Phase**	
• Base		Solid
• Caustic		Liquid
	✓	Gas/Vapor
	Sensitivity	
	Qualitative only. No specific sensitivity listed.	
Directions and Comments		
Amines and other reactive basic gases can be detected with a pH indicator. Compare to Drager Acid Test. The package insert may contain additional information.		
References	**Manufacturer**	
Drager-Tube/CMS Handbook, 13th Edition. Drager Safety AG & Co. KGaA, Lubeck, Germany, 2004.	Drager Safety AG & Co. KGaA, Lubeck, Germany	

Drager Ammonia 2/a		
Detects	**Test Principle**	
• Amine	This test is a pH indicator.	
• Ammonia	**Test Phase**	
• Base		Solid
• Caustic		Liquid
	✓	Gas/Vapor
	Sensitivity	
	2 ppm ammonia	
Directions and Comments		
Ammonia, amines, and other basic gases and vapors are detected with a pH indicator. The package insert may contain additional information.		
References	**Manufacturer**	
Drager-Tube/CMS Handbook, 13th Edition. Drager Safety AG & Co. KGaA, Lubeck, Germany, 2004.	Drager Safety AG & Co. KGaA, Lubeck, Germany	

Drager Aniline 0.5/a	
Detects	**Test Principle**
• Aniline • Aromatic compound • Ester • Ether • Hydrocarbon (petroleum) • Ketone • Methyl aniline	Aniline reduces orange chromium(VI) to green chromium(III).

Test Phase	
	Solid
	Liquid
✓	Gas/Vapor

Sensitivity
0.5 ppm aniline

Directions and Comments
This test is sensitive to aniline and methyl aniline. Some ethers, ketones, aromatics, petroleum hydrocarbons, and esters are indicated at varying sensitivities. The package insert may contain additional information.

References	**Manufacturer**
Drager-Tube/CMS Handbook, 13th Edition. Drager Safety AG & Co. KGaA, Lubeck, Germany, 2004.	Drager Safety AG & Co. KGaA, Lubeck, Germany

Drager Arsine 0.05/a	
Detects	**Test Principle**
• Antimony hydride • Arsine • Phosphine	Arsine and phosphine react with gold(III) to form a gray-violet color (colloidal gold).

Test Phase	
	Solid
	Liquid
✓	Gas/Vapor

Sensitivity
Arsine - 0.05 ppm

Directions and Comments
This test is sensitive for arsine, phosphine, and antimony hydride in air. The package insert may contain additional information.

References	**Manufacturer**
Drager-Tube/CMS Handbook, 13th Edition. Drager Safety AG & Co. KGaA, Lubeck, Germany, 2004.	Drager Safety AG & Co. KGaA, Lubeck, Germany

Drager Benzene 0.5/a	
Detects	**Test Principle**
• Aromatic • Benzene • Ethyl benzene • Toluene • Xylene	Two aromatic molecules are linked by formaldehyde in a condensation reaction. The linked aromatics react with sulfuric acid and a pale brown color is produced.

(continued)

Drager Benzene 0.5/a (*Continued*)

	Test Phase	
		Solid
		Liquid
✓		Gas/Vapor
	Sensitivity	
	0.5 ppm benzene	

Directions and Comments

This test is a screen for many aromatic compounds. Benzene as well as toluene, ethyl benzene, xylene, and others are detected. The package insert may contain additional information.

References	Manufacturer
Drager-Tube/CMS Handbook, 13th Edition. Drager Safety AG & Co. KGaA, Lubeck, Germany, 2004.	Drager Safety AG & Co. KGaA, Lubeck, Germany

Drager Carbon Dioxide 5%/A

Detects	Test Principle	
• Carbon dioxide • Sulfur dioxide	Carbon dioxide reacts with hydrazine to produce a purple color.	
	Test Phase	
		Solid
		Liquid
	✓	Gas/Vapor
	Sensitivity	
	<1% carbon dioxide	

Directions and Comments

This test is designed for the measurement of carbon dioxide in percent amounts. A conventional "four-gas" air monitor will be more practical monitoring depressed oxygen level. This tube may be used to verify carbon dioxide in the case of other inert, oxygen displacing gases. Sulfur dioxide reacts with approximately the same sensitivity within the measuring range of 5–60%. The package insert may contain additional information.

References	Manufacturer
Drager-Tube/CMS Handbook, 13th Edition. Drager Safety AG & Co. KGaA, Lubeck, Germany, 2004.	Drager Safety AG & Co. KGaA, Lubeck, Germany

Drager Carbon Monoxide 2/a

Detects	Test Principle	
• Carbon monoxide	Carbon monoxide liberates free iodine from iodine pentoxide to produce a brown color.	
	Test Phase	
		Solid
		Liquid
	✓	Gas/Vapor
	Sensitivity	
	2 ppm carbon monoxide	

(*continued*)

Drager Carbon Monoxide 2/a (*Continued*)
Directions and Comments
This reaction is sensitive to many compounds, but this tube contains a prefilter that removes many cross-sensitive compounds. Petroleum hydrocarbons, benzene, halogenated hydrocarbons, and hydrogen sulfide are blocked in low concentration. Higher concentration can overload the filter and produce a brown color in the layer. The package insert may contain additional information.

References	Manufacturer
Drager-Tube/CMS Handbook, 13th Edition. Drager Safety AG & Co. KGaA, Lubeck, Germany, 2004.	Drager Safety AG & Co. KGaA, Lubeck, Germany

Drager Carbon Tetrachloride 5/c		
Detects	**Test Principle**	
• Carbon tetrachloride • Chlorohydrocarbon • Phosgene	Carbon tetrachloride and disulfuric acid react to produce phosgene. Phosgene and an aromatic amine react to produce a blue-green reaction product. This is probably a reaction of phosgene with p-dimethylaminobenzaldehyde in the presence of N,N-dimethylaniline, which produces a green to deep blue color.	
	Test Phase	
		Solid
		Liquid
	✓	Gas/Vapor
	Sensitivity	
	5 ppm carbon tetrachloride	
Directions and Comments		
This test is not specific to carbon tetrachloride and does not indicate all chlorinated hydrocarbons. Phosgene is indicated with approximately the same sensitivity. The package insert may contain additional information.		

References	Manufacturer
Drager-Tube/CMS Handbook, 13th Edition. Drager Safety AG & Co. KGaA, Lubeck, Germany, 2004. Pitschmann, V. et al., A simple in situ visual and tristimulus colorimetric method for the determination of diphosgene in air, *J. Serb. Chem. Soc.,* 72, 1031–1037 (2007).	Drager Safety AG & Co. KGaA, Lubeck, Germany

Drager CDS Simultaneous Test Set I - Tube 1		
Detects	**Test Principle**	
• Sulfur mustard • Thioether	Gold chloride reacts with thioether (R-S-R') and chloramide to form an orange complex.	
	Test Phase	
		Solid
		Liquid
	✓	Gas/Vapor
	Sensitivity	
	1 mg/m³ sulfur mustard.	

(continued)

Drager CDS Simultaneous Test Set I - Tube 1 (*Continued*)	
Directions and Comments	
This test is a screen for all thioethers and is not specific to sulfur mustard. Compare to Drager Thioether tube. The package insert may contain additional information.	
References	**Manufacturer**
Drager-Tube/CMS Handbook, 13th Edition. Drager Safety AG & Co. KGaA, Lubeck, Germany, 2004. *Drager Civil Defense Simultest Kit Accessory Guide*, Drager Safety, Inc. Pittsburgh, PA, USA, April 2002.	Drager Safety AG & Co. KGaA, Lubeck, Germany

Drager CDS Simultaneous Test Set I - Tube 2		
Detects	**Test Principle**	
• Acetyl chloride • Carbonyl bromide • Phosgene	Phosgene, ethylaniline and dimethylaminobenzaldehyde react to form a blue-green reaction product.	
	Test Phase	
		Solid
		Liquid
	✓	Gas/Vapor
	Sensitivity	
	0.2 ppm phosgene	
Directions and Comments		
This method is different than the phosgene test in Simultaneous Test Set II. Low amounts of hydrochloric acid (<100 ppm) do not interfere. The test is cross sensitive to acetyl chloride and carbonyl bromide. Compare to Drager Phosgene 0.25/c tube. The package insert may contain additional information.		
References	**Manufacturer**	
Drager-Tube/CMS Handbook, 13th Edition. Drager Safety AG & Co. KGaA, Lubeck, Germany, 2004. *Drager Civil Defense Simultest Kit Accessory Guide*, Drager Safety, Inc., Pittsburgh, PA, USA, April 2002.	Drager Safety AG & Co. KGaA, Lubeck, Germany	

Drager CDS Simultaneous Test Set I - Tube 3		
Detects	**Test Principle**	
• Hydrocyanic acid	Mercuric chloride reacts with hydrogen cyanide to produce hydrogen chloride. Hydrogen chloride reacts with methyl red to form a red color.	
	Test Phase	
		Solid
		Liquid
	✓	Gas/Vapor
	Sensitivity	
	1 ppm hydrocyanic acid	

(continued)

Drager CDS Simultaneous Test Set I - Tube 3 (*Continued*)
Directions and Comments
This test is cross sensitive to phosphine. Do not use for hydrogen chloride detection as 1000 ppm hydrogen chloride does not interfere. High concentrations of other gases may interfere. Compare to Drager Hydrocyanic Acid 2/a tube. The package insert may contain additional information.

References	Manufacturer
Drager-Tube/CMS Handbook, 13th Edition. Drager Safety AG & Co. KGaA, Lubeck, Germany, 2004. *Drager Civil Defense Simultest Kit Accessory Guide*, Drager Safety, Inc., Pittsburgh, PA, USA, April 2002.	Drager Safety AG & Co. KGaA, Lubeck, Germany

Drager CDS Simultaneous Test Set I - Tube 4		
Detects	**Test Principle**	
• Arsine • Lewisite • Organic arsenic compound • Phosphine	Step one collects a sample in the reaction media. Arsine separates gold from a complex to form a gray to violet color. In step two, zinc and hydrochloric acid reduce organic arsenic to arsine and with more pump strokes is indicated by the reaction from step one.	
	Test Phase	
		Solid
		Liquid
	✓	Gas/Vapor
	Sensitivity	
	0.1 ppm arsine. 3 mg/m³ of organic arsenic.	
Directions and Comments		
Step one is sensitive to arsine and phosphine. Step two detects organic arsenic (arsenic[III]) compounds. Both results appear in the same colorimetric area of the tube. Arsine and phosphine can produce a gray color and colloidal gold can form rose, purple, or yellow color depending on concentration. Compare to Drager Organic Arsenic Compounds and Arsine tube. The package insert may contain additional information.		

References	Manufacturer
Drager-Tube/CMS Handbook, 13th Edition. Drager Safety AG & Co. KGaA, Lubeck, Germany, 2004. *Drager Civil Defense Simultest Kit Accessory Guide*, Drager Safety, Inc., Pittsburgh, PA, USA, April 2002.	Drager Safety AG & Co. KGaA, Lubeck, Germany

Drager CDS Simultaneous Test Set I - Tube 5	
Detects	**Test Principle**
• Nitrogen mustard • Organic basic nitrogen compound	Organic basic nitrogen compounds (mostly amines) react with potassium bismuth iodide to form an orange-red indication.

(continued)

Drager CDS Simultaneous Test Set I - Tube 5 (*Continued*)

	Test Phase	
		Solid
		Liquid
	✓	Gas/Vapor
	Sensitivity	
	1 mg/m^3 unspecified organic basic nitrogen compound	

Directions and Comments

This is a general screening test for organic amines. Individual compounds cannot be identified. Compare to Drager Organic Basic Nitrogen Compounds tube and Dragendorff's Reagent. The package insert may contain additional information.

References	Manufacturer
Drager-Tube/CMS Handbook, 13th Edition. Drager Safety AG & Co. KGaA, Lubeck, Germany, 2004. *Drager Civil Defense Simultest Kit Accessory Guide,* Drager Safety, Inc., Pittsburgh, PA, USA, April 2002.	Drager Safety AG & Co. KGaA, Lubeck, Germany

Drager CDS Simultaneous Test Set V - Tube 1

Detects	Test Principle	
• Cyanogen bromide • Cyanogen chloride	Cyanogen chloride and pyridine will form glutaconaldehyde cyanamide. Glutaconaldehyde in the presence of barbituric acid forms a pink reaction product.	
	Test Phase	
		Solid
		Liquid
	✓	Gas/Vapor
	Sensitivity	
	0.25 ppm cyanogen chloride	

Directions and Comments

This test is sensitive to 0.25 ppm cyanogen chloride and is cross sensitive to cyanogen bromide. This test could be used to verify cyanogen chloride from other halogenated hydrocarbons. Compare to Drager Cyanogen Chloride 0.25/a tube. The package insert may contain additional information.

References	Manufacturer
Drager-Tube/CMS Handbook, 13th Edition. Drager Safety AG & Co. KGaA, Lubeck, Germany, 2004. *Drager Civil Defense Simultest Kit Accessory Guide,* Drager Safety, Inc., Pittsburgh, PA, USA, April 2002.	Drager Safety AG & Co. KGaA, Lubeck, Germany

Drager CDS Simultaneous Test Set V - Tube 2

Detects	Test Principle
• Sulfur mustard • Thioether	Gold chloride and thioether (R-S-R´) react to form a bright yellow to orange complex.

(*continued*)

Drager CDS Simultaneous Test Set V - Tube 2 (*Continued*)		
	Test Phase	
		Solid
		Liquid
	✓	Gas/Vapor
	Sensitivity	
	1 mg/m³ sulfur mustard.	

Directions and Comments
The test is a screen for all thioethers and is not specific to sulfur mustard. This is the same test as CDS Set I. Compare to Drager Thioether tube. The package insert may contain additional information.

References	Manufacturer
Drager-Tube/CMS Handbook, 13th Edition. Drager Safety AG & Co. KGaA, Lubeck, Germany, 2004. *Drager Civil Defense Simultest Kit Accessory Guide*, Drager Safety, Inc., Pittsburgh, PA, USA, April 2002.	Drager Safety AG & Co. KGaA, Lubeck, Germany

Drager CDS Simultaneous Test Set V - Tube 3		
Detects	**Test Principle**	
• Acetyl chloride • Carbonyl bromide • Phosgene	Phosgene, ethylaniline and dimethylaminobenzaldehyde form a blue-green reaction product.	
	Test Phase	
		Solid
		Liquid
	✓	Gas/Vapor
	Sensitivity	
	0.2 ppm phosgene	

Directions and Comments
This method is different than the phosgene test in Test Set II. Low amounts of hydrochloric acid (<100 ppm) do not interfere. Test is cross sensitive to acetyl chloride and carbonyl bromide. This is the same test as CDS Set I. Compare to Drager Phosgene 0.25/c tube. The package insert may contain additional information.

References	Manufacturer
Drager-Tube/CMS Handbook, 13th Edition. Drager Safety AG & Co. KGaA, Lubeck, Germany, 2004. *Drager Civil Defense Simultest Kit Accessory Guide*, Drager Safety, Inc. Pittsburgh, PA, USA, April 2002.	Drager Safety AG & Co. KGaA, Lubeck, Germany

Drager CDS Simultaneous Test Set V - Tube 4	
Detects	**Test Principle**
• Bromine • Chlorine • Chlorine dioxide • Nitrogen dioxide	Chlorine gas and o-tolidine produce an orange reaction product.

(continued)

Drager CDS Simultaneous Test Set V - Tube 4 (*Continued*)		
	Test Phase	
		Solid
		Liquid
	✓	Gas/Vapor
	Sensitivity	
	0.2 ppm chlorine	

Directions and Comments

This test is cross sensitive to bromine, chlorine dioxide, and nitrogen dioxide. This is the same as the test contained in Simultaneous Test Set II. Compare to Drager Chlorine 0.2/a tube. The package insert may contain additional information.

References	**Manufacturer**
Drager-Tube/CMS Handbook, 13th Edition. Drager Safety AG & Co. KGaA, Lubeck, Germany, 2004. *Drager Civil Defense Simultest Kit Accessory Guide,* Drager Safety, Inc., Pittsburgh, PA, USA, April 2002.	Drager Safety AG & Co. KGaA, Lubeck, Germany

Drager CDS Simultaneous Test Set V - Tube 5		
Detects	**Test Principle**	
• G-nerve agent • Pesticide • Phosphoric acid ester • V-nerve agent	If phosphoric acid ester is present, cholinesterase in the tube is inactivated after a sample is drawn into the tube and 1 minute passes. Butylcholine iodide and water are added and if the cholinesterase is active, it will cleave butylcholine iodide and form butyric acid. Butyric acid and phenol red pH indicator produces a yellow color (negative result). If the cholinesterase has been inactivated, butyric acid will not form and the phenol red pH indicator will produce a red color (positive result).	
	Test Phase	
		Solid
		Liquid
	✓	Gas/Vapor
	Sensitivity	
	0.025 ppm dichlorvos	

Directions and Comments

This is a sensitive test for airborne phosphoric acid esters (dichlorvos 0.05 ppm) but detects others at varying sensitivity. The reference does not clearly state if other cholinesterase-inhibiting materials (carbamates, thiophosphates, phosphonates and others) are detected at similar sensitivity. The package insert may contain additional information.

References	**Manufacturer**
Drager-Tube/CMS Handbook, 13th Edition. Drager Safety AG & Co. KGaA, Lubeck, Germany, 2004. *Drager Civil Defense Simultest Kit Accessory Guide*, Drager Safety, Inc., Pittsburgh, PA, USA, April 2002.	Drager Safety AG & Co. KGaA, Lubeck, Germany

Drager Chlorine 0.2/a	
Detects	**Test Principle**
• Bromine • Chlorine • Chlorine dioxide • Nitrogen dioxide	Chlorine gas and o-tolidine produce an orange reaction product.

	Test Phase	
		Solid
		Liquid
	✓	Gas/Vapor

Sensitivity	
0.2 ppm chlorine	

Directions and Comments	
This test is cross sensitive to bromine, chlorine dioxide, and nitrogen dioxide. This is the same as the test contained in Simultaneous Test Set II and V. Compare to Drager Chlorine 0.2/a tube. The package insert may contain additional information.	
References	**Manufacturer**
Drager-Tube/CMS Handbook, 13th Edition. Drager Safety AG & Co. KGaA, Lubeck, Germany, 2004.	Drager Safety AG & Co. KGaA, Lubeck, Germany

Drager Clan Lab Simultaneous Test Set - Tube 1	
Detects	**Test Principle**
• Antimony hydride • Arsine • Phosphine	Arsine and phosphine react with gold(III) to form a gray-violet color (colloidal gold).

	Test Phase	
		Solid
		Liquid
	✓	Gas/Vapor

Sensitivity	
0.3 ppm phosphine	

Directions and Comments	
Test is sensitive for phosphine but is cross sensitive to arsine and antimony hydride. The package insert may contain additional information.	
References	**Manufacturer**
Drager-Tube/CMS Handbook, 13th Edition. Drager Safety AG & Co. KGaA, Lubeck, Germany, 2004. *Drager Clan Lab Simultest Kit Illustrated Accessory Guide,* Drager Safety, Inc., Pittsburgh, PA, USA, January 2002.	Drager Safety AG & Co. KGaA, Lubeck, Germany

Drager Clan Lab Simultaneous Test Set - Tube 2	
Detects	**Test Principle**
• Phosgene	Phosgene and an aromatic amine produce a red color.

(continued)

Drager Clan Lab Simultaneous Test Set - Tube 2 (*Continued*)

	This is probably a reaction of phosgene with p-nitrobenzylpyridine (4-[4-nitrobenzyl]pyridine, CAS 1083-48-3) to form a derivative of glutaconic aldehyde. N-benzylaniline (CAS 103-32-2) then condenses the aldehyde to form a pink to red color.

Test Phase

	Solid
	Liquid
✓	Gas/Vapor

Sensitivity

0.05 ppm phosgene

Directions and Comments

This test may be used to verify phosgene in the presence of chlorinated hydrocarbons. High concentration of phosgene will bleach the color indicator (>30 ppm). Chlorine and hydrochloric acid interfere. Compare to Drager Phosgene 0.02/a tube. The package insert may contain additional information.

References	Manufacturer
Drager-Tube/CMS Handbook, 13th Edition. Drager Safety AG & Co. KGaA, Lubeck, Germany, 2004. *Drager Clan Lab Simultest Kit Illustrated Accessory Guide*, Drager Safety, Inc., Pittsburgh, PA, USA, January 2002. Feigl, F. and Anger, V., *Spot Tests in Inorganic Analysis,* Sixth Edition, Elsevier Publishing Company, Amsterdam, 1972.	Drager Safety AG & Co. KGaA, Lubeck, Germany

Drager Clan Lab Simultaneous Test Set - Tube 3

Detects	Test Principle
• Amine • Ammonia • Bases • Caustics	Yellow pH indicator turns blue in the presence of basic gas or vapor.

Test Phase

	Solid
	Liquid
✓	Gas/Vapor

Sensitivity

20 ppm ammonia

Directions and Comments

This is a sensitive pH test for air. All basic gases produce a yellow to blue color. Compare to Drager Acid Test. The package insert may contain additional information.

References	Manufacturer
Drager-Tube/CMS Handbook, 13th Edition. Drager Safety AG & Co. KGaA, Lubeck, Germany, 2004. *Drager Clan Lab Simultest Kit Illustrated Accessory Guide,* Drager Safety, Inc., Pittsburgh, PA, USA, January 2002.	Drager Safety AG & Co. KGaA, Lubeck, Germany

Drager Clan Lab Simultaneous Test Set - Tube 4	
Detects	**Test Principle**
• Acid • Hydrochloric acid	Bromophenol blue pH indicator and hydrochloric acid produces a yellow color.

	Test Phase	
		Solid
		Liquid
	✓	Gas/Vapor

Sensitivity
2 ppm hydrochloric acid

Directions and Comments

This is a sensitive pH test for air. All acid gases produce a blue to yellow color. Compare to acid gas tube in Set I. Hydrochloric acid cannot be discerned from other mineral acids. Compare to Drager Hydrochloric Acid 1/a tube. The package insert may contain additional information.

References	**Manufacturer**
Drager-Tube/CMS Handbook, 13th Edition. Drager Safety AG & Co. KGaA, Lubeck, Germany, 2004. *Drager Clan Lab Simultest Kit Illustrated Accessory Guide*, Drager Safety, Inc., Pittsburgh, PA, USA, January 2002.	Drager Safety AG & Co. KGaA, Lubeck, Germany

Drager Clan Lab Simultaneous Test Set - Tube 5	
Detects	**Test Principle**
• Iodine	Reaction principle not available. Possibly the formation of hydroiodic acid alters a pH indicator from yellow to pink.

	Test Phase	
		Solid
		Liquid
	✓	Gas/Vapor

Sensitivity
0.5 ppm iodine

Directions and Comments

A yellow compound changes to pale-pink in the presence of iodine. The reaction principle is not available. Possibly the formation of hydroiodic acid alters a pH indicator from yellow to pink. The package insert may contain additional information.

References	**Manufacturer**
Drager-Tube/CMS Handbook, 13th Edition. Drager Safety AG & Co. KGaA, Lubeck, Germany, 2004. *Drager Clan Lab Simultest Kit Illustrated Accessory Guide*, Drager Safety, Inc., Pittsburgh, PA, USA, January 2002.	Drager Safety AG & Co. KGaA, Lubeck, Germany

Drager Cyanide 2/a		
Detects	**Test Principle**	
• Acid • Cyanide • Phosphine	Inorganic cyanide compounds in air react with sulfuric acid to form hydrogen cyanide. Hydrogen cyanide reacts with mercuric chloride to form hydrogen chloride, which reacts with a pH indicator to form a red color.	
	Test Phase	
		Solid
		Liquid
	✓	Gas/Vapor
	Sensitivity	
	2 mg/m^3	
Directions and Comments		
Acid in air moves directly to the indicating layer and produces a red result. Phosphine will also produce a red result. The package insert may contain additional information.		
References	**Manufacturer**	
Drager-Tube/CMS Handbook, 13th Edition. Drager Safety AG & Co. KGaA, Lubeck, Germany, 2004.	Drager Safety AG & Co. KGaA, Lubeck, Germany	

Drager Cyanogen Chloride 0.25/a		
Detects	**Test Principle**	
• Cyanogen bromide • Cyanogen chloride	Cyanogen chloride reacts with pyridine to form glutaconaldehyde cyanamide, which then reacts with barbituric acid to produce a pink reaction product.	
	Test Phase	
		Solid
		Liquid
	✓	Gas/Vapor
	Sensitivity	
	0.25 ppm cyanogen chloride	
Directions and Comments		
This reaction is used in some aqueous cyanide tests. Many cyanide compounds can be converted to cyanogen chloride, which go on to be indicated through this method. Hydrogen cyanide is also detected, but at a different sensitivity. The package insert may contain additional information.		
References	**Manufacturer**	
Drager-Tube/CMS Handbook, 13th Edition. Drager Safety AG & Co. KGaA, Lubeck, Germany, 2004.	Drager Safety AG & Co. KGaA, Lubeck, Germany	

Drager Ethylene 0.1/a	
Detects	**Test Principle**
• Alkene • Carbon monoxide • Ethylene	Alkene (C=C) compounds and palladium molybdate complex react to form a blue reaction product.

Test Phase		
		Solid
		Liquid
	✓	Gas/Vapor

Sensitivity
0.2 ppm ethylene

Directions and Comments
This test screens for material containing carbon-carbon double bonds. The test does not discern material. Carbon monoxide produces a gray-blue color through the entire layer. The package insert may contain additional information.

References	**Manufacturer**
Drager-Tube/CMS Handbook, 13th Edition. Drager Safety AG & Co. KGaA, Lubeck, Germany, 2004.	Drager Safety AG & Co. KGaA, Lubeck, Germany

Drager Halogenated Hydrocarbon 100/a	
Detects	**Test Principle**
• Acid • Freon • Halogen gas • Halohydrocarbon • Halon® • Hydrogen halide	Tube contains a heat source that will pyrolize halogenated hydrocarbons. Chlorine radical forms hydrogen chloride and the pH indicator produces a yellow color from purple.

Test Phase		
		Solid
		Liquid
	✓	Gas/Vapor

Sensitivity
200 ppm R113 (trichlorotrifluoroethane)

Directions and Comments
Test is an ignition source and is not intrinsically safe. The test only works while the metal in the tube is hot. The test detects all halogenated hydrocarbons and is cross sensitive to halogens and hydrogen halides such as hydrogen fluoride, hydrogen chloride, hydrogen bromide, and hydrogen iodide. The package insert may contain additional information.

References	**Manufacturer**
Drager-Tube/CMS Handbook, 13th Edition. Drager Safety AG & Co. KGaA, Lubeck, Germany, 2004.	Drager Safety AG & Co. KGaA, Lubeck, Germany

Drager Hydrazine 0.2/a		
Detects	**Test Principle**	
• Hydrazine • Hydrogen sulfide • Organic hydrazine	Hydrazine reacts with a silver salt to produce colloidal silver.	
	Test Phase	
		Solid
		Liquid
	✓	Gas/Vapor
	Sensitivity	
	0.5 ppm hydrazine	
Directions and Comments		
Silver salts can also be reduced by organic hydrazine and hydrogen sulfide. The package insert may contain additional information.		
References	**Manufacturer**	
Drager-Tube/CMS Handbook, 13th Edition. Drager Safety AG & Co. KGaA, Lubeck, Germany, 2004.	Drager Safety AG & Co. KGaA, Lubeck, Germany	

Drager Hydrocyanic Acid 2/a		
Detects	**Test Principle**	
• Arsine • Hydrocyanic acid • Phosphine	Mercuric chloride reacts with hydrogen cyanide to produce hydrogen chloride. Hydrogen chloride reacts with methyl red to form a red color.	
	Test Phase	
		Solid
		Liquid
	✓	Gas/Vapor
	Sensitivity	
	2 ppm hydrocyanic acid	
Directions and Comments		
The test is cross sensitive to phosphine and possibly antimony hydride. Do not use for hydrogen chloride detection as 1000 ppm hydrogen chloride does not interfere. High concentrations of other gases may interfere. The package insert may contain additional information.		
References	**Manufacturer**	
Drager-Tube/CMS Handbook, 13th Edition. Drager Safety AG & Co. KGaA, Lubeck, Germany, 2004.	Drager Safety AG & Co. KGaA, Lubeck, Germany	

Drager Hydrogen Fluoride 1.5/b	
Detects	**Test Principle**
• Hydrogen fluoride	Fluoride reacts with zirconium hydroxide - quinalizarin complex to produce a pale pink color.

(*continued*)

Drager Hydrogen Fluoride 1.5/b (*Continued*)

	Test Phase	
		Solid
		Liquid
	✓	Gas/Vapor
	Sensitivity	
	1.5 ppm hydrogen fluoride	

Directions and Comments

Test is specific to fluoride ion and is effective only for hydrogen fluoride in dry air. Moisture causes the formation of hydrofluoric acid which cannot be indicated within the range of the test; higher concentrations of hydrofluoric acid will be indicated. The package insert may contain additional information.

References	Manufacturer
Drager-Tube/CMS Handbook, 13th Edition. Drager Safety AG & Co. KGaA, Lubeck, Germany, 2004.	Drager Safety AG & Co. KGaA, Lubeck, Germany

Drager Hydrogen Peroxide 0.1/a

Detects	Test Principle	
• Hydrogen peroxide • Oxidizer	Hydrogen peroxide (and many other oxidizers) will produce free iodine from potassium iodide. Iodine in the presence of a starch compound in the layer forms a brown color.	
	Test Phase	
		Solid
		Liquid
	✓	Gas/Vapor
	Sensitivity	
	0.1 ppm hydrogen peroxide	

Directions and Comments

This test uses the same reaction as potassium iodide-starch test strips. It is not expected to detect nitrate, organic peroxide, or perchlorate mists in air. The package insert may contain additional information.

References	Manufacturer
Drager-Tube/CMS Handbook, 13th Edition. Drager Safety AG & Co. KGaA, Lubeck, Germany, 2004.	Drager Safety AG & Co. KGaA, Lubeck, Germany

Drager Hydrogen Sulfide 0.2/a

Detects	Test Principle	
• Hydrogen sulfide • Hydrogen cyanide	Hydrogen sulfide reacts with a lead salt to form brown lead sulfide.	
	Test Phase	
		Solid
		Liquid
	✓	Gas/Vapor
	Sensitivity	
	0.2 ppm hydrogen sulfide	

(*continued*)

Drager Hydrogen Sulfide 0.2/a (*Continued*)	
Directions and Comments	
This reaction is similar to the lead acetate test strip. However, a wet lead acetate test strip will not detect hydrogen sulfide gas in air unless exposed for an extended time. This test would be much faster and more efficient at detecting hydrogen sulfide gas in air. The package insert may contain additional information.	
References	**Manufacturer**
Drager-Tube/CMS Handbook, 13th Edition. Drager Safety AG & Co. KGaA, Lubeck, Germany, 2004.	Drager Safety AG & Co. KGaA, Lubeck, Germany

Drager Hydrogen Sulfide 0.2/b		
Detects	**Test Principle**	
<ul><li>Arsine</li><li>Hydrogen cyanide</li><li>Hydrogen sulfide</li><li>Mercaptan</li><li>Nitrogen dioxide</li><li>Phosphine</li></ul>	Hydrogen sulfide reacts with mercuric chloride to produce hydrogen chloride, which converts a pH indicator from yellow to pink.	
	Test Phase	
		Solid
		Liquid
	✓	Gas/Vapor
	Sensitivity	
	0.2 ppm hydrogen sulfide	
Directions and Comments		
This test could be useful in detecting several compounds that react with mercuric chloride. Hydrogen cyanide turns the entire layer from yellow to light orange. The package insert may contain additional information.		
References	**Manufacturer**	
Drager-Tube/CMS Handbook, 13th Ed. Drager Safety AG & Co. KGaA, Lubeck, Germany, 2004.	Drager Safety AG & Co. KGaA, Lubeck, Germany	

Drager Mercaptan 0.5/a		
Detects	**Test Principle**	
<ul><li>Mercaptan</li></ul>	Mercaptans react with palladium(II) to form a yellow palladium-mercaptan compound.	
	Test Phase	
		Solid
		Liquid
	✓	Gas/Vapor
	Sensitivity	
	0.5 ppm mercaptan	
Directions and Comments		
Hydrogen sulfide colors the prelayer black, but does not affect the detection layer. Higher-molecular-weight mercaptans are detected with approximately the same sensitivity. The package insert may contain additional information.		
References	**Manufacturer**	
Drager-Tube/CMS Handbook, 13th Edition. Drager Safety AG & Co. KGaA, Lubeck, Germany, 2004.	Drager Safety AG & Co. KGaA, Lubeck, Germany	

Drager Mercury Vapour 0.1/b	
Detects	**Test Principle**
• Mercury	Elemental mercury reacts with copper iodide to form a pale orange copper-mercury complex.

	Test Phase	
		Solid
		Liquid
	✓	Gas/Vapor

Sensitivity
0.05 mg/m^3 mercury

Directions and Comments	
This test is designed to detect elemental mercury. Bound inorganic mercury and organic mercury compounds will not be detected without prior reduction of mercury. The package insert may contain additional information.	
References	**Manufacturer**
Drager-Tube/CMS Handbook, 13th Edition. Drager Safety AG & Co. KGaA, Lubeck, Germany, 2004.	Drager Safety AG & Co. KGaA, Lubeck, Germany

Drager Methyl Bromide 0.5/a	
Detects	**Test Principle**
• Halogenated hydrocarbon • Methyl bromide	Methyl bromide reacts with disulfuric acid to form hydrogen bromide. Chromium(VI) reacts with hydrogen bromide to form free bromine. Bromine reacts with o-tolidine to form a blue reaction product.

	Test Phase	
		Solid
		Liquid
	✓	Gas/Vapor

Sensitivity
0.5 ppm methyl bromide

Directions and Comments	
Chlorine, bromine, and iodine may cause a similar result. The package insert may contain additional information.	
References	**Manufacturer**
Drager-Tube/CMS Handbook, 13th Edition. Drager Safety AG & Co. KGaA, Lubeck, Germany, 2004.	Drager Safety AG & Co. KGaA, Lubeck, Germany

Drager Natural Gas Odorization Tertiary Butylmercaptan	
Detects	**Test Principle**
• Arsine • Butylmercaptan • Hydrogen sulfide	Mercaptans react with mercury dichlorine heptoxide to liberate hydrogen chloride, which causes a pH indicator to change to pink.

(continued)

Drager Natural Gas Odorization Tertiary Butylmercaptan (*Continued*)		
• Mercaptan • Nitrogen dioxide • Sulfur dioxide	**Test Phase**	
		Solid
		Liquid
	✓	Gas/Vapor
	Sensitivity	
	$1\ mg/m^3$ tertiary butylmercaptan	
Directions and Comments		
This test also detects phosphine, hydrogen sulfide, mercaptans, arsine, phosphine, and nitrogen dioxide at various sensitivities. The package insert may contain additional information.		
References	**Manufacturer**	
Drager-Tube/CMS Handbook, 13th Edition. Drager Safety AG & Co. KGaA, Lubeck, Germany, 2004.	Drager Safety AG & Co. KGaA, Lubeck, Germany	

Drager Nickel Tetracarbonyl 0.1/a		
Detects	**Test Principle**	
• Iron tetracarbonyl • Nickel tetracarbonyl	Nickel tetracarbonyl reacts with iodine to produce nickel iodide. Nickel iodide reacts with dimethylglyoxime to form a pink color.	
	Test Phase	
		Solid
		Liquid
	✓	Gas/Vapor
	Sensitivity	
	0.1 ppm nickel tetracarbonyl	
Directions and Comments		
This test uses dimethylglyoxime to detect nickel. Iron tetracarbonyl is also detected but forms a brown color. Substances that can react with iodine, such as hydrogen sulfide or sulfur dioxide can suppress the indication due to competition with the nickel or iron. Test instructions describe a way to differentiate. Other color indications may be available—review the Dimethylglyoxime Reagent for nickel. The package insert may contain additional information.		
References	**Manufacturer**	
Drager-Tube/CMS Handbook, 13th Edition. Drager Safety AG & Co. KGaA, Lubeck, Germany, 2004.	Drager Safety AG & Co. KGaA, Lubeck, Germany	

Drager Nitrogen Dioxide 0.5/c	
Detects	**Test Principle**
• Nitrogen dioxide • Chlorine • Ozone • Oxidizer	Nitrogen dioxide reacts with diphenylbenzidine to form a blue-gray color.

(*continued*)

Drager Nitrogen Dioxide 0.5/c *(Continued)*		
	Test Phase	
		Solid
		Liquid
	✓	Gas/Vapor
	Sensitivity	
	0.5 ppm nitrogen dioxide	

Directions and Comments	
Chlorine and ozone react similar to nitrogen dioxide but at different sensitivity. Other oxidizers would be expected to react, but a complete list is not available. High concentration may bleach the color from the indicating layer. The package insert may contain additional information.	
References	**Manufacturer**
Drager-Tube/CMS Handbook, 13th Edition. Drager Safety AG & Co. KGaA, Lubeck, Germany, 2004.	Drager Safety AG & Co. KGaA, Lubeck, Germany

Drager Nitrous Fumes 0.5/a		
Detects	**Test Principle**	
• Chlorine • Nitrogen dioxide • Nitrogen monoxide • Nitrous fume • Oxidizer • Ozone	A prelayer oxidizes nitrogen monoxide with chromium(VI) to form nitrogen dioxide. Nitrogen dioxide reacts with diphenylbenzidine to form a blue-gray color.	
	Test Phase	
		Solid
		Liquid
	✓	Gas/Vapor
	Sensitivity	
	0.5 ppm nitrogen monoxide	

Directions and Comments	
Chlorine, ozone, and other oxidizers are indicated, but a complete list is not available. High concentration of nitrogen dioxide (>300 ppm) may bleach the indicator color to white. The package insert may contain additional information.	
References	**Manufacturer**
Drager-Tube/CMS Handbook, 13th Edition. Drager Safety AG & Co. KGaA, Lubeck, Germany, 2004.	Drager Safety AG & Co. KGaA, Lubeck, Germany

Drager Nitrous Fumes 20/a	
Detects	**Test Principle**
• Chlorine • Nitrogen dioxide • Nitrogen monoxide • Nitrous fume • Oxidizer • Ozone	A prelayer oxidizes nitrogen monoxide with chromium(VI) to form nitrogen dioxide. Nitrogen dioxide is detected with o-dianisidine to form a red-brown color.

(continued)

Drager Nitrous Fumes 20/a (Continued)		
	Test Phase	
		Solid
		Liquid
	✓	Gas/Vapor
	Sensitivity	
	20 ppm nitrous fume	

Directions and Comments	
This test uses a different method, and is less sensitive to chlorine and ozone, than the Drager Nitrous Fumes 0.5/a tube. The package insert may contain additional information.	
References	**Manufacturer**
Drager-Tube/CMS Handbook, 13th Edition. Drager Safety AG & Co. KGaA, Lubeck, Germany, 2004.	Drager Safety AG & Co. KGaA, Lubeck, Germany

Drager Olefine 0.05%/a		
Detects	**Test Principle**	
• Alkene • Olefin • Olefine	Alkene bonds react with permanganate to produce manganese(IV). The original purple color of permanganate is changed to pale brown.	
	Test Phase	
		Solid
		Liquid
	✓	Gas/Vapor
	Sensitivity	
	0.06% (600 ppm) propylene	

Directions and Comments	
The test detects many compounds with a carbon-carbon double bond. The test also reacts to dialkyl sulfides and other substances that can be oxidized by permanganate. Compare to Permanganate Test. The package insert may contain additional information.	
References	**Manufacturer**
Drager-Tube/CMS Handbook, 13th Edition. Drager Safety AG & Co. KGaA, Lubeck, Germany, 2004.	Drager Safety AG & Co. KGaA, Lubeck, Germany

Drager Organic Arsenic Compounds and Arsine	
Detects	**Test Principle**
• Arsenic, organic • Arsine • Phosphine	Organic arsenic compounds react with zinc and hydrogen chloride to form arsine. Arsine or phosphine is detected when a reaction with gold-mercury complex forms a rose to purple colored colloidal gold.

(continued)

Drager Organic Arsenic Compounds and Arsine (*Continued*)

	Test Phase	
		Solid
		Liquid
	✓	Gas/Vapor
	Sensitivity	
	0.1 ppm arsine. 3 mg/m^3 organic arsenic.	

Directions and Comments

This multistep test detects arsine and phosphine first. Additional steps indicate the presence of organic arsenic compounds. This test could detect lewisite as well as arsine formed as lewisite hydrolyzes in humid air. The package insert may contain additional information.

References	Manufacturer
Drager-Tube/CMS Handbook, 13th Edition. Drager Safety AG & Co. KGaA, Lubeck, Germany, 2004.	Drager Safety AG & Co. KGaA, Lubeck, Germany

Drager Organic Basic Nitrogen Compounds

Detects	Test Principle	
• Amines • Nitrogen mustard • Organic basic nitrogen compounds	Organic basic nitrogen compounds (amines) react with potassium bismuth iodide to form an orange-red indication.	
	Test Phase	
		Solid
		Liquid
	✓	Gas/Vapor
	Sensitivity	
	1 mg/m3 unspecified organic basic nitrogen compound	

Directions and Comments

This test detects various organic basic nitrogen compounds but cannot differentiate them. The reaction principle would allow differentiation from other basic airborne material. Compare to Dragendorff's Reagent. The package insert may contain additional information.

References	Manufacturer
Drager-Tube/CMS Handbook, 13th Edition. Drager Safety AG & Co. KGaA, Lubeck, Germany, 2004.	Drager Safety AG & Co. KGaA, Lubeck, Germany

Drager Ozone 0.05/b

Detects	Test Principle
• Chlorine • Nitrogen dioxide • Oxidizer • Ozone	Ozone reacts with indigo to form isatine. The original pale blue indigo layer is bleached white.

(*continued*)

Drager Ozone 0.05/b (*Continued*)

	Test Phase	
		Solid
		Liquid
	✓	Gas/Vapor
	Sensitivity	
	0.05 ppm ozone	

Directions and Comments

This is a sensitive test for ozone, but any material that can bleach indigo might be indicated similarly. Chlorine and nitrogen dioxide can be differentiated from ozone by the pale gray color imparted to the indicating layer. This test is helpful in circumstances probably caused by ozone, such as a stinging odor in a building following a lightning strike or other massive electrical arc. The package insert may contain additional information.

References	Manufacturer
Drager-Tube/CMS Handbook, 13th Edition. Drager Safety AG & Co. KGaA, Lubeck, Germany, 2004.	Drager Safety AG & Co. KGaA, Lubeck, Germany

Drager Perchloroethylene 0.1/a

Detects	Test Principle	
• Bromine • Chlorine • Fluorine • Halogen • Halohydrocarbon • Perchloroethylene	Halogenated hydrocarbons react with permanganate to form free halogen. Chlorine (or other strong oxidizers) reacts with diphenylbenzidine to form a blue reaction product.	
	Test Phase	
		Solid
		Liquid
	✓	Gas/Vapor
	Sensitivity	
	0.5 ppm perchloroethylene	

Directions and Comments

Diphenylbenzidine produces a blue color upon reaction with a strong oxidizer. Halogenated hydrocarbons react with permanganate to release halogen. Free halogen would be detected directly. Other strong oxidizers in air, such as hydrogen peroxide vapor would most likely be indicated. Vapors that produce a positive result that come from a heavier-than-water, insoluble liquid are most likely chlorine. The package insert may contain additional information.

References	Manufacturer
Drager-Tube/CMS Handbook, 13th Edition. Drager Safety AG & Co. KGaA, Lubeck, Germany, 2004.	Drager Safety AG & Co. KGaA, Lubeck, Germany

Drager Phenol 1/b

Detects	Test Principle
• Aromatic alcohol • Cresol • Phenol	Phenol reacts with cerium sulfate and sulfuric acid to produce a brown-gray reaction product.

(*continued*)

Drager Phenol 1/b (*Continued*)		
	Test Phase	
		Solid
		Liquid
	✓	Gas/Vapor
	Sensitivity	
	1 ppm phenol	

Directions and Comments
This test is useful in the detection of aromatic alcohols. Other aromatics without a hydroxyl group are not indicated. Aliphatic hydrocarbons and alcohols are not indicated. The package insert may contain additional information.

References	**Manufacturer**
Drager-Tube/CMS Handbook, 13th Edition. Drager Safety AG & Co. KGaA, Lubeck, Germany, 2004.	Drager Safety AG & Co. KGaA, Lubeck, Germany

Drager Phosgene 0.02/a		
Detects	**Test Principle**	
• Chlorine • Hydrochloric acid • Phosgene	Phosgene and an aromatic amine produce a red reaction product. This is probably a reaction of phosgene with p-nitrobenzylpyridine (4-[4-nitrobenzyl]pyridine, CAS 1083-48-3) to form a derivative of glutaconic aldehyde. N-benzylaniline (CAS 103-32-2) then condenses the aldehyde to form a pink to red color.	
	Test Phase	
		Solid
		Liquid
	✓	Gas/Vapor
	Sensitivity	
	0.02 ppm phosgene	

Directions and Comments
This test may be used to verify phosgene in the presence of chlorinated hydrocarbons. High concentration of phosgene will bleach the color indicator (>30 ppm). Chlorine and hydrochloric acid interfere. The package insert may contain additional information.

References	**Manufacturer**
Drager-Tube/CMS Handbook, 13th Edition. Drager Safety AG & Co. KGaA, Lubeck, Germany, 2004. Feigl, F. and Anger, V., *Spot Tests in Inorganic Analysis*, Sixth Edition, Elsevier Publishing Company, Amsterdam, 1972.	Drager Safety AG & Co. KGaA, Lubeck, Germany

Drager Phosgene 0.05/a	
Detects	**Test Principle**
• Acetyl chloride • Carbonyl bromide • Phosgene	Phosgene reacts with ethylaniline and dimethylaminobenzaldehyde to form a blue-green reaction product.

	Test Phase	
		Solid
		Liquid
	✓	Gas/Vapor

Sensitivity
0.04 ppm phosgene

Directions and Comments
Carbonyl bromide and acetyl chloride are also indicated but with differing sensitivity. The package insert may contain additional information.

References	**Manufacturer**
Drager-Tube/CMS Handbook, 13th Edition. Drager Safety AG & Co. KGaA, Lubeck, Germany, 2004.	Drager Safety AG & Co. KGaA, Lubeck, Germany

Drager Phosphine 0.1/a	
Detects	**Test Principle**
• Antimony hydride • Arsine • Phosphine	Phosphine and gold(III) form a gray-violet colloid.

	Test Phase	
		Solid
		Liquid
	✓	Gas/Vapor

Sensitivity
0.01 ppm phosphine

Directions and Comments
The test is sensitive for phosphine (0.1 ppm) but is cross sensitive to arsine and antimony hydride. It might be sensitive to sulfur dioxide and hydrochloric acid. The package insert may contain additional information.

References	**Manufacturer**
Drager-Tube/CMS Handbook, 13th Edition. Drager Safety AG & Co. KGaA, Lubeck, Germany, 2004.	Drager Safety AG & Co. KGaA, Lubeck, Germany

Drager Phosphoric Acid Esters 0.05/a	
Detects	**Test Principle**
• G-nerve agent • Pesticide • Phosphoric acid ester • V-nerve agent	If phosphoric acid ester is present, cholinesterase in the tube is inactivated after a sample is drawn into the tube and 1 minute passes. Butylcholine iodide and water are added and if the cholinesterase is active, will cleave butylcholine

(continued)

Drager Phosphoric Acid Esters 0.05/a (*Continued*)		
	iodide and form butyric acid. Butyric acid and phenol red pH indicator produces a yellow color (negative result). If the cholinesterase has been inactivated, butyric acid will not form and the phenol red pH indicator will produce a red color (positive result).	
	Test Phase	
		Solid
		Liquid
	✓	Gas/Vapor
	Sensitivity	
	0.05 ppm dichlorvos	

Directions and Comments	
This is a sensitive test for airborne phosphoric acid esters (dichlorvos 0.05 ppm) but detects others at varying sensitivity. The reference does not clearly state if other cholinesterase-inhibiting materials (carbamates, thiophosphates, phosphonates, and others) are detected at similar sensitivity. The package insert may contain additional information.	
References	**Manufacturer**
Drager-Tube/CMS Handbook, 13th Edition. Drager Safety AG & Co. KGaA, Lubeck, Germany, 2004.	Drager Safety AG & Co. KGaA, Lubeck, Germany

Drager Simultaneous Test Set I - Tube 1		
Detects	**Test Principle**	
• Acid	This test is a pH indicator.	
	Test Phase	
		Solid
		Liquid
	✓	Gas/Vapor
	Sensitivity	
	Not available. Qualitative result only.	

Directions and Comments	
This test is designed as a presumptive screening test. This is a sensitive pH test in air. All acid gases produce a blue color from yellow. The package insert may contain additional information.	
References	**Manufacturer**
Drager-Tube/CMS Handbook, 13th Edition. Drager Safety AG & Co. KGaA, Lubeck, Germany, 2004.	Drager Safety AG & Co. KGaA, Lubeck, Germany

Drager Simultaneous Test Set I - Tube 2	
Detects	**Test Principle**
• Arsine • Hydrocyanic acid • Phosphine	Mercuric chloride produces hydrogen chloride when hydrogen cyanide is present. Methyl red reacts with hydrogen chloride to produce a red color.

(continued)

Drager Simultaneous Test Set I - Tube 2 (*Continued*)		
	Test Phase	
		Solid
		Liquid
	✓	Gas/Vapor
	Sensitivity	
	Not available. Qualitative result only.	

Directions and Comments
This test is designed as a presumptive screening test. It is cross sensitive to phosphine. Do not use for hydrogen chloride detection as 1000 ppm hydrogen chloride does not interfere. Tube probably contains a filter to remove some interfering gases, such as hydrogen sulfide or sulfur dioxide. Does not detect organic nitriles (cyanides). Compare to Drager Hydrocyanic Acid 2/a tube. The package insert may contain additional information.

References	**Manufacturer**
Drager-Tube/CMS Handbook, 13th Edition. Drager Safety AG & Co. KGaA, Lubeck, Germany, 2004.	Drager Safety AG & Co. KGaA, Lubeck, Germany

Drager Simultaneous Test Set I - Tube 3		
Detects	**Test Principle**	
• Acetylene • Carbon monoxide • Lewisite	Carbon monoxide reacts with iodine pentoxide and disulfuric acid in tube to produce a brown iodine stain.	
	Test Phase	
		Solid
		Liquid
	✓	Gas/Vapor
	Sensitivity	
	Not available. Qualitative result only.	

Directions and Comments
This test is designed as a presumptive screening test. A conventional "four-gas" air monitor will be more practical for CO monitoring. This tube may be used to verify CO in the case of gases that produce cross-sensitive results in the electronic monitor. This tube also detects acetylene, a Lewisite degradation product. Compare to Carbon Monoxide 2/a tube. High concentration of alkene will interfere. The package insert may contain additional information.

References	**Manufacturer**
Drager-Tube/CMS Handbook, 13th Edition. Drager Safety AG & Co. KGaA, Lubeck, Germany, 2004.	Drager Safety AG & Co. KGaA, Lubeck, Germany

Drager Simultaneous Test Set I - Tube 4	
Detects	**Test Principle**
• Base • Ammonia • Amine	This test is a pH indicator.

(*continued*)

Drager Simultaneous Test Set I - Tube 4 (*Continued*)		
	Test Phase	
		Solid
		Liquid
	✓	Gas/Vapor
	Sensitivity	
	Not available. Qualitative result only.	
Directions and Comments		
This test is designed as a presumptive screening test. This is a sensitive pH test for compounds in air. All basic gases produce a blue color from yellow. Compare to Ammonia 2/a tube. The package insert may contain additional information.		
References	**Manufacturer**	
Drager-Tube/CMS Handbook, 13th Edition. Drager Safety AG & Co. KGaA, Lubeck, Germany, 2004.	Drager Safety AG & Co. KGaA, Lubeck, Germany	

Drager Simultaneous Test Set I - Tube 5		
Detects	**Test Principle**	
• Chlorine • Nitrogen dioxide • Nitrogen monoxide • Nitrous gas • Oxidizer • Ozone	Nitrogen monoxide and chromium react to produce nitrogen dioxide. Nitrogen dioxide and diphenylbenzidine produce a blue-gray color.	
	Test Phase	
		Solid
		Liquid
	✓	Gas/Vapor
	Sensitivity	
	Not available. Qualitative result only.	
Directions and Comments		
This test is designed as a presumptive screening test. The test detects nitrogen monoxide and nitrogen dioxide. Greater than 300 ppm nitrogen dioxide will bleach the indication. Not effective in the presence of more than 1 ppm chlorine or 0.1 ppm ozone. Compare to Nitrous Fumes 0.5/a tube. The package insert may contain additional information.		
References	**Manufacturer**	
Drager-Tube/CMS Handbook, 13th Edition. Drager Safety AG & Co. KGaA, Lubeck, Germany, 2004.	Drager Safety AG & Co. KGaA, Lubeck, Germany	

Drager Simultaneous Test Set II - Tube 1	
Detects	**Test Principle**
• Sulfur dioxide • Hydrogen sulfide	Sulfur dioxide reacts with iodine and starch to produce sulfuric acid and hydrogen iodide.

(continued)

Drager Simultaneous Test Set II - Tube 1 (*Continued*)		
	Test Phase	
		Solid
		Liquid
	✓	Gas/Vapor
	Sensitivity	
	Not available. Qualitative result only.	

Directions and Comments	
This test is designed as a presumptive screening test. This is the complementary reaction to potassium iodide starch strips. Sulfur dioxide (and hydrogen sulfide) is able to reduce iodine (but not other halogens), which removes the blue color and forms white as a positive indication. The package insert may contain additional information.	
References	**Manufacturer**
Drager-Tube/CMS Handbook, 13th Edition. Drager Safety AG & Co. KGaA, Lubeck, Germany, 2004.	Drager Safety AG & Co. KGaA, Lubeck, Germany

Drager Simultaneous Test Set II - Tube 2		
Detects	**Test Principle**	
• Bromine • Chlorine • Chlorine dioxide • Nitrogen dioxide • Oxidizer	Chlorine gas and o-tolidine produce an orange reaction product.	
	Test Phase	
		Solid
		Liquid
	✓	Gas/Vapor
	Sensitivity	
	Not available. Qualitative result only.	

Directions and Comments	
This test is designed as a presumptive screening test. The test is cross sensitive to bromine, chlorine dioxide, nitrogen dioxide, and other strong oxidizers. Compare to Chlorine 0.2/a tube. The package insert may contain additional information.	
References	**Manufacturer**
Drager-Tube/CMS Handbook, 13th Edition. Drager Safety AG & Co. KGaA, Lubeck, Germany, 2004.	Drager Safety AG & Co. KGaA, Lubeck, Germany

Drager Simultaneous Test Set II - Tube 3	
Detects	**Test Principle**
• Hydrogen sulfide	The production of mercuric sulfide (or possibly lead sulfide) forms a pale brown color.

(*continued*)

Drager Simultaneous Test Set II - Tube 3 (*Continued*)		
	Test Phase	
		Solid
		Liquid
	✓	Gas/Vapor
	Sensitivity	
	Not available. Qualitative result only.	

Directions and Comments
This test is designed as a presumptive screening test. A conventional "four-gas" air monitor will be more practical for hydrogen sulfide monitoring. This tube may be used to verify hydrogen sulfide in the case of gases that produce cross-sensitive results in the electronic monitor. The package insert may contain additional information.

References	Manufacturer
Drager-Tube/CMS Handbook, 13th Edition. Drager Safety AG & Co. KGaA, Lubeck, Germany, 2004.	Drager Safety AG & Co. KGaA, Lubeck, Germany

Drager Simultaneous Test Set II - Tube 4		
Detects	**Test Principle**	
• Carbon dioxide • Sulfur dioxide	Carbon dioxide and hydrazine react to form a purple reaction product.	
	Test Phase	
		Solid
		Liquid
	✓	Gas/Vapor
	Sensitivity	
	Not available. Qualitative result only.	

Directions and Comments
This test is designed as a presumptive screening test. This test is designed for the measurement of carbon dioxide in percent amounts. A conventional "four-gas" air monitor will be more practical monitoring depressed oxygen level. This tube may be used to verify carbon dioxide in the case of other inert, oxygen-displacing gases. Sulfur dioxide produces a similar result with about one-third the sensitivity. The package insert may contain additional information.

References	Manufacturer
Drager-Tube/CMS Handbook, 13th Edition. Drager Safety AG & Co. KGaA, Lubeck, Germany, 2004.	Drager Safety AG & Co. KGaA, Lubeck, Germany

Drager Simultaneous Test Set II - Tube 5	
Detects	**Test Principle**
• Phosgene	Phosgene and an aromatic amine produce a red color. This is probably a reaction of phosgene with p-nitrobenzylpyridine (4-[4-Nitrobenzyl]pyridine,

(continued)

Drager Simultaneous Test Set II - Tube 5 (*Continued*)	
	CAS 1083-48-3) to form a derivative of glutaconic aldehyde. N-benzylaniline (CAS 103-32-2) then condenses the aldehyde to form a pink to red color.

Test Phase	
	Solid
	Liquid
✓	Gas/Vapor

Sensitivity
Not available. Qualitative result only.

Directions and Comments

This test is designed as a presumptive screening test. This test may be used to verify phosgene in the presence of chlorinated hydrocarbons. High concentration of phosgene will bleach the color indicator (>30 ppm). Chlorine and hydrochloric acid interfere. Compare to Phosgene 0.02/a tube. The package insert may contain additional information.

References	Manufacturer
Drager-Tube/CMS Handbook, 13th Edition. Drager Safety AG & Co. KGaA, Lubeck, Germany, 2004.	Drager Safety AG & Co. KGaA, Lubeck, Germany

Drager Simultaneous Test Set III - Tube 1	
Detects	**Test Principle**
• Ketone • Aldehyde	Probably a reaction of ketones or aldehydes with 2,4-dinitrophenylhydrazine to form various intensity of yellow hydrazone.

Test Phase	
	Solid
	Liquid
✓	Gas/Vapor

Sensitivity
Not available. Qualitative result only.

Directions and Comments

This test is designed as a presumptive screening test. The most practical detection of these compounds at higher concentration will be with a combustible gas sensor. Esters are not indicated. Compare to Acetone 100/b tube. The package insert may contain additional information.

References	Manufacturer
Drager-Tube/CMS Handbook, 13th Edition. Drager Safety AG & Co. KGaA, Lubeck, Germany, 2004.	Drager Safety AG & Co. KGaA, Lubeck, Germany

Drager Simultaneous Test Set III - Tube 2	
Detects	**Test Principle**
• Aromatic compounds	Probably a reaction of an aromatic compound with iodine pentoxide and disulfuric acid to produce iodine.

(*continued*)

Drager Simultaneous Test Set III - Tube 2 (*Continued*)		
	Test Phase	
		Solid
		Liquid
	✓	Gas/Vapor
	Sensitivity	
	Not available. Qualitative result only.	

Directions and Comments
This test is designed as a presumptive screening test. The most practical detection of these compounds at higher concentration will be with a combustible gas sensor. There is a pretreatment area or filter in the tube that may help differentiate aromatics from aliphatics. Compare to the Aliphatic Hydrocarbon tube (fourth tube in the Simultaneous Tube Set III). The package insert may contain additional information.

References	**Manufacturer**
Drager-Tube/CMS Handbook, 13th Edition. Drager Safety AG & Co. KGaA, Lubeck, Germany, 2004.	Drager Safety AG & Co. KGaA, Lubeck, Germany

Drager Simultaneous Test Set III - Tube 3		
Detects	**Test Principle**	
• Alcohol	Alcohols convert yellow-orange chromium(VI) to green chromium(III).	
• Aldehyde	**Test Phase**	
• Ester		Solid
• Ether		Liquid
• Ketone	✓	Gas/Vapor
	Sensitivity	
	Not available. Qualitative result only.	

Directions and Comments
This test is designed as a presumptive screening test. The most practical detection of these compounds at higher concentration will be with a combustible gas sensor. More complex alcohols give a reduced sensitivity. Some, but not all, aldehydes, ethers, ketones, and esters are indicated. The package insert may contain additional information.

References	**Manufacturer**
Drager-Tube/CMS Handbook, 13th Edition. Drager Safety AG & Co. KGaA, Lubeck, Germany, 2004.	Drager Safety AG & Co. KGaA, Lubeck, Germany

Drager Simultaneous Test Set III - Tube 4	
Detects	**Test Principle**
• Aliphatic compound	Probably a reaction of most hydrocarbons with iodine pentoxide and disulfuric acid to produce brown iodine.
• Alkene	
• Alkyne	
• Alkane	

(*continued*)

Drager Simultaneous Test Set III - Tube 4 (*Continued*)		
	Test Phase	
		Solid
		Liquid
	✓	Gas/Vapor
	Sensitivity	
	Not available. Qualitative result only.	

Directions and Comments
This test is designed as a presumptive screening test. The most practical detection of these compounds at higher concentration will be with a combustible gas sensor. The test also detects aliphatic alkanes, alkenes, and alkynes. It may indicate aromatics but the tube appears to have a pretreatment area and the reaction principle is not clear. Compare to the aromatics tube (second tube) in Simultaneous Test Set III. The package insert may contain additional information.

References	**Manufacturer**
Drager-Tube/CMS Handbook, 13th Edition. Drager Safety AG & Co. KGaA, Lubeck, Germany, 2004.	Drager Safety AG & Co. KGaA, Lubeck, Germany

Drager Simultaneous Test Set III - Tube 5		
Detects	**Test Principle**	
• Chlorohydrocarbon • Freon • Halon*	Halogenated hydrocarbons form chlorine gas after oxidation by permanganate. Diphenylbenzidine forms a blue color in the presence of chlorine gas and possibly other halogen gases.	
	Test Phase	
		Solid
		Liquid
	✓	Gas/Vapor
	Sensitivity	
	Not available. Qualitative result only.	

Directions and Comments
This test is designed as a presumptive screening test. Another method of detecting these compounds, even at low concentration, is a refrigerant gas detector. Compare to Perchloroethylene 0.1/a tube. The package insert may contain additional information.

References	**Manufacturer**
Drager-Tube/CMS Handbook, 13th Edition. Drager Safety AG & Co. KGaA, Lubeck, Germany, 2004.	Drager Safety AG & Co. KGaA, Lubeck, Germany

Drager Thioether	
Detects	**Test Principle**
• Thioether • Sulfur mustard	Gold chloride, thioether (R-S-R´), and chloramide react to form an orange complex.

(*continued*)

Drager Thioether (*Continued*)		
	Test Phase	
		Solid
		Liquid
	✓	Gas/Vapor
	Sensitivity	
	1 mg/m³ unspecified thioether.	

Directions and Comments	
This test detects all thioethers and is not specific to sulfur mustard. The package insert may contain additional information.	
References	**Manufacturer**
Drager-Tube/CMS Handbook, 13th Edition. Drager Safety AG & Co. KGaA, Lubeck, Germany, 2004.	Drager Safety AG & Co. KGaA, Lubeck, Germany

Drager Toluene Diisocyanate 0.02/A		
Detects	**Test Principle**	
• Toluene diisocyanate • 2,4-TDI • 2,6-TDI	The first step of the test prepares fresh glutaconaldehyde. In the second step, 2,4-toluene diisocyanate reacts hydrogen chloride to produce an aromatic amine. The resulting aromatic amine reacts with the glutaconaldehyde to form an orange reaction product.	
	Test Phase	
		Solid
		Liquid
	✓	Gas/Vapor
	Sensitivity	
	0.02 ppm 2,4-TDI	

Directions and Comments	
This test indicates 2,4-TDI and 2,6-TDI. Other isocyanates are not indicated. Mercaptans will discolor the indicating layer, but it is not clear if they can be discerned from TDI. The test takes about 20 minutes. The package insert may contain additional information.	
References	**Manufacturer**
Drager-Tube/CMS Handbook, 13th Edition. Drager Safety AG & Co. KGaA, Lubeck, Germany, 2004.	Drager Safety AG & Co. KGaA, Lubeck, Germany

Drop-Ex Plus™	
Detects	**Test Principle**
• Bromate • Chlorate • Inorganic nitrate • Nitramine • Nitrate ester	The test is a proprietary kit of presumptive, sequential colorimetric tests for the detection of explosive materials.

(continued)

Drop-Ex Plus™ (*Continued*)		
• Nitroaromatic • Peroxide	**Test Phase**	
	✓	Solid
	✓	Liquid
		Gas/Vapor
	Sensitivity	
	20 nanograms	

Directions and Comments	
Drop-Ex Plus™ is a proprietary kit of presumptive, sequential colorimetric test reactions for the detection of explosive material. The test uses wipe samples or other methods to collect a very small sample. Spray bottles contain mixtures of pretreatment chemicals, colorimetric reagents, solvents, buffers, and so on. The sprays are applied to the sample and filter paper in a specified order because sequential tests utilize reaction products from previous sprays. False positive results are possible from interfering material.	

References	**Manufacturer**
Explosives Detection Field Test Kits, The Mistral Group, http://www.mistralgroup.com/SEC_explosives.asp#drop, April 16, 2008.	The Mistral Group, Bethesda, Maryland, USA (www.mistralgroup.com).

Duquenois-Levine Reagent, Modified		
Detects	**Test Principle**	
• Cannabis • Hashish • Marijuana • Tetrahydrocannabinol • THC	Tetrahydrocannabinol (THC) reacts with vanillin under test conditions to form a reaction product of varying shades of gray to red to blue.	
	Test Phase	
	✓	Solid
	✓	Liquid
		Gas/Vapor
	Sensitivity	
	5 µg THC	

Directions and Comments	
The test is packaged as a presumptive field test for marijuana and other material containing THC. Vanillin forms colors with many other compounds. The test is modified by solvent extractions and uses three solutions. Solution A is 2.5 ml of acetaldehyde (CAS 75-07-0) and 2.0 g of vanillin (CAS 121-33-5) in 100 ml of 95% ethanol. Solution B is concentrated hydrochloric acid. Solution C contains chloroform (CAS 67-66-3). The test is performed by placing a small amount of sample in a spot plate or test tube and adding one volume of solution A to the drug and then agitating for 1 minute. Next, add one volume of Solution B, agitate gently and determine the color. Lastly, add three volumes of Solution C and note whether the color is extracted from the mixture to A and B. Consult the package insert for specific details. Refer to the appendix for drug screen results.	

References	**Manufacturer**
U.S. Department of Justice. Color Test Reagents/Kits for Preliminary Identification of Drugs of Abuse, NIJ Standard–0604.01, National Institute of Justice, National Law Enforcement and Corrections Technology Center, Rockville, MD, July 2000.	Several commercial test kits are available.

Ethyl-8-Hydroxytetrahydroquinoline HCl Reagent		
Detects	**Test Principle**	
• Arsenite • Arsenic(III)	N-Ethyl-8-hydroxytetrahydroquinoline HCl reacts with arsenite to form a red-brown precipitate.	
	Test Phase	
	✓	Solid
	✓	Liquid
		Gas/Vapor
	Sensitivity	
	0.0000006 mg arsenite	

Directions and Comments

Mix a 0.5% (w/v) aqueous solution of N-ethyl-8-hydroxytetrahydroquinoline HCl (1-ethyl-1,2,3,4-tetrahydro-quinolin-8-ol hydrochloride, no CAS listing). Place a drop of the test solution on a filter paper. When dry, moisten with a drop of hydrochloric acid followed by a drop of the test solution. Finally, add a drop of ferric chloride (CAS 10025-77-1) solution and warm the spot. A red-brown color indicates arsenite ion. Lead, mercury, and copper interfere.

References	**Manufacturer**
Welcher, F., *Chemical Solutions: Reagents Useful to the Chemist, Biologist, and Bacteriologist*, D. Van Nostrand Company, Inc., New York, 1942. *Chemical Abstracts* 29, 5037 (1935).	

Fehling's Solution		
Detects	**Test Principle**	
• Aldehyde • Fructose • Glucose • Reducing Sugar	Potassium sodium tartrate holds cupric ions in basic solution. Reducing sugars react with blue cupric ions to form red cuprous ions. Tartrate cannot complex cuprous ions and red cuprous oxide precipitate forms.	
	Test Phase	
	✓	Solid
	✓	Liquid
		Gas/Vapor
	Sensitivity	
	Not available	

Directions and Comments

The test consists of two solutions that must be stored separately. Solution A is 0.35 g cupric sulfate pentahydrate (CAS 7758-99-8) in 5 ml water. Solution B is 1.7 g of Rochelle salt (potassium sodium tartrate tetrahydrate, CAS 6381-59-5) and 1.3 g of potassium hydroxide in 5 ml water (0.7 g sodium hydroxide is acceptable [Shriner]; do not use ammonium hydroxide). Mix 1 ml of each solution and add a few drops of test solution or a few milligrams of solid sample and gently heat to boiling. A red precipitate that forms at anytime before or during heating indicates a reducing sugar. The color change may be gradual from blue to green to yellow to red depending on concentration and reactivity. Reducing sugars contain either a free aldehyde group (CH=O) or a free ketose group (C=O). Aldoses and ketoses are indicated, but through separate pathways. Not all sugars are reducing. For example, glucose, xylose, lactose, pentose and fructose are reducing sugars; sucrose is not a reducing sugar. Sucrose can be detected by first boiling the sample with

(continued)

Fehling's Solution (*Continued*)

two drops of concentrated hydrochloric acid, neutralizing, adding Fehling's solution, and boiling. Most, but not all, disaccharides are reducing. The test will indicate some aldehydes but not ketones. Compare to Luff's Reagent.

References	Manufacturer
Welcher, F., *Chemical Solutions: Reagents Useful to the Chemist, Biologist, and Bacteriologist*, D. Van Nostrand Company, Inc., New York, 1942. Jacobs, M., *The Chemical Analysis of Foods and Food Products*, D. Van Nostrand, New York, 1938. Keusch, P. http://www.uni-regensburg.de/Fakultaeten/nat_Fak_IV/Organische_Chemie/Didaktik/Keusch/index_e.html, Institut für Organische Chemie, Universität Regensburg, Regensburg, Germany, January 2008. Shriner, R., Fuson, R. and Curtin, D., *The Systematic Identification of Organic Compounds, A Laboratory Manual*, Fifth Edition, John Wiley and Sons, Inc., New York, 1964. *J. Chem. Educ.*, 37, 205 (1960).	

Ferric Chloride Reagent

Detects	Test Principle	
• Acetaminophen • Chlorpromazine HCl • Ferrocyanides • Illegal drugs • Morphine • Phenols • Sodium bicarbonate	Ferric chloride reacts with various compounds to form various colored reaction products.	
	Test Phase	
	✓	Solid
	✓	Liquid
		Gas/Vapor
	Sensitivity	
	200 µg morphine monohydrate	

Directions and Comments

This is a presumptive test for illegal drugs. The test solution is 2% anhydrous ferric chloride or 3.3% ferric chloride hexahydrate (CAS 10025-77-1) in water (w/v). It is most helpful to compare any color formed in this test with results from other presumptive drug tests (Marquis Reagent, etc.) in a matrix to determine a pattern of unique results. Many, but not all, phenols, enols, oximes, and hydrooxamic acids produce a positive result. Some colors are weak or fleeting. Consider performing the test simultaneously on a drop of water to detect light changes in color. Watch the test closely to observe color changes that occur and immediately clear. Ferrocyanides produce a blue color. Refer to the appendix for drug screen results. See color images 2.20 and 2.21.

References	Manufacturer
U.S. Department of Justice. Color Test Reagents/Kits for Preliminary Identification of Drugs of Abuse, NIJ Standard–0604.01, National Institute of Justice, National Law Enforcement and Corrections Technology Center, Rockville, MD, July 2000. Shriner, R., Fuson, R. and Curtin, D., *The Systematic Identification of Organic Compounds, A Laboratory Manual*, Fifth Edition, John Wiley and Sons, Inc., New York, 1964. *J. Chem. Educ.*, 37, 205 (1960).	Several commercial test kits are available.

Ferric Ferricyanide Test		
Detects	**Test Principle**	
• Alloys • Elemental metal • Metals	Metals reduce either ferric or ferricyanide in the ferric ferricyanide test solution to form Turnbull's blue or Prussian blue respectively.	
	Test Phase	
	✓	Solid
	✓	Liquid
		Gas/Vapor
	Sensitivity	
	Not available but described as sensitive	
Directions and Comments		
This is a sensitive test for unbound metal or alloy in solid or liquid sample. The reagent may be purchased or mixed fresh by adding equal volumes of 0.4% ferric chloride hexahydrate (CAS 10025-77-1) and 0.8% potassium ferricyanide (CAS 13746-66-2). The sample is placed in a spot plate and mixed with a drop of the test reagent. A blue (likely) or green (dilute) color indicates unbound metal or alloy is present. The test is sensitive and works with many metals, including silver and platinum.		
References	**Manufacturer**	
Feigl, F., *Spot Tests in Inorganic Analysis*, Fifth Edition, Elsevier Publishing Company, New York, 1958.		

Ferric Solution Test		
Detects	**Test Principle**	
• Thiocyanate	Thiocyanate compounds react with ferric solution to form a red blood color.	
	Test Phase	
	✓	Solid
	✓	Liquid
		Gas/Vapor
	Sensitivity	
	Not available	
Directions and Comments		
Mix a 2% solution of ferric chloride (CAS 10025-77-1). Add a drop of the test solution to the sample. A deep red color indicates thiocyanate. This test forms a ferric thiocyanide complex that may appear dark in color. It may be necessary to dilute the solution with water to observe the blood red color. Iodine interferes by covering the red color with a dark brown.		
References	**Manufacturer**	
Engelder, C., Dunkelberger, T. and Schiller, W., *Semi-Micro Qualitative Analysis*, Second Edition, John Wiley & Sons, Inc., New York, May 1946.		

Ferric Thiocyanate Reagent	
Detects	**Test Principle**
<ul><li>Hydrocarbon</li><li>Organic nitrogen</li><li>Organic oxygen</li><li>Organic sulfur</li></ul>	Red ferric thiocyanate is soluble in organic molecules containing polar coordination centers, such as oxygen, nitrogen, and sulfur. Ferric thiocyanate is not soluble in organic molecules containing only carbon, hydrogen, or halogen.

	Test Phase	
	✓	Solid
	✓	Liquid
		Gas/Vapor

Sensitivity
Not available

Directions and Comments
This test is suggestive of hydrocarbon compounds containing oxygen, nitrogen, or sulfur containing hydrocarbons; however, a negative result can occur with higher complexity structures such as nitro compounds and large ethers. The test must be performed on paper in order to eliminate interference from the alcohol solvent. The papers do not store well and must be made just prior to use. Add 1 g of ferric chloride (CAS 10025-77-1) to 10 ml of methanol (CAS 67-56-1). Separately, add 1 g of potassium thiocyanate (CAS 333-20-0) to 10 ml of methanol. Mix the two solutions and let stand for a few hours. Filter the precipitate and then soak filter papers in the liquid and dry. Liquid sample is added directly to the dry paper; a red color indicates oxygen, nitrogen, or sulfur in the organic compound. Solid samples should be dissolved in a hydrocarbon solvent, such as hexane or toluene and then tested on the paper. Acids and oxidizing compounds interfere.

References	**Manufacturer**
Feigl, F. and Anger, V., *Spot Tests in Inorganic Analysis*, Sixth Edition, Elsevier Publishing Company, Amsterdam, 1972.	

Ferricyanide Test	
Detects	**Test Principle**
<ul><li>Zinc</li></ul>	Ferricyanide will form a yellow precipitate when mixed with a zinc solution.

	Test Phase	
	✓	Solid
	✓	Liquid
		Gas/Vapor

Sensitivity
Not available

Directions and Comments
Mix a 1% solution (w/v) of potassium ferricyanide (CAS 13746-66-2) and add to an equal volume of sample. Insoluble compounds may require acidification or heating to form a colored precipitate. A yellow to brown indicates zinc; a red precipitate may indicate cobalt but is not dependable.

References	**Manufacturer**
Engelder, C., Dunkelberger, T. and Schiller, W., *Semi-Micro Qualitative Analysis,* Second Edition, John Wiley & Sons, Inc., New York, May 1946.	

Ferrocyanide Test (Ferric Chloride)

Detects	Test Principle		
• Ferrocyanide	The addition of an aqueous ferric solution (ferric chloride) to a ferrocyanide compound will produce Prussian blue in acid solution.		
	Test Phase		
	✓	Solid	
	✓	Liquid	
		Gas/Vapor	
	Sensitivity		
	Not available		

Directions and Comments

Mix a 1% solution of ferric chloride hexahydrate (CAS 10025-77-1). Add a drop of test solution to a drop of sample solution. The development of Prussian blue indicates ferrocyanide. Alternatively, ferrocyanide can be oxidized to ferricyanide by hydrogen peroxide, permanganate, or other oxidizers. The resulting ferricyanide can be confirmed with the appropriate test. Addition of acid to a cyanide compound can produce hydrogen cyanide gas; work with small amounts and adequate ventilation. Compare to Ferrocyanide Test (Gutzeit Scheme).

References	Manufacturer
Engelder, C., Dunkelberger, T. and Schiller, W., *Semi-Micro Qualitative Analysis,* Second Edition, John Wiley & Sons, Inc., New York, May 1946.	

Ferrocyanide Test (Gutzeit Scheme)

Detects	Test Principle		
• Ferrocyanide • Thiocyanate • Thiosulfate	Ferric chloride solution with hydrochloric acid will react with ferrocyanide to form a blue color; thiosulfate to form a white color; and thiocyanate to form a red color.		
	Test Phase		
		Solid	
	✓	Liquid	
		Gas/Vapor	
	Sensitivity		
	Not available		

Directions and Comments

The sample is first reacted with a soluble carbonate compound such as sodium carbonate (CAS 497-19-8), and acetic acid (CAS 64-19-7) is added until no more carbon dioxide is produced. A drop of ferric chloride solution (CAS 10025-77-1) is placed on filter paper and almost dried. The ferric chloride is spotted with a drop of dilute hydrochloric acid and then the sample solution. The paper provides a chromatographic method to separate the three target substances through the relative mobility of the ions. Rings will form with a blue ferrocyanide ring being the least mobile and in the center. A white thiosulfate ring is intermediate. A red thiocyanate ring is the outermost ring.

References	Manufacturer
Gutzeit, G., *Helv. Chim. Acta,* 12, 829 (1929).	

Ferrous Hydroxide Reagent	
Detects	**Test Principle**
• Alkyl nitrate • Alkyl nitrite • Explosive • Hydroxylamine • Nitro • Nitroso	Nitro compounds oxidize ferrous hydroxide to form a brown to red-brown precipitate of ferric hydroxide.

	Test Phase	
	✓	Solid
	✓	Liquid
		Gas/Vapor

Sensitivity
Not available

Directions and Comments
Solution A is formed by boiling 50 ml of distilled water to eliminate dissolved oxygen. 2.5 g of ferrous ammonium sulfate (CAS 7783-85-9) is added followed by 0.2 g of concentrated sulfuric acid. If the solution will not be used immediately, a piece of scrap iron metal is added to reduce oxidation by air. Solution B is 3 g of potassium hydroxide (CAS 1310-58-3) dissolved in 3 ml of water and then added to 20 ml of 95% ethanol (CAS 64-17-5). Place 10 mg of solid or one drop of liquid sample in a test tube and then add 1 ml (10 drops) of Solution A and 0.7 ml (7 drops) of Solution B. Immediately displace all oxygen in the tube by adding another gas, such as nitrogen, through a tube that extends to the bottom of the test tube. Cover and shake the tube for up to 30 seconds. A brown to red-brown precipitate indicates nitro, nitroso, organic nitrate and nitrite, quinones or hydroxylamines. Practically all nitro compounds produce a positive result in 30 seconds. The speed is based on solubility of the sample. A green precipitate is a negative result. A light brown color can be caused by atmospheric oxygen in the tube, which can also oxidize ferrous hydroxide. If inert gas cannot be added, perform the test alongside a sample of water and compare the colors; a marked difference in brown color is a positive result.

References	**Manufacturer**
Shriner, R., Fuson, R. and Curtin, D., *The Systematic Identification of Organic Compounds, A Laboratory Manual,* Fifth Edition, John Wiley and Sons, Inc., New York, 1964. *J. Chem. Educ.,* 37, 205 (1960).	

Firearm Discharge Residue Test	
Detects	**Test Principle**
• Antimony • Barium • Lead	Three individual tests are combined as a screen for trace materials in firearm residue.

	Test Phase	
	✓	Solid
	✓	Liquid
		Gas/Vapor

Sensitivity
Detects trace amounts from cloth wipes used to remove firearm residue

Firearm Discharge Residue Test (*Continued*)
Directions and Comments
The sample is placed in a small amount of water and one drop is then dried on a white cotton cloth. A drop of 10% alcoholic solution of triphenylmethylarsonium iodide (CAS 1499-33-8) is added and will form an orange ring in about 30 seconds to 2 minutes if antimony is present. The cloth is dried with a hair dryer and two drops of freshly prepared 0.2% sodium rhodizonate (CAS 523-21-7) solution is added; a red to red-brown color within the orange ring indicates barium and/or lead. The cloth is dried again and one or two drops of dilute (1:20) hydrochloric acid are added to the red area; blue indicates lead and a brighter red indicates barium. Avoid strong indoor lights or sunlight as this will cause decomposition of the rhodizonate solution.

References	Manufacturer
Jungreis, E., *Spot Test Analysis: Clinical, Environmental, Forensic, and Geochemical Applications,* Second Edition, John Wiley & Sons, Inc., New York, 1997.	

Fizz Test	
Detects	**Test Principle**
• Bicarbonate • Carbonate	Carbonate reacts with acid to form carbon dioxide gas.

	Test Phase	
	✓	Solid
	✓	Liquid
		Gas/Vapor

Sensitivity
0.2% calcium carbonate equivalent (dry soil basis)

Directions and Comments
Place a small amount of material in a spot plate and add a drop of 3N hydrochloric acid or other suitable acid on the sample. Effervescence indicates the presence of carbonate or bicarbonate. This test is commonly used in soil testing on solid samples. It is not quantitative, but greater concentration of carbonate or bicarbonate will produce more aggressive effervescence. To rule out water-reactive materials, repeat the test using water instead of acid. Consider other materials that will produce gas from acid such as cyanides and sulfides; analyze the gas as necessary. For example, the gas in the tube neck can be tested with a colorimetric tube for carbon dioxide. See color images 2.22 and 2.23.

References	Manufacturer
Gavlak, R. et al., *Soil, Plant and Water Reference Methods for the Western Region,* Second Edition, Western Coordinating Committee on Nutrient Management, 2003.	

Fluoride Test Strip	
Detects	**Test Principle**
• Fluoride	Fluoride added to a red complex of zirconium-alizarin in hydrochloric acid will decompose the complex and produce a yellow color.

(continued)

Fluoride Test Strip (*Continued*)		
	Test Phase	
	✓	Solid
	✓	Liquid
		Gas/Vapor
	Sensitivity	
	10 ppm fluoride	

Directions and Comments

Alizarin (CAS 72-48-0) impregnated pad detects fluoride in acidic solution. The solution being tested must be acidified with hydrochloric acid to a pH of 0.5. Dip the strip in the sample and compare the results to the color chart. A red color that becomes pink to yellow indicates fluoride.

References	**Manufacturer**
Kimel, H., Notes from correspondence, Precision Laboratories, Inc., Cottonwood, Arizona, USA, December 26, 2007.	Precision Laboratories, Inc., Cottonwood, Arizona, USA

Fluorescamine Test for Protein		
Detects	**Test Principle**	
• Protein	Fluorescamine (CAS 38183-12-9) attaches to the amine portion of protein to produce a glowing blue color when illuminated with UV light.	
	Test Phase	
	✓	Solid
	✓	Liquid
		Gas/Vapor
	Sensitivity	
	0.75% dried egg white (albumin); 0.9% soy protein.	

Directions and Comments

Fluorescamine and many other fluorophores are used in sensitive laboratory assays to detect primary amino acids and amines, peptides, and proteins. Typically, a target protein is tagged with a fluorophore and quantitatively measured with an optical reader that can detect a specific wavelength of light caused by the tagged protein. If the protein concentration is high enough, usually about 1% to 2% concentration for most proteins, the fluorescence is visible. The test solution is freshly mixed by dissolving a few milligrams of fluorescamine (CAS 38183-12-9) in 0.25 ml acetone and then applying to the sample. At room temperature results should be visible within 3 minutes and longer at cooler temperatures. Protein is indicated by a fluorescent blue color when illuminated with a UV light source. Visibility is improved if room lights are darkened. A premixed solution is commercially available (Sigma-Aldrich F5928).

References	**Manufacturer**
Houghton, R. Unpublished studies, St. Johns, MI, 2003.	

Fluorescein Reagent (Bromides)	
Detects	**Test Principle**
• Bromide • Hydrobromic acid • Hydroiodic acid • Iodide	Bromides or hydrogen bromide is oxidized to free bromine by lead peroxide and acetic acid. Yellow fluorescein reacts with bromine or iodine to produce red or red-yellow reaction products.

Test Phase	
✓	Solid
✓	Liquid
	Gas/Vapor

Sensitivity
2 µg/ml bromide

Directions and Comments
Yellow fluorescein will react with bromine or iodine to produce a red color. The paper test strips for bromine and iodine are made by soaking paper in a saturated aqueous solution of fluorescein (2-[6-hydroxy-3-oxo-(3H)-xanthen-9-yl]benzoic acid, CAS 2321-07-5) and dried. Dried paper strips of both types store indefinitely. The sample is treated to oxidize any bromide or iodide to free halogen by adding one or two drops of sample to a test tube followed by a few milligrams of lead peroxide (CAS 1309-60-0) and one or two drops of acetic acid (CAS 64-19-7). Chromic acid (CAS 1333-82-0) may be substituted for acetic acid. Suspend a fluorescein test strip in the test tube and gently heat. A red spot on the yellow paper indicates bromide or iodide. Chloride compounds do not interfere.

References	**Manufacturer**
Feigl, F. and Anger, V., *Spot Tests in Inorganic Analysis*, Sixth Edition, Elsevier Publishing Company, Amsterdam, 1972.	

Fluorescein Reagent (Halogens)	
Detects	**Test Principle**
• Bromine • Chlorine • Iodine	Yellow fluorescein reacts with bromine, iodine, or chlorine (the latter in the presence of potassium bromide) to produce red or red-yellow colored reaction products.

Test Phase	
✓	Solid
✓	Liquid
	Gas/Vapor

Sensitivity
1 µg/ml chlorine; 2 µg/ml bromine; 6 µg/ml iodine.

(*continued*)

Fluorescein Reagent (Halogens) (*Continued*)
Directions and Comments
Yellow fluorescein will react with bromine or iodine to produce a red color; chlorine will produce the red color indirectly if potassium bromide is present, thus chlorine and bromine produce red tetrabromofluorescein and iodine produces yellow-red tetraiodofluorescein. The paper test strips for bromine and iodine are made by soaking paper in a saturated aqueous solution of fluorescein (2-[6-hydroxy-3-oxo-(3H)-xanthen-9-yl]benzoic acid, CAS 2321-07-5) and dried. The test strip also becomes sensitive to chlorine if bathed in a weakly alkaline solution of fluorescein and potassium bromide (CAS 7758-02-3) in a 1:8 proportion respectively and dried. Alternately, use a dried paper strip containing fluorescein and add a drop of the potassium bromide solution at the time of the test. Dried paper strips of both types store indefinitely. This test can selectively identify halogens among other oxidizers that pass a potassium iodide-starch test strip screen.

References	Manufacturer
Frehden, O., and Huang, C., *Mikrochem. ver. Mikrochim. Acta*, 26, 41 (1939). Feigl, F., *Spot Tests in Inorganic Analysis*, Fifth Edition, Elsevier Publishing Company, New York, 1958.	

Fluorescein Reagent (Uranyl)			
Detects	**Test Principle**		
• Uranium • Uranyl	Uranyl (UO_2^{2+}) ions react with fluorescein and ammonium chloride to produce a red color.		
	Test Phase		
		Solid	
	✓	Liquid	
		Gas/Vapor	
	Sensitivity		
	0.12 µg/ml uranyl ion		

Directions and Comments
Mix a 0.12% aqueous solution of fluorescein (2-[6-hydroxy-3-oxo-(3H)-xanthen-9-yl]benzoic acid, CAS 2321-07-5). Separately, mix a 5% solution of ammonium chloride. Dissolve the sample in water and add a drop to a spot plate or test tube. Add a drop of the two test solutions. A red color indicates uranyl (UO_2^{2+}) ions.

References	Manufacturer
Feigl, F. and Anger, V., *Spot Tests in Inorganic Analysis*, Sixth Edition, Elsevier Publishing Company, Amsterdam, 1972. *Z. Anal. Chem.*, 142, 161 (1954).	

Formaldoxime Reagent	
Detects	**Test Principle**
• Cobalt • Copper(I) • Iron(I) • Manganese(I) • Nickel	Formaldoxime reacts with metal ions to form various colors.

(*continued*)

Formaldoxime Reagent (*Continued*)

	Test Phase	
		Solid
	✓	Liquid
		Gas/Vapor
	Sensitivity	
	0.05 mg/l manganese; 0.10 mg/l copper, nickel or iron; 0.20 mg/l cobalt.	

Directions and Comments

Mix a 50% (w/v) aqueous solution of formadloxime (CAS 62479-72-5) in water. Add one drop of the test solution to 10 ml of sample solution and then add two drops of 1N sodium hydroxide solution. Green to yellow indicates nickel; yellow indicates cobalt; violet indicates cupric; red to violet indicates ferric; orange-red indicates manganous ions. Formaldoxime is moisture sensitive and must be stored under inert gas, such as nitrogen.

References	Manufacturer
Welcher, F., *Chemical Solutions: Reagents Useful to the Chemist, Biologist, and Bacteriologist*, D. Van Nostrand Company, Inc., New York, 1942. *Chemical Abstracts* 26, 2935 (1932).	

Froede Reagent

Detects	Test Principle
• Alkaloid • Aspirin • Codeine • LSD Morphine • Mace	Molybdate anion reacts with alkaloids in the presence of concentrated sulfuric acid to form various colored reaction products.

	Test Phase	
	✓	Solid
	✓	Liquid
		Gas/Vapor
	Sensitivity	
	50 μg LSD; 100 μg mescaline HCl; 25 μg morphine monohydrate.	

Directions and Comments

This is a presumptive test for illegal drugs. The test solution is 0.5 g of molybdic acid or sodium molybdate (CAS 10102-40-6) in 100 ml of hot concentrated sulfuric acid. The test solution is applied to the sample and any color production is noted. It is most helpful to compare any color formed in this test with results from other presumptive drug tests (Marquis Reagent, etc.) in a matrix to determine a pattern of unique results. Refer to the appendix for drug screen results.

References	Manufacturer
U.S. Department of Justice. Color Test Reagents/Kits for Preliminary Identification of Drugs of Abuse, NIJ Standard–0604.01, National Institute of Justice, National Law Enforcement and Corrections Technology Center, Rockville, MD, July 2000.	Several commercial test kits are available.

Gas Production with Sulfuric Acid	
Detects	**Test Principle**
• Acetate • Azide • Cyanide • Halide • Nitrite • Sulfide	Dilute sulfuric acid reacts with certain compounds to produce various gases.

Test Phase	
✓	Solid
✓	Liquid
	Gas/Vapor

Sensitivity
Various

Directions and Comments
Colorless gases produced by contact with dilute sulfuric acid include carbon dioxide from carbonates; hydrogen cyanide from cyanides; hydrogen sulfide from sulfides; sulfur dioxide from sulfites and thiosulfates; acetic acid from acetates; oxygen from peroxides; hydrogen from base metals; hydrazoic acid from azides. Brown, yellow, or violet gases may be produced by nitrogen oxides from nitrites; chlorine, bromine, or iodine from halides in the presence of oxidizers. The use of concentrated sulfuric acid will add other detectable gases to the test: hydrogen chloride from chlorides; oxygen from chromates, permanganates and peroxides; hydrogen fluoride from fluorides and fluosilicates. Generated gases may be confirmed by a colorimetric tube or other suitable test.

References	**Manufacturer**
Feigl, F., *Spot Tests in Inorganic Analysis,* Fifth Edition, Elsevier Publishing Company, New York, 1958.	

Gastec Tube #180	
Detects	**Test Principle**
• Amine	Amines react with sulfuric acid. The neutralization reaction probably induces a color change in a pH indicator from pink to yellow. Some amine sulfate salts formed in the neutralization reaction impart additional colors.

Test Phase	
	Solid
	Liquid
✓	Gas/Vapor

Sensitivity
>0.5 ppm amine

Directions and Comments
This test detects amines in general. The test instructions include a table of 29 specific amines detected, listing eight groups that produced uniquely colored results. See the test instruction for details.

References	**Manufacturer**
Gastec Corporation website, http://www.gastec.co.jp/english/products/frame.php?place=seihin/dtube/short/index.htm, December 13, 2007.	Gastec Corporation, Ayase-shi, Kanagawa, Japan

Gold Chloride Reagent		
Detects	**Test Principle**	
• Bromide • Iodide	Gold reacts with iodide and starch to form a blue color; with bromide to form an orange-red color.	
	Test Phase	
	✓	Solid
	✓	Liquid
		Gas/Vapor
	Sensitivity	
	100 ppm bromide; 20 ppm iodide.	

Directions and Comments

Heat 8 ml of water to 90°C and add 1 g soluble starch (CAS 9005-25-8; corn starch may be substituted) dissolved in 2 ml cool water. After mixing, cool the solution, allow it to settle, and decant the clear starch solution. Add to the starch solution an equal volume of 2% (w/v) gold chloride (gold chloride trihydrate, CAS 16961-25-4) solution. The gold chloride solution will be a clear yellow color. The test is pH dependent and is overwhelmed by higher concentrations of bromide and iodide. Neutralize the sample as necessary. Test increasingly dilute sample solutions if muddy precipitate forms. Solid samples should be diluted in water to about 0.1% to 1.0% and neutralized. Add equal volumes of test solution to sample solution in a test tube. Bromide will produce an orange-red to yellow-red color. Iodide will produce a blue color due to the formation of the starch complex. If both bromide and iodide are present, a blue color will form but upon heating orange-yellow will appear; the blue color will return on cooling.

References	**Manufacturer**
Welcher, F., *Chemical Solutions: Reagents Useful to the Chemist, Biologist, and Bacteriologist,* D. Van Nostrand Company, Inc., New York, 1942. *Chemical Abstracts* 33, 4904 (1939).	

Gold Chloride Reagent (Thiol)		
Detects	**Test Principle**	
• H • HD • Mercaptan • Mustard • Thiol	Clear, yellow gold ion solution reacts with thiol to form an opaque yellow or orange solution.	
	Test Phase	
	✓	Solid
	✓	Liquid
		Gas/Vapor
	Sensitivity	
	Low ppm range. The test is adaptable to detection of thiol in air.	

Directions and Comments

Mix a 0.5% (w/v) gold chloride (gold chloride trihydrate, CAS 16961-25-4) solution. The gold chloride solution will be a clear yellow color. Place 0.5 ml of test solution in a test tube or use a spot plate. Add a drop of sample solution or a few milligrams of solid sample to the test solution. A thick, opaque yellow or orange color indicates thiol. Neutralize strongly acidic or basic sample as necessary. Test increasingly dilute sample solutions if muddy precipitate forms. Solid samples can be diluted in water or alcohol to about 0.1% to 1.0% and neutralized as necessary. Compare to Drager Thioether test. See color images 2.24 and 2.25.

(continued)

Gold Chloride Reagent (Thiol) (*Continued*)	
References	**Manufacturer**
Jacobs, M. *War Gases: Their Identification and Decontamination*, Interscience Publishers, Inc., New York, 1942. Schroter, G. *Method of Detecting the Presence of Mustard Gas (Yperite)*, United States Patent 2054885, 1936.	

Griess Test		
Detects	**Test Principle**	
• Nitrite • Nitrous acid	Nitrate reacts with primary aromatic amines in acid solution to form diazonium compounds, which in turn react with other or the same aromatic amine to produce intensely colored amino azo compounds.	
	Test Phase	
		Solid
	✓	Liquid
		Gas/Vapor
	Sensitivity	
	0.01 µg/ml nitrous acid	
Directions and Comments		
The sample is mixed with water as needed. If the solution is basic, acetic acid is added until pH <7. One gram of sulfanilic acid (CAS 121-57-3; other primary amines may be substituted with decreased sensitivity) is mixed with 100 ml of 30% acetic acid (CAS 64-19-7) and warmed until dissolved; 0.03 g of a-naphthylamine (CAS 134-32-7) is boiled in 70 ml of water and allowed to settle before removing the colorless portion of liquid and mixing with 30 ml acetic acid. The three solutions are combined in equal volume in the order of the sample first, sulfanilic acid and then naphthylamine. A red color indicates nitrite.		
References	**Manufacturer**	
Griess, P., *Ber.*, 12, 427 (1879). Feigl, F., *Spot Tests in Inorganic Analysis,* Fifth Edition, Elsevier Publishing Company, New York, 1958.		

Griess' Paper (Yellow)		
Detects	**Test Principle**	
• Nitrous Acid	Nitrous acid reacts with sulfanilic acid and metaphenylenediamine dried on paper to form a yellow-brown color.	
	Test Phase	
	✓	Solid
	✓	Liquid
		Gas/Vapor
	Sensitivity	
	Not available, but described as a "delicate" test	

(continued)

Griess' Paper (Yellow) (*Continued*)

Directions and Comments

Mix an aqueous solution of sulfanilic acid (CAS 121-57-3) and metaphenylenediamine (CAS 108-45-2). Soak filter papers in the solution and dry. Solid sample should be placed in a water solution; liquid solutions may be tested directly or diluted as necessary. Place a drop of sample solution on the paper. A yellow-brown color indicates nitrous acid.

References	Manufacturer
Welcher, F., *Chemical Solutions: Reagents Useful to the Chemist, Biologist, and Bacteriologist,* D. Van Nostrand Company, Inc., New York, 1942.	

Guerin's Reagent

Detects	Test Principle	
• Selenium	Selenium and soluble selenites react with mercuric nitrate in nitric acid to form crystalline precipitates.	
	Test Phase	
	✓	Solid
	✓	Liquid
		Gas/Vapor
	Sensitivity	
	Not available	

Directions and Comments

Dissolve 1 g mercuric nitrate monohydrate (CAS 7783-34-8) in 10 ml of 2N nitric acid. Mix equal volumes of test solution with sample solution. A crystalline precipitate indicates selenium or soluble selenite.

References	Manufacturer
Welcher, F., *Chemical Solutions: Reagents Useful to the Chemist, Biologist, and Bacteriologist,* D. Van Nostrand Company, Inc., New York, 1942. *Chemical Abstracts* 4235 (1930).	

Hearon-Gustavson's Reagents

Detects	Test Principle	
• Nitro • Organic nitro • Oxidizer	Many oxidizers, including atmospheric oxygen, react with ferrous ammonium sulfate in basic solution to form a red-brown ferric precipitate. The organic nitro group is also detected.	
	Test Phase	
	✓	Solid
	✓	Liquid
		Gas/Vapor
	Sensitivity	
	Not available	

(*continued*)

Hearon-Gustavson's Reagents (*Continued*)	
Directions and Comments	
Solution A consists of 50 ml of distilled water that is cooled and mixed with 2.5 g of ferrous ammonium sulfate hexahydrate (CAS 7783-85-9) and 0.2 ml (about 4 drops) of concentrated sulfuric acid. Solution B is 3 g of potassium hydroxide (sodium hydroxide may be substituted) in 3 ml of water diluted to 20 ml with 95% alcohol. The test is performed by placing 0.7 ml of Solution A in a test tube followed by a few drops of liquid test or a few milligrams of finely ground solid. Add 0.5 ml of Solution B and then inject a stream of inert gas such as nitrogen or argon into the head space of the test tube (the test is sensitive to atmospheric oxygen). Immediately stopper and shake. A red-brown precipitate indicates an organic nitro group.	
References	**Manufacturer**
Welcher, F., *Chemical Solutions: Reagents Useful to the Chemist, Biologist, and Bacteriologist*, D. Van Nostrand Company, Inc., New York, 1942. *Ind. Eng. Chem., Anal. Ed.* 9, 352 (1937).	

Hydrogen Peroxide Reagent		
Detects	**Test Principle**	
• Chromate • Iron salt • Molybdate • Titanium(IV) • Vanadate	Hydrogen peroxide reacts with various metal ions to produce colored compounds.	
	Test Phase	
	✓	Solid
	✓	Liquid
		Gas/Vapor
	Sensitivity	
	2 µg/ml titanium(IV)	
Directions and Comments		
The test reagent is 3% hydrogen peroxide (medical grade, CAS 7722-84-1). Acidify the sample with hydrochloric or sulfuric acid and then add a drop of 3% hydrogen peroxide. A yellow color indicates titanium (compare to Sensidyne Tube #247S). A violet color indicates iron, which can be cleared with phosphoric acid. Molybdate produces a yellow color. Chromates can produce a variety of colors: light yellow, gray, violet-blue, violet brown. Brown-pink to brown-red to bloodred indicates vanadate. Fluoride interferes but can be demasked by adding a beryllium salt. Large amounts of acetates, bromides, chlorides, nitrates, or colored compounds decrease sensitivity. See color images 2.26 and 2.27.		
References	**Manufacturer**	
Feigl, F. and Anger, V., *Spot Tests in Inorganic Analysis*, Sixth Edition, Elsevier Publishing Company, Amsterdam, 1972.		

Hydroxylamine Hydrochloride Reagent	
Detects	**Test Principle**
• Amide • Cyanate • Cyanic acid • Nitrile	Nitriles and amides react with hydroxylamine and ferric chloride in base with heating to produce red to violet colors.

(*continued*)

Hydroxylamine Hydrochloride Reagent (*Continued*)		
	Test Phase	
		Solid
	✓	Liquid
		Gas/Vapor
	Sensitivity	
	Various, but less than 30 µg/ml cyanic acid	

Directions and Comments

Mix 0.2 g of hydroxylamine hydrochloride (CAS 5470-11-1) in 2 ml of propylene glycol (CAS 57-55-6). Separately, dissolve one drop of liquid or about 30 mg of solid sample in the least amount of propylene glycol necessary to dissolve it and then add 1 ml of 1N potassium hydroxide (CAS 1310-58-30). Mix the two solutions, boil gently for 2 minutes and allow it to cool. Mix an 8% (w/v) ferric chloride hexahydrate (CAS 10025-77-1) solution and add 0.5 to 1.0 ml to the test solution. A red to violet color indicates an amine or amide. A violet color that clears quickly indicates a cyanate or cyanic acid. Yellow is a negative result and brown is indeterminate. All common nitriles and amides produce a positive result. Benzanilide and diacetylbenzidine are not detected.

References	Manufacturer
Shriner, R., Fuson, R. and Curtin, D., *The Systematic Identification of Organic Compounds, A Laboratory Manual,* Fifth Edition, John Wiley and Sons, Inc., New York, 1964. *J. Chem. Educ.,* 37, 205 (1960). *J. Chem. Educ.,* 17, 81 (1940). *Anal. Chem.,* 22, 676 (1950). *Anal. Chem.,* 24, 898 (1952). Feigl, F. and Anger, V., *Spot Tests in Inorganic Analysis,* Sixth Edition, Elsevier Publishing Company, Amsterdam, 1972. *Chemist-Analyst,* 50, 77 (1961).	

Hypohalogenite Differentiation Test		
Detects	**Test Principle**	
• Hypobromite • Hypochlorite • Hypohalogenite • Hypoiodite	Hypohalogenite reacts with excess zinc chloride to release free halogen; chlorine, bromine, and iodine are indicated in unique ways.	
	Test Phase	
	✓	Solid
	✓	Liquid
		Gas/Vapor
	Sensitivity	
	2 µg/ml as sodium hypochlorite; 2.5 µg/ml as sodium hypobromite or sodium hypoiodite.	

(*continued*)

Hypohalogenite Differentiation Test (*Continued*)

Directions and Comments

The first test indicates hypohalogenites in general and no inorganic oxidizers interfere. Prepare an aqueous solution of 20% zinc chloride (CAS 7646-85-7), 0.1 g potassium iodide (CAS 7681-11-0), and 0.2 g thiodene (CAS 9005-84-9; Feigl calls for thiodene to be used as a starch indicator; other starches may be used, however, sensitivities are listed using thiodene). Add a drop of the test solution to a drop of liquid sample or a few milligrams of solid sample; a blue color indicates the presence of a hypohalogenite. To differentiate hypohalogenites perform the following three tests on as little as one drop of test solution. Prepare an aqueous solution of 20% zinc chloride and 0.2 g thiodene (starch) and mix with the sample; a blue color indicates hypoiodite. Prepare an aqueous solution of 20% zinc chloride and saturated alcoholic solution of fluorescein (CAS 2321-07-5) and mix with the sample; a red color indicates hypobromite. Hypochlorite is unreactive in the two previous tests. To confirm hypochlorite, prepare an aqueous solution of 20% zinc chloride, saturated alcoholic solution of fluorescein which is then saturated with potassium bromide and mix with the sample; a red color indicates hypochlorite. Hypochlorite, hypobromite, and hypoiodite ions are unstable in mixtures that include more than one hypohalogenite; only a single hypohalogenite is expected to be found in a sample.

References	Manufacturer
Feigl, F., *Spot Tests in Inorganic Analysis,* Fifth Edition, Elsevier Publishing Company, New York, 1958.	

Hypohalogenite Test

Detects	Test Principle		
<ul><li>Hypobromite</li><li>Hypochlorite</li><li>Hypohalogenite</li><li>Hypoiodite</li></ul>	Hypohalogenites react with ammonium to form free nitrogen. Reduced concentration of ammonium is detected.		
	Test Phase		
			Solid
		✓	Liquid
			Gas/Vapor
	Sensitivity		
	0.1 µg/ml hypohalogenite (Feigl)		

Directions and Comments

Ammonia from an ammonium salt is immediately oxidized by hypohalogenites to form free nitrogen. Hypohalogenites, even in the presence of other oxidizers, are indicated by measuring the consumption of ammonium with an appropriate test. The test is performed by mixing a 0.003% solution of ammonium chloride (CAS 12125-02-9) solution and placing two drops side by side on a spot plate. To one drop, add a drop of the test solution; to the other, add a drop of sodium hydroxide solution. Add a drop of Nessler's solution (CAS 7783-33-7 or see Merckoquant® Ammonium Test) to each. If the test solution drop is less red than the control drop, hypohalogenite is indicated. A distinct red color that remains indicates no ammonium was consumed and no hypohalogenite is present. An alternate test at only slightly lowered sensitivity involves the use of Merckoquant® Ammonium Test in place of the Nessler's solution. Stronger solutions of hypohalogenite may be indicated by overt bubbling of nitrogen from the solution upon addition of concentrated ammonium chloride. If the sample is colored or may contain chromate, ferricyanide or permanganate, refer to the Hypohalogenate (Colored Solution) Test.

References	Manufacturer
Feigl, F., *Spot Tests in Inorganic Analysis,* Fifth Edition, Elsevier Publishing Company, New York, 1958.	

Indigo Carmine Reagent (Chlorate)	
Detects	**Test Principle**
• Chlorate	Indigo carmine dye is discolored by chlorate.
	Test Phase
	✓ Solid
	✓ Liquid
	Gas/Vapor
	Sensitivity
	5 ppm chlorate in 5 ml of test solution

Directions and Comments

Indigo carmine dye (indigo-5,5′-disulfonic acid disodium salt, CAS 860-22-0) is mixed as a 0.1% (w/v) solution. Add 1 ml of indigo carmine solution to 5 ml of concentrated hydrochloric acid and boil. Add up to 5 ml of sample solution. The blue test solution is discolored by chlorate. Perchlorate is not detected.

References	**Manufacturer**
U.S. Patent No. 2,392,769	

Indigo Carmine Reagent (Chromate)	
Detects	**Test Principle**
• Chromate	Chromium(VI) ions catalyze the oxidation of indigo carmine by hydrogen peroxide.
	Test Phase
	Solid
	✓ Liquid
	Gas/Vapor
	Sensitivity
	0.02 µg/ml chromate

Directions and Comments

The test will compare the rate of decolorization of a blank to the test solution. Chromate will catalyze the reaction and the sample solution will become clear faster than the blank if chromate is present. The test uses a 0.04% solution of indigo carmine (CAS 860-22-0) in water. Add a drop of indigo carmine solution to a test tube followed by a drop of 0.2N hydrochloric acid and a drop of 2% hydrogen peroxide (medical grade 3% hydrogen peroxide may be diluted). Make a blank in a second test tube using identical amounts. Add a drop of sample solution to the test solution and a drop of water to the blank. Compare the rate of decolorization of the test solution to the blank solution. If the test solution clears more rapidly than the blank, chromate is indicated.

References	**Manufacturer**
Feigl, F. and Anger, V., *Spot Tests in Inorganic Analysis,* Sixth Edition, Elsevier Publishing Company, Amsterdam, 1972. *Anal. Abstracts*, 16 (1969).	

Iodic Acid Test 1	
Detects	**Test Principle**
• Iodate • Iodic Acid	Iodates are reduced quickly by hypophosphorous acid to form iodide and phosphorous acid. Iodine and iodates react to form free iodine. Free iodine reacts slowly with phosphorous acid to produce phosphoric acid. The free, but transient, iodine is detected with starch.

	Test Phase	
		Solid
✓		Liquid
		Gas/Vapor

Sensitivity

1 μg/ml iodic acid

Directions and Comments

If the sample is strongly acidic or basic, adjust the pH near neutral and then add a drop of starch solution (CAS 9005-25-8; corn starch may be used) and a drop of hypophosphorous acid (CAS 6303-21-5). A transient blue color indicates iodate.

References	**Manufacturer**
Feigl, F., *Spot Tests in Inorganic Analysis,* Fifth Edition, Elsevier Publishing Company, New York, 1958.	

Iodic Acid Test 2	
Detects	**Test Principle**
• Iodic Acid • Iodate	Iodates react with alkali thiocyanates in acidic solution to form free iodine, which is detected with starch.

	Test Phase	
		Solid
✓		Liquid
		Gas/Vapor

Sensitivity

4 μg/ml sodium iodate

Directions and Comments

A drop of 5% potassium thiocyanate solution (CAS 333-20-0) is placed on starch paper (not potassium iodide-starch paper) or on a little corn starch. A drop of acid suspected of containing iodic acid is added (toxic gas may be released from the drop). A blue color indicates iodate.

References	**Manufacturer**
Feigl, F., *Spot Tests in Inorganic Analysis,* Fifth Edition, Elsevier Publishing Company, New York, 1958.	

Iodine-Sulfide Reagent	
Detects	**Test Principle**
• Azide • Hydrazoic acid	Azides react with iodine to evolve nitrogen gas when the reaction is catalyzed by sulfide.

(continued)

Iodine-Sulfide Reagent (*Continued*)

	Test Phase	
	✓	Solid
	✓	Liquid
		Gas/Vapor
	Sensitivity	
	5 µg/ml sodium azide	

Directions and Comments

Many azides are explosive. Place a few milligrams of solid sample in a test tube containing five drops of water. If the sample dissolves, add a crystal of iodine (CAS 7553-56-2) or a drop of 0.1N iodine solution and mix. If the sample does not dissolve, dissolve a few milligrams of potassium iodide (CAS 7681-11-0) into the liquid, mix, and then add a crystal of iodine or a drop of 0.1N iodine solution and mix again. Allow the solution to settle and add a few milligrams of sodium sulfide (CAS 1313-84-4). The appearance of small bubbles of nitrogen within 2 minutes indicates an azide. Sodium azide is water soluble; many others are insoluble.

References	Manufacturer
Feigl, F. and Anger, V., *Spot Tests in Inorganic Analysis,* Sixth Edition, Elsevier Publishing Company, Amsterdam, 1972.	

Isatin Reagent (Mercaptan)

Detects	Test Principle	
• Mercaptan • Thioalcohol	Mercaptan (R-C-SH) reacts with isatin in sulfuric acid to form a green color.	
	Test Phase	
		Solid
	✓	Liquid
		Gas/Vapor
	Sensitivity	
	Not available	

Directions and Comments

Dissolve 2 mg of isatin (CAS 91-56-5) in 1 ml concentrated sulfuric acid. Dissolve the sample in alcohol. Slowly and carefully add the sample solution to the test solution. A green color indicates mercaptan.

References	Manufacturer
Welcher, F., *Chemical Solutions: Reagents Useful to the Chemist, Biologist, and Bacteriologist,* D. Van Nostrand Company, Inc., New York, 1942. *J. Pharm. Chim.* 276 (1889).	

ITS Cyanide ReagentStrip™

Detects	Test Principle
• Cyanide • Thiocyanate • Cyanogen chloride	Cyanide reacts with a chlorinating agent (such as chloramine-T or sodium hypochlorite) to form cyanogen chloride. The cyanogen chloride reacts with isonicotinic acid and barbituric acid which in the presence of cyanide to produce a blue color. The test uses a buffer to maintain optimal pH of 6 to 7.

(*continued*)

ITS Cyanide ReagentStrip™ (*Continued*)		
	Test Phase	
	✓	Solid
	✓	Liquid
		Gas/Vapor
	Sensitivity	
	0.1 mg/l free cyanide	

Directions and Comments
The test detects free cyanide in water. Generally, cyanide dissociates easily from smaller cations (e.g., sodium, potassium) and is more tightly bound by larger cations (e.g., iron, silver). If the cyanide anion is free in water, this test will detect it qualitatively in 1 minute. Thiocyanate yields a positive result that is about 60% of the corresponding cyanide concentration. All reagents are dry and captive on test strip pads until immersed in the aqueous sample. Sulfides, aldehydes, and heavy metals interfere, but will not prevent detection of cyanide in higher concentrations. Test directions are for quantitative results. For faster qualitative results of higher concentrations of free cyanide, simply drop Test Strip 1 into a test tube containing the aqueous sample and shake for a few seconds. Next, drop Test Strip 2 into the solution and watch for the appearance of a blue color that indicates cyanide. See color images 2.28 through 2.30.

References	**Manufacturer**
James, R. et al., *Industrial Test Systems, Inc. Cyanide ReagentStrip™ Test Kit ETV Report*, Environmental Technology Verification Advanced Monitoring Systems Center, Columbus, Ohio, April 2005.	Industrial Test Systems, Inc., Rock Hill, South Carolina, USA (www.sensafe.com)

ITS SenSafe™ Water Metal Test		
Detects	**Test Principle**	
• Arsenic • Cadmium • Chromium • Cobalt • Copper • Heavy metal • Lead • Mercury • Metal • Nickel • Zinc	The test uses a colorimetric change at near neutral pH to indicate a broad range of heavy metals.	
	Test Phase	
		Solid
	✓	Liquid
		Gas/Vapor
	Sensitivity	
	10 ppb	

Directions and Comments
The ITS SenSafe™ Water Metal Test is a rapid field screen for heavy metals in water. Various colors are possible depending on the metal and concentration. See color image 2.31.

References	**Manufacturer**
ITS SenSafe™ Water Metal Test product insert.	Industrial Test Systems, Inc., Rock Hill, South Carolina, USA (www.sensafe.com)

Jaffe's Reagent

Detects	Test Principle		
• Bismuth • Antimony	Iodine and triethanolamine react with bismuth to form a scarlet precipitate; antimony forms a yellow-gold precipitate.		
	Test Phase		
	✓	Solid	
	✓	Liquid	
		Gas/Vapor	
	Sensitivity		

Directions and Comments

The test solution is 8 g of iodine (CAS 7553-56-2) in 100 ml of triethanolamine (CAS 102-71-6). Add hydrochloric acid to the sample and add a few drops of test solution. A scarlet precipitate indicates bismuth; a yellow-gold precipitate indicates antimony.

References	Manufacturer
Welcher, F., *Chemical Solutions: Reagents Useful to the Chemist, Biologist, and Bacteriologist,* D. Van Nostrand Company, Inc., New York, 1942. *Chemical Abstracts* 28, 6382 (1934).	

Janovsky Test

Detects	Test Principle		
• 2,4,6-Trinitrotoluene • 2,4-Dinitrotoluene • DNT • TNT	Di- and tri-nitroaromatic compounds develop colors in alkaline solution.		
	Test Phase		
	✓	Solid	
	✓	Liquid	
		Gas/Vapor	
	Sensitivity		
	Not available		

Directions and Comments

A few milligrams of sample is dissolved in a drop of acetone and then mixed with a drop of strong potassium hydroxide solution. A characteristic red-violet color indicates di- and tri-nitroaromatic compounds.

References	Manufacturer
Janovsky, J. and Erb, L., *Ber. Dtsch. Chem. Ges.,* 19, 2156 (1886). Jungreis, E., *Spot Test Analysis: Clinical, Environmental, Forensic, and Geochemical Applications,* Second Edition, John Wiley & Sons, Inc., New York, 1997.	

Lactase Test	
Detects	**Test Principle**
• Dairy • Lactose • Milk	Lactose is cleaved by beta-galactosidase (lactase) to form glucose and galactose. The resulting glucose is detected with a glucose test strip.

	Test Phase	
	✓	Solid
	✓	Liquid
		Gas/Vapor

Sensitivity	
0.5 mg/ml dried milk in 3 minutes; 5 mg/ml dried milk in 30 seconds.	

Directions and Comments

Dairy products can be detected if sugar is not present by the use of an enzyme, beta-galactosidase (Lactaid®). Milk contains the carbohydrate lactose. Lactase will quickly cleave lactose into glucose and galactose. The increase in glucose can be detected by a glucose urine test strip (Bayer® Uristix™ Glucose Strips or Merckoquant® Glucose Test).

Perform the test by placing two drops of liquid sample or a few milligrams of solid sample in a test tube and mix with two drops of water. Use a glucose test strip to measure the glucose present in the solution (split the sample and expose this test strip simultaneously with the one below). Crush a few milligrams of beta-galactosidase into a powder and mix it in the solution. Use a second glucose test strip to measure the glucose present in the solution. An increase in glucose indicates lactose is present and a dairy product is indicated.

If the sample already contains a high amount of glucose (flavored milk) before the test, the test strip will be overloaded and the small increase in glucose from the lactase will not be detected. The glucose test strip requires no more than 30 seconds for a quantitative result for glucose in urine. If the strip is allowed to react for about 3 minutes, the sensitivity improves about 10 times. See color images 2.32 and 2.33.

References	**Manufacturer**
Houghton, R., *Emergency Characterization of Unknown Materials,* CRC Press, 2007.	McNeil Nutritionals, LLC, Ft. Washington, Pennsylvania (www.lactaid.com). Bayer AG, 51368 Leverkusen, Germany (www.bayer.com).

Lead Acetate Test Strip (JD175)	
Detects	**Test Principle**
• Hydrogen sulfide • Sulfide	Hydrogen sulfide reacts with lead acetate to produce black lead sulfide.

	Test Phase	
	✓	Solid
	✓	Liquid
	✓	Gas/Vapor

Sensitivity	
2 ppm hydrogen sulfide	

(*continued*)

Lead Acetate Test Strip (JD175) (*Continued*)

Directions and Comments

The sensitivity of the test is dependent on exposure time. Lead sulfide begins to tint the paper to a brown color at very low concentrations. At 300 to 400 ppm the strip turns black quickly. Fleeting exposure to hydrogen sulfide may not provide sufficient time for the hydrogen sulfide to be absorbed by the water and transmitted to the lead acetate on the strip. Liquids and solids suspected of containing sulfide may be tested directly if the test strip is moistened with water.

References	Manufacturer
Kimel, H., Notes from correspondence, Precision Laboratories, Inc., Cottonwood, Arizona, USA, December 26, 2007.	Precision Laboratories, Inc., Cottonwood, Arizona, USA.

Lead(II) Test

Detects	Test Principle
• Barium • Cadmium • Lead • Silver • Thallium • Tin	Lead reacts with rhodizonate to form a yellow/orange to pink/red color under acidic conditions.

Test Phase	
✓	Solid
✓	Liquid
	Gas/Vapor

Sensitivity
5–15 µg lead as a wipe sample

Directions and Comments

Lead is dissolved into a solution of 1% nitric acid or household vinegar and mixed with a freshly prepared solution of approximately 0.13% sodium or potassium rhodizonate (CAS 523-21-7 or 13021-40-4, respectively). Lead is indicated by a bright red color in slightly acidic solution or a violet color in neutral solution. The rhodizonate solution should be orange when mixed; a solution that is yellow after mixing should be discarded. The test detects lead and lead compounds but not alkyl lead. Thallium(I) and organothallium compounds (brown), silver(I), cadmium(II), barium(II), strontium, and tin(II) also form colored compounds with rhodizonate ion at a pH of about 3, but with less sensitivity than that of lead(II), and only the lead-rhodizonate complex gives the characteristic pink or red color. If the test is performed with a neutral solution of 0.2% sodium or potassium rhodizonate, the test will produce a red-brown precipitate from a neutral salt of barium or strontium. To differentiate barium from strontium, add a dilute (1:20) solution of hydrochloric acid, which will dissolve the strontium complex but the barium complex will convert to an insoluble, bright red acid salt.

References	Manufacturer
Esswein, E. and Ashley, K., Lead in Dust Wipes by Chemical Spot Test (Colorimetric Screening Method), *NIOSH Manual of Analytical Methods (NMAM),* Fourth Edition, March 15, 2003. Jungreis, E., *Spot Test Analysis: Clinical, Environmental, Forensic, and Geochemical Applications,* Second Edition, John Wiley & Sons, Inc., New York, 1997. Feigl, F. and Anger, V., *Spot Tests in Inorganic Analysis*, Sixth Edition, Elsevier Publishing Company, Amsterdam, 1972. *Anal. Abstracts*, 16 (1969).	

Lead Sulfide Paper	
Detects	**Test Principle**
• Hydrogen peroxide • Alkali peroxide	Black-brown lead sulfide reacts with peroxide to form white lead sulfate.

Test Phase	
✓	Solid
✓	Liquid
	Gas/Vapor

Sensitivity
0.04 µg/ml hydrogen peroxide

Directions and Comments
Mix a 0.5% aqueous solution of lead sulfide (CAS 1314-87-0). Soak filter papers in the solution and dry. The paper test strips will store indefinitely in a sealed bottle. Adjust the pH of the test solution to neutral or just slightly acidic. Place a drop of the test solution on the test strip. A white spot forming on the light brown paper indicates peroxide. Alternatively and with less sensitivity, a lead acetate test strip can be suspended in a test tube over a grain of sulfide material (e.g., sodium sulfide, CAS 1313-84-4) in a drop of acid. The lead acetate contains lead sulfide when it turns brown/black. Use this strip to directly test a drop of the pH adjusted sample solution. See color images 2.34 and 2.35.

References	**Manufacturer**
Feigl, F. and Anger, V., *Spot Tests in Inorganic Analysis*, Sixth Edition, Elsevier Publishing Company, Amsterdam, 1972.	

Lowry Protein Test	
Detects	**Test Principle**
• Protein	Copper(II) complexes with a protein under alkaline condition and is reduced to copper(I). Copper(I) and phosphomolybdotungstate cause a color change to an intense blue-green.

Test Phase	
✓	Solid
✓	Liquid
	Gas/Vapor

Sensitivity
About 3% to 4% as a visually determined field test. This is a sensitive test for total protein when performed with an optical reader.

Directions and Comments
The Lowry method combines sodium carbonate, sodium potassium tartrate, cupric sulfate, and phosphomolybdotungstate in water. The change is proportional to the amount of protein present in the sample. Proportions have been recommended in several modifications. A commercially available Lowry Reagent (Sigma-Aldrich L3540) is a dry mixture of sodium dodecyl sulfate, sodium carbonate, lithium hydroxide, and cupric sulfate that is reconstituted with 40 ml of water.

(continued)

Lowry Protein Test (*Continued*)	
References	**Manufacturer**
Lowry, O.H., Rosebrough, N.J., Farr, A.L. and Randall, R.J., *J. Biol. Chem.* 193, 265–275 (1951). Peterson, G.L., A simplification of the protein assay method of Lowry et al. which is more generally applicable. *Anal. Biochem.* 83, 346–356 (1977). Material Safety Data Sheet Lowry Reagent, Powder, Version 1.6, www.sigmaaldrich.com.	Sigma-Aldrich, St. Louis, Missouri (www.sigmaaldrich.com)

Luff's Reagent		
Detects	**Test Principle**	
• Fructose • Glucose • Reducing sugar	Reducing sugars react with cupric citrate and citric acid to produce red cuprous oxide.	
	Test Phase	
	✓	Solid
	✓	Liquid
		Gas/Vapor
	Sensitivity	
	Not available	
Directions and Comments		
Add enough water to dissolve 63 g of citric acid (CAS 5949-29-1) and then add 35.9 g cupric citrate (CAS 10402-15-0). Warm as necessary. Cool and add 67.2 g potassium hydroxide (CAS 1310-58-3). Add a few drops of test solution to the sample. A red color indicates a reducing sugar. Compare to Fehling's Solution.		
References	**Manufacturer**	
Welcher, F., *Chemical Solutions: Reagents Useful to the Chemist, Biologist, and Bacteriologist*, D. Van Nostrand Company, Inc., New York, 1942.		

Lugol Starch Test		
Detects	**Test Principle**	
• Polysaccharide • Starch	Iodine will interact with the coil structure of a polysaccharide and produce a black color.	
	Test Phase	
	✓	Solid
	✓	Liquid
		Gas/Vapor
	Sensitivity	

(continued)

Lugol Starch Test (*Continued*)
Directions and Comments
Lugol Solution is an aqueous solution of various proportions up to 5% iodine and 10% potassium iodide used as a disinfectant among other uses. The potassium iodide makes the volatile iodine water soluble by formation of I_3^- in solution. As a starch test, this solution is a deep, dark brown and obscures test results. A lower concentration is adequate. Sigma Aldrich offers a solution of 0.34% iodine and 0.68% potassium iodide, which is sufficient for starch detection. A solution of about 0.5% iodine (CAS 7553-56-2) in water stabilized with potassium iodide (CAS 7681-11-0) may last a year in a tightly capped amber glass bottle. Iodine will stain starches due to its interaction with the coil structure of the polysaccharide (starch) but will not detect simple sugars such as glucose or fructose. Compare to Povidone Test for Starch.

References	Manufacturer
Material Safety Data Sheet Iodine Potassium Iodide Solution According to Lugol for Microscopy, Sigma-Aldrich, St. Louis, Missouri, January 4, 2008.	Sigma-Aldrich, St. Louis, Missouri (www.sigmaaldrich.com)

M256 Series Chemical Agent Detector Kit		
Detects	**Test Principle**	
• Blister agent • Blood agent • Chemical warfare agent • CX • Lewisite • Nerve agent	Multiple reactions. See specific tests immediately below.	
	Test Phase	
	✓	Solid
	✓	Liquid
	✓	Gas/Vapor
	Sensitivity	
	See specific test immediately below.	
Directions and Comments		
The M256 series chemical agent detector kits include the M256A1 and the obsolete M256. The only difference between the two kits is that the M256A1 kit will detect lower levels of nerve agent. The kits are intended for the detection of vapors in air in about 20 minutes but can be modified to test solids and liquids more quickly. Each test is listed separately below as M256 Nerve Agent Test, M256 Blister Agent Test, M256 Blood Agent Test, and M256 Lewisite Test.		

References	Manufacturer
Rostke, B., Information Paper - M256 Series Chemical Agent Detector Kit, http://www.gulflink.osd.mil/m256/index.htm, United States Department of Defense, Washington, DC, July 12, 1999.	Anachemia Canada Inc., Lachine, Quebec, Canada

M256A1 Blister Agent Test	
Detects	**Test Principle**
• Blister agent • CX • H-series blister agents • Nitrogen mustard • Phosgene oxime • Sulfur mustard	DB3 and mercuric cyanide react with sulfur mustard and CX, which is then exposed to potassium carbonate. Sulfur mustard produces a blue color and CX produces a red color.

(*continued*)

M256A1 Blister Agent Test (*Continued*)

	Test Phase	
	✓	Solid
	✓	Liquid
	✓	Gas/Vapor
	Sensitivity	
	Minimum detectable amount 2 mg as H or HD; 3 mg CX.	

Directions and Comments

Prepackaged reagents consisting of 2.25 mg of DB3 (4-[4-nitrobenzyl] pyridine, CAS 1083-48-3), and 2.64 mg of mercuric cyanide (CAS 592-04-1) in 0.2 ml of methanol are released onto an absorbent surface where it is heated by steaming for 2 minutes. The reagents are then exposed to air for 10 minutes or exposed to a small solid or liquid sample immediately. The reagents are heated by steam again for 1 minute and then treated with a solution of potassium carbonate (CAS 584-08-7). Sulfur mustard and similar organic sulfur compounds will cause a blue to purple spot to form. CX (dichloroformoxime or phosgene oxime [CAS 1794-86-1]) will form a pink to red color. The kit is presumptive but fairly specific for chemical warfare agents while maintaining the ability to detect very small amounts of agent. Nitrogen mustards (HN-1, HN-2 and HM-3) are indicated as blister agents in the test.

References	Manufacturer
Rostke, B., Information Paper - M256 Series Chemical Agent Detector Kit, http://www.gulflink.osd.mil/m256/index.htm, United States Department of Defense, Washington, DC, July 12, 1999. Chemical Agent Detector Kit, M256A1, U.S. Army Fact Files, http://www.army.mil/factfiles/equipment/nbc/m256a1.html, accessed January 3, 2009.	Anachemia Canada Inc., Lachine, Quebec, Canada

M256A1 Blood Agent Test

Detects	Test Principle	
• AC • Blood agent • CK • Cyanogen chloride • Hydrogen cyanide	Hydrogen cyanide is converted by sodium hypochlorite to cyanogen chloride. Either hydrogen cyanide or cyanogen chloride is detected when cyanogen chloride reacts with barbituric acid and 4-benzyl pyridine in 2-methoxy ethanol. A pink to purple to blue color is formed.	
	Test Phase	
	✓	Solid
	✓	Liquid
	✓	Gas/Vapor
	Sensitivity	
	9 mg as hydrogen cyanide; 8 mg as cyanogen chloride.	

Directions and Comments

An absorbent surface is impregnated with a solution of barbituric acid (CAS 67-52-7) and allowed to dry. At the time of the test, the absorbent surface is soaked with 0.79% solution of sodium hypochlorite (CAS 7681-52-9), which will convert any hydrogen cyanide to cyanogen chloride. About 10 minutes is necessary to collect agent vapor on the test surface. Next, a 4% solution of 4-benzyl pyridine (CAS 2116-65-6) in 2-methoxy ethanol (CAS 109-86-4) is placed on the test surface. A pink to purple to blue color forms

(continued)

M256A1 Blood Agent Test (*Continued*)	
immediately depending on the concentration of detected cyanogen chloride. Some ink solvents will give a weak false positive for blood agents. Compare to ITS Cyanide ReagentStrip™ or Merckoquant Cyanide Test.	

References	Manufacturer
Rostke, B., Information Paper - M256 Series Chemical Agent Detector Kit, http://www.gulflink.osd.mil/m256/index.htm, United States Department of Defense, Washington, DC, July 12, 1999.	Anachemia Canada Inc., Lachine, Quebec, Canada

M256A1 Lewisite Test		
Detects	**Test Principle**	
• Lewisite • Organometal	Arsenic compounds react with Michler's thioketone to change a red-clay color to olive green.	
	Test Phase	
		Solid
		Liquid
	✓	Gas/Vapor
	Sensitivity	
	9 mg Lewisite	

Directions and Comments
Michler's thioketone (4,4′-Bis[dimethylamino]thiobenzophenone, CAS 1226-46-6) is suspended in an inert, clay-like mixture and used like a crayon to make a clay-colored mark on a white surface. The mark is allowed to absorb vapors for 10 to 20 minutes. Lewisite will cause a color change to olive green. The presence of several organometallic compounds will produce a color change; the test is presumptive but sensitive for lewisite. The lewisite crayon is prone to false positives from many other chemicals, including some common organic solvents such as acetone. The lewisite crayon will not detect decomposition products formed when lewisite is exposed to water or humid air. Michler's thioketone is a carcinogen. Compare to Drager Organic Arsenic Compounds and Arsine.

References	Manufacturer
Rostke, B., Information Paper - M256 Series Chemical Agent Detector Kit, http://www.gulflink.osd.mil/m256/index.htm, United States Department of Defense, Washington, DC, July 12, 1999. New Jersey Department of Health and Senior Services, "Hazardous Substance Fact Sheet for Michler's Ketone," www.state.nj.us/health/eoh/rtkweb/1305.pdf, September 1994.	Anachemia Canada Inc., Lachine, Quebec, Canada

M256A1 Nerve Agent Test	
Detects	**Test Principle**
• Carbamate pesticide • G-nerve agent • Nerve agent • Novichok agent • Organophosphate pesticide • V-nerve agent	Acetylcholinesterase-inhibiting materials deactivate acetylcholinesterase, blocking the conversion of indoxyl acetate to a blue-colored reaction product.

(continued)

M256A1 Nerve Agent Test (*Continued*)		
	Test Phase	
	✓	Solid
	✓	Liquid
	✓	Gas/Vapor
	Sensitivity	
	0.005 mg as GA, GB, GD, GF; 0.02 mg as VX.	

Directions and Comments
An absorbent surface is coated with stabilized eel acetylcholinesterase (CAS 9000-81-1). The surface is soaked in a buffer solution and allowed to absorb vapor for 10 minutes. Finally, the surface is soaked in a solution containing indoxyl acetate (CAS 608-08-2). If active acetylcholinesterase remains, the indoxyl acetate will be converted to a blue reaction product. If the acetylcholinesterase on the pad has been poisoned by a nerve agent or acetylcholinesterase-inhibiting pesticide, the blue reaction product will not form and the test surface color will remain. Compare to the Agri-Screen Ticket. Note that if the test kit expired or has been stored improperly, the eel acetylcholinesterase may degrade and cause a false positive result.

References	**Manufacturer**
Rostke, B., Information Paper - M256 Series Chemical Agent Detector Kit, http://www.gulflink.osd.mil/m256/index.htm, United States Department of Defense, Washington, DC, July 12, 1999.	Anachemia Canada Inc., Lachine, Quebec, Canada

Mandelin Reagent		
Detects	**Test Principle**	
• Acetaminophen • Alkaloid • Aspirin • Codeine • Illegal drug • Mace	Vanadate anion reacts with various compounds in acidic condition to form various colored reaction products.	
	Test Phase	
	✓	Solid
	✓	Liquid
		Gas/Vapor
	Sensitivity	
	20 µg codeine; 5 µg morphine monohydrate.	

Directions and Comments
Mandelin Reagent is used as a presumptive drug test. The test solution is 1% ammonium vanadate (CAS 7803-55-6) in concentrated sulfuric acid (w/v); the solution will be orange. The test solution is applied to the sample and any color change noted. Many compounds will form colors with this presumptive test for illegal drugs. It is most helpful to compare any color formed in this test with results from other presumptive drug tests (Marquis Reagent, etc.) in a matrix to determine a pattern of unique results. Aspirin and acetaminophen form shades of olive green color, as does codeine and mace. Refer to the appendix for drug screen results.

(*continued*)

Mandelin Reagent (*Continued*)	
References	**Manufacturer**
U.S. Department of Justice. Color Test Reagents/Kits for Preliminary Identification of Drugs of Abuse, NIJ Standard–0604.01, National Institute of Justice, National Law Enforcement and Corrections Technology Center, Rockville, MD, July 2000.	Several commercial test kits are available.

Marquis Reagent		
Detects	**Test Principle**	
• Amphetamine • Aspirin • Codeine • Illegal drug • Mace • Salt	Formaldehyde reacts with various compounds in the presence of concentrated sulfuric acid to form various colored reaction products.	
	Test Phase	
	✓	Solid
	✓	Liquid
		Gas/Vapor
	Sensitivity	
	10 µg d-amphetamine HCl; 1 µg codeine; 5 µg LSD.	
Directions and Comments		
This is a presumptive test for illegal drugs. To make the test solution, carefully add 100 ml of concentrated sulfuric acid to 5 ml of 40% formaldehyde (CAS 50-00-0) in water. To mix a smaller volume, mix 10 ml of concentrated sulfuric acid with 8 to 10 drops of 40% formaldehyde solution. Apply the test solution to the sample and note any color formed. It is most helpful to compare any color formed in this test with results from other presumptive drug tests (Marquis Reagent, etc.) in a matrix to determine a pattern of unique results. Refer to the appendix for drug screen results.		
References	**Manufacturer**	
U.S. Department of Justice. Color Test Reagents/Kits for Preliminary Identification of Drugs of Abuse, NIJ Standard–0604.01, National Institute of Justice, National Law Enforcement and Corrections Technology Center, Rockville, MD, July 2000.	Several commercial test kits are available.	

Mecke Reagent		
Detects	**Test Principle**	
• Codeine • Hydrocodone tartrate • Illegal drug • LSD • Morphine	Selenous acid reacts with various compounds in the presence of concentrated sulfuric acid to form various colored reaction products.	
	Test Phase	
	✓	Solid
	✓	Liquid
		Gas/Vapor
	Sensitivity	
	25 µg codeine; 50 µg LSD.	

(continued)

Mecke Reagent (*Continued*)

Directions and Comments

This test is a presumptive field test for illegal drugs. The test solution is 1% selenious acid (CAS 7783-00-8) in concentrated sulfuric acid (w/v). It is most helpful to compare any color formed in this test with results from other presumptive drug tests (Marquis Reagent, etc.) in a matrix to determine a pattern of unique results. Refer to the appendix for drug screen results.

References	Manufacturer
U.S. Department of Justice. Color Test Reagents/ Kits for Preliminary Identification of Drugs of Abuse, NIJ Standard–0604.01, National Institute of Justice, National Law Enforcement and Corrections Technology Center, Rockville, MD, July 2000.	Several commercial test kits are available.

Melibiose Test

Detects	Test Principle	
• Bean • Melibiose • Vegetable	Melibiose is cleaved by alpha-galactosidase to form glucose and galactose. The resulting glucose is detected by a glucose test strip.	
	Test Phase	
	✓	Solid
	✓	Liquid
		Gas/Vapor
	Sensitivity	
	100 mg/ml wet hummus (contains chickpeas and less than 1 g sugar per 30 g)	

Directions and Comments

Melibiose is the carbohydrate in beans and other vegetables that causes dietary gas. Melibiose will react to form glucose and galactose when cleaved by the enzyme alpha-galactosidase (Beano®). Alpha-galactosidase is the stereoisomer (mirror image) of beta-galactosidase. The resulting increase in glucose can be detected by a glucose urine test strip (Bayer® Uristix™ Glucose Strips or Merckoquant® Glucose Test) or other suitable glucose test.

Perform the test by placing two drops of liquid sample or a few milligrams of solid sample in a test tube or spot plate and mix with two drops of water. Use a glucose test strip to measure the glucose present in the solution (split the sample and expose this test strip simultaneously with the one below). Add a drop of liquid alpha-galactosidase or a few milligrams of powdered tablet to the solution. Use a second glucose test strip to measure the glucose present in the solution. An increase in glucose indicates melibiose is present and a vegetable product is indicated.

If the sample already contains a high amount of glucose before the test, the test strip will be overloaded and the small increase in glucose from the melibiose will not be detected. The glucose test strip requires no more than 30 seconds for a quantitative result for glucose in urine. If the strip is allowed to react for about 3 minutes, the sensitivity improves about 10 times.

References	Manufacturer
Houghton, R., *Emergency Characterization of Unknown Materials*, CRC Press, 2007.	GlaxoSmithKline, Brentford, Middlesex, United Kingdom (www.beanogas.com). Bayer AG, 51368 Leverkusen, Germany (www.bayer.com).

Merckoquant® Aluminum Test	
Detects	**Test Principle**
• Aluminum	Aluminum is converted to aluminate, which then reacts with aurin tricarboxylic acid (aluminon) to form a red color.

Test Phase	
	Solid
✓	Liquid
	Gas/Vapor

Sensitivity
10 mg/l aluminum

Directions and Comments
The sample is treated with potassium hydroxide to form aluminate and to precipitate other metals (aluminum is soluble in strong base). Any precipitate is filtered. This test is largely interference free, but copper and iron interfere in low amounts. The test strip is immersed in the basic sample for 1 second and then wetted with aurin tricarboxylic acid to form the red color. Results are read after 1 minute. Other metals such as copper, iron, cobalt and others can react with aurin tricarboxylic acid to form colors if the basic precipitation step is omitted. Colors formed by omitting the precipitation step indicate further testing is necessary.

References	**Manufacturer**
Merckoquant® Aluminum Test 1.10015.0001, Merck KGaA, Darmstadt, Germany, October 2005. Merck Safety Data Sheet 1.10015.0001, Merck KGaA, Darmstadt, Germany, October 27, 2004.	Merck KGaA, Darmstadt, Germany

Merckoquant® Ammonium Test	
Detects	**Test Principle**
• Ammonia • Ammonium salt	Ammonium reacts with Nessler's reagent (a solution of mercury(II) iodide in potassium iodide and potassium hydroxide) to form a yellow-brown color from $Hg_2I_3NH_2$.

Test Phase	
	Solid
✓	Liquid
	Gas/Vapor

Sensitivity
10 mg/l ammonium

Directions and Comments
The sample is treated with sodium hydroxide, which forces ammonium to produce ammonia. Ammonia immediately reacts with the sensitive Nessler's reagent. The test strip is immersed in the treated sample for 3 seconds and results are read after 15 seconds.

(continued)

Merckoquant® Ammonium Test (*Continued*)	
References	Manufacturer
Merckoquant® Ammonium Test 1.10024.000, Merck KGaA, Darmstadt, Germany, December 2007. Merck Safety Data Sheet 1.10024.0001, Merck KGaA, Darmstadt, Germany, September 7, 2007. Tannanaeff, N. and Budkewitsch, A., *Chem. Abstracts*, 30, 5905 (1936).	Merck KGaA, Darmstadt, Germany

Merckoquant® Arsenic Test		
Detects	**Test Principle**	
• Arsenic • Arsenic(III) • Arsenic(V)	Arsenic(III) and arsenic(V) compounds are processed with test reagents to form arsine, which is then detected on the test strip with imbedded mercury(II) bromide. Various yellow-brown arsenic-mercury halogenides are formed.	
	Test Phase	
		Solid
	✓	Liquid
		Gas/Vapor
	Sensitivity	
	0.005 mg/l As(III)/As(V)	
Directions and Comments		
The liquid sample containing arsenic(III) or arsenic(V) is treated with potassium permanganate to eliminate sulfide interference. Malonic acid is dissolved into the solution in the second step. Sodium tungstate and zinc powder (used to eliminate sulfide interference) is added to the solution and arsine is generated in a reduction reaction. The test strip is suspended in the vapor space of the test vessel and sealed for 20 minutes. The test strip is removed, dipped in water, and observed for a yellow-brown color indicating arsenic. The test is designed for arsenic in drinking water; other forms of arsenic may be insoluble and undetectable by this method. Other metals may interfere; consult test instructions.		
References	**Manufacturer**	
Merckoquant® Arsenic Test (highly sensitive) 1.17927.0001, Merck KGaA, Darmstadt, Germany, February 2003. Merck Safety Data Sheet 1.17927.0001, Merck KGaA, Darmstadt, Germany, October 26, 2005.	Merck KGaA, Darmstadt, Germany	

Merckoquant® Ascorbic Acid Test	
Detects	**Test Principle**
• Ascorbic acid	Ascorbic acid reduces yellow molybdophosphoric acid to phosphomolybdenum blue.

(*continued*)

Merckoquant® Ascorbic Acid Test (*Continued*)

		Test Phase	
			Solid
		✓	Liquid
			Gas/Vapor
		Sensitivity	
		50 mg/l ascorbic acid	

Directions and Comments
This test is designed for use on food products; many other nonfood products can reduce yellow molybdophosphoric acid to phosphomolybdenum blue. pH must be adjusted a range of 2 to 7.

References	Manufacturer
Merckoquant® Ascorbic Acid Test 1.10023.0001, Merck KGaA, Darmstadt, Germany, October 2007.	Merck KGaA, Darmstadt, Germany

Merckoquant® Calcium Test

Detects	Test Principle		
• Calcium • Cadmium • Copper • Lead	Calcium in pH adjusted solution reacts with hydrogen peroxide and glyoxal-bis(2-hydroxyanil) to form a red color.		
	Test Phase		
			Solid
		✓	Liquid
			Gas/Vapor
	Sensitivity		
	10 mg/l calcium		

Directions and Comments
pH is adjusted to 4 to 10 and the test strip is immersed for 1 second and then set aside. An aqueous solution is prepared with urea hydrogen peroxide and sodium hydroxide. The test strip is immersed in the solution for 45 seconds and observed for a red color. This test can also be used to detect some heavy metals; cadmium is indicated by a blue-violet color; copper is indicated by a gray color; lead is indicated by a green color.

References	Manufacturer
Merckoquant® Calcium Test 1.10083.0001, Merck KGaA, Darmstadt, Germany, August 2007. Merck Safety Data Sheet 1.10083.0001, Merck KGaA, Darmstadt, Germany, August 22, 2007.	Merck KGaA, Darmstadt, Germany

Merckoquant® Chloride Test

Detects	Test Principle
• Chloride	Chloride reacts with silver ions to de-color red-brown silver chromate.

(*continued*)

Merckoquant® Chloride Test (*Continued*)		
	Test Phase	
	✓	Solid
	✓	Liquid
		Gas/Vapor
	Sensitivity	
	500 mg/l chloride	

Directions and Comments
Optimal pH for this test is 5 to 8. Solid samples may be wiped and then wetted with deionized water. The test strip has four reaction zones for quantitative measure. Any positive result indicates the presence of chloride. A few compounds may interfere but are not likely. See the package insert for more detail.

References	**Manufacturer**
Merckoquant® Chloride Test 1.10079.0001, Merck KGaA, Darmstadt, Germany, June 2007.	Merck KGaA, Darmstadt, Germany

Merckoquant® Chlorine Tests 1.17924.0001, 1.17925.0001		
Detects	**Test Principle**	
• Chlorine • Oxidizer	Chlorine oxidizes an organic compound to a violet dye.	
	Test Phase	
		Solid
	✓	Liquid
		Gas/Vapor
	Sensitivity	
	0.5 and 25 mg/l chlorine, respectively	

Directions and Comments
Both tests are designed for use in systems known to contain chlorine. Strong oxidizers partially interfere.

References	**Manufacturer**
Merckoquant® Chlorine Tests 1.17924.0001, 1.17925.0001, Merck KGaA, Darmstadt, Germany, November 2007.	Merck KGaA, Darmstadt, Germany

Merckoquant® Chromate Test	
Detects	**Test Principle**
• Chromate • Chromium(III) • Chromium(VI) • Chromium, bound • Dichromate	Chromate reacts with diphenylcarbazide in phosphoric solution to form chromium(III) and red-violet diphenylcarbazone.

(*continued*)

Merckoquant® Chromate Test		
	Test Phase	
		Solid
	✓	Liquid
		Gas/Vapor
	Sensitivity	
	3 mg/l chromate	

Directions and Comments

Sulfuric acid is added to the sample until pH <1. The test strip is immersed for 1 second and the result is read in 15 seconds. This test measures chromium(VI) present as chromate or dichromate ions. Complex-bound chromium, including chromium(III), can be detected if the sample is digested before testing.

References	Manufacturer
Merckoquant® Chromate Test 1.10012.0001, Merck KGaA, Darmstadt, Germany, November 2007. Merck Safety Data Sheet 1.10012.0001, Merck KGaA, Darmstadt, Germany, May 22, 2004.	Merck KGaA, Darmstadt, Germany

Merckoquant® Cobalt Test		
Detects	**Test Principle**	
• Cobalt • Iron(III) • Copper(II) • Mercury(I)	Cobalt(II) reacts with thiocyanate ions to form a blue cobalt thiocyanate, which tints the yellow pad to various shades of green.	
	Test Phase	
	✓	Solid
	✓	Liquid
		Gas/Vapor
	Sensitivity	
	10 mg/l cobalt	

Directions and Comments

The test requires pH adjustment to a range of 1 to 7. The test strip is immersed for 1 second and read in 15 seconds. Cyanide, ferric cyanide, and ferrous cyanide interfere in low concentration. Other colors can be caused by other cations that can be determined by the addition of masking agents and retesting. Brown indicates >3500 mg/l iron(III) or >1000 mg/l copper(II), which can be masked by the addition of potassium fluoride or sodium thiosulfate, respectively. Gray indicates >300 mg/l mercury(I), which can be masked with the addition of sodium chloride. Blue cobalt thiocyanate is used as a reagent in other tests; if a pink result occurs, cobalt is present in the sample and the resulting cobalt thiocyanate has reacted with an unknown component of the sample. This test strip may be used for nondestructive surface testing for cobalt.

References	Manufacturer
Merckoquant® Cobalt Test 1.10002.0001, Merck KGaA, Darmstadt, Germany, October 2007.	Merck KGaA, Darmstadt, Germany

Merckoquant® Copper Test	
Detects	**Test Principle**
• Copper • Copper(I) • Copper(II)	Copper(I) ions react with 2,2′-biquinoline (cuproin, CAS 119-91-5) to form a violet color. Copper(II) is reduced to copper(I) by a reducing agent so that total copper is detected.

	Test Phase	
	✓	Solid
	✓	Liquid
		Gas/Vapor

Sensitivity
10 mg/l copper

Directions and Comments
The pH of the sample must be adjusted to 2 to 6. The test strip is immersed for 1 second and read in 30 seconds. Cyanide, ferrocyanide, and ferricyanide interfere in low concentration. This test strip may be used for nondestructive surface testing for copper, including copper wood preservative.

References	**Manufacturer**
Merckoquant® Copper Test 1.10003.0001, Merck KGaA, Darmstadt, Germany, October 2007.	Merck KGaA, Darmstadt, Germany

Merckoquant® Cyanide Test	
Detects	**Test Principle**
• Cyanide	Cyanide reacts with chloramine-T to form cyanogen chloride, which in turn reacts with 1,3-dimethylbarbituric acid and pyridine to form a violet color. This is also known as the König reaction.

	Test Phase	
		Solid
	✓	Liquid
		Gas/Vapor

Sensitivity
1 mg/l cyanide

Directions and Comments
Optimum pH is 6 to 7 (adding too much acid will produce hydrogen cyanide gas), but qualitative results are possible at higher pH. The sample is reacted with sodium triphosphate and chloramine-T to convert soluble cyanide to cyanogen chloride. The test strip is immersed for 30 seconds. Pyridine (included in the test solution with chloramine-T) reacts with cyanogen chloride and 1,3-dimethylbarbituric acid in the test pad to form a violet color. This test measures only cyanide ions (free cyanide) and cannot detect insoluble cyanide. Some metals interfere in low concentration; consult the test instructions.

(continued)

Merckoquant® Cyanide Test (*Continued*)	
References	**Manufacturer**
Merckoquant® Cyanide Test 1.10044.0001, Merck KGaA, Darmstadt, Germany, September 2007. Merck Safety Data Sheet 1.10044.0001, Merck KGaA, Darmstadt, Germany, February 10, 2004.	Merck KGaA, Darmstadt, Germany

Merckoquant® Fixing Bath Test		
Detects	**Test Principle**	
• Silver	Silver reacts with cadmium sulfide to form brown to black silver sulfide.	
	Test Phase	
		Solid
	✓	Liquid
		Gas/Vapor
	Sensitivity	
	0.5 mg/l silver	

Directions and Comments

This test is quantitative up to 10 g/l (1%) silver. High concentration of other metals may interfere. Simply immerse the strip for 1 second and read results in 30 seconds.

References	**Manufacturer**
Merckoquant® Fixing Bath Test 1.10008.0001, Merck KGaA, Darmstadt, Germany, May 2007.	Merck KGaA, Darmstadt, Germany

Merckoquant® Formaldehyde Test		
Detects	**Test Principle**	
• Formaldehyde • Aldehyde	Formaldehyde reacts with 4-amino-3-hydrazino-5-mercapto-1,2,4-triazole to form a purple-red color.	
	Test Phase	
		Solid
	✓	Liquid
		Gas/Vapor
	Sensitivity	
	10 mg/l formaldehyde	

Directions and Comments

The sample is mixed with strong sodium hydroxide and the test strip is immersed for 1 second. The test strip must be read at exactly 60 seconds for a quantitative result. The test also responds to other aldehydes at higher concentrations and with a different color, but the colors are not specified. Ketones, esters, amides, hydrazines, hydroxylamines, quinones, aminophenol, uric acid, and formic acid interfere. Strong oxidizing and reducing agents reduce sensitivity.

References	**Manufacturer**
Merckoquant® Formaldehyde Test 1.10036.0001, Merck KGaA, Darmstadt, Germany, April 2007. Merck Safety Data Sheet 1.10036.0001, Merck KGaA, Darmstadt, Germany, November 19, 2007.	Merck KGaA, Darmstadt, Germany

Merckoquant® Glucose Test	
Detects	**Test Principle**
• Glucose	Glucose is converted by the enzyme glucose oxidase to gluconic acid lactone and hydrogen peroxide. A peroxidase enzyme transfers peroxide oxygen to an organic redox indicator, which then produces a blue color.

	Test Phase	
		Solid
	✓	Liquid
		Gas/Vapor

Sensitivity
10 mg/l glucose

Directions and Comments
Sample pH must be adjusted to 2 to 10 for quantitative measurement. The test strip is simply immersed for 1 second and read in 1 minute. Ascorbic acid, sulfate, sulfite, hydrogen sulfide, and peracetic acid interfere. The test is specific to glucose and can be used to discern glucose among other sugars.

References	**Manufacturer**
Merckoquant® Glucose Test 1.17866.0001, Merck KGaA, Darmstadt, Germany, April 2006.	Merck KGaA, Darmstadt, Germany

Merckoquant® Iron Test	
Detects	**Test Principle**
• Iron • Iron(II) • Iron(III)	Iron(II) reacts with 2,2′-bipyridine to form a red color.

	Test Phase	
	✓	Solid
	✓	Liquid
		Gas/Vapor

Sensitivity
3 mg/l iron(II)

Directions and Comments
Sample pH is must be acidic. The test strip is immersed for 1 second and read in 15 seconds. Iron(III) may be detected by first reducing to iron(II) with ascorbic acid (total iron is measured). This test strip may be used for nondestructive surface testing of iron. Ferric cyanide and ferrous cyanide interfere in low concentration.

References	**Manufacturer**
Merckoquant® Iron Test 1.10004.0001, Merck KGaA, Darmstadt, Germany, April 2007.	Merck KGaA, Darmstadt, Germany

Merckoquant® Lead Test	
Detects	**Test Principle**
• Lead	Lead(II) reacts with rhodizonic acid to form a red color.

(continued)

Merckoquant® Lead Test (*Continued*)		
	Test Phase	
	✓	Solid
	✓	Liquid
		Gas/Vapor
	Sensitivity	
	20 mg/l lead	

Directions and Comments	
The optimal pH of a sample should be 2 to 5. This test measures lead(II), but not complex-bound lead or organolead compounds. The test strip may be used as a wipe for solid material. Calcium, strontium, barium, and other materials may be detected. Compare to Sodium Rhodizonate Solutions.	
References	**Manufacturer**
Merckoquant® Lead Test 1.10077.0001, Merck KGaA, Darmstadt, Germany, December 2006.	Merck KGaA, Darmstadt, Germany

Merckoquant® Manganese Test		
Detects	**Test Principle**	
• Manganese	Manganese(II) is oxidized on the test strip to manganese(IV) oxide, which reacts with an organic redox indicator to form a blue color.	
	Test Phase	
		Solid
	✓	Liquid
		Gas/Vapor
	Sensitivity	
	2 mg/l manganese(II)	

Directions and Comments	
The sample pH must be acidic. The test strip is immersed in the sample and then a drop of sodium hydroxide is placed on the test strip. Next a drop of acetic acid is applied to the test strip and if a blue color indicates manganese. The redox indicator may be tetramethyl-p-diaminodiphenylmethane. All steps are timed. Strong oxidizers interfere; the test pad will turn green before the addition of the test reagents.	
References	**Manufacturer**
Merckoquant® Manganese Test 1.10080.0001, Merck KGaA, Darmstadt, Germany, September 17, 2001. Merck Safety Data Sheet 1.10080.0001, Merck KGaA, Darmstadt, Germany, November 16, 2006.	Merck KGaA, Darmstadt, Germany

Merckoquant® Molybdenum Test	
Detects	**Test Principle**
• Molybdenum	Molybdate reacts with toluene-3,4-dithiol in acidic solution to form a green color.

(*continued*)

Merckoquant® Molybdenum Test (*Continued*)

	Test Phase	
		Solid
	✓	Liquid
		Gas/Vapor
	Sensitivity	
	5 mg/l molybdenum; 8 mg/l molybdate.	

Directions and Comments
Sample pH must be adjusted to 4 to 6 for quantitative results. Sulfamic acid reacts with molybdenum (or molybdate) and sulfuric acid is added if necessary until pH <1. The test strip is immersed for 1 minute and read. Copper, tin, and sulfur interfere.

References	Manufacturer
Merckoquant® Molybdenum Test 1.10049.0001, Merck KGaA, Darmstadt, Germany, July 2007. Merck Safety Data Sheet 1.10049.0001, Merck KGaA, Darmstadt, Germany, June 10, 2004.	Merck KGaA, Darmstadt, Germany

Merckoquant® Nickel Test

Detects	Test Principle	
• Cobalt • Copper • Mercury • Molybdenum • Nickel	Nickel(II) ions react with dimethylglyoxime to form a red color.	
	Test Phase	
	✓	Solid
	✓	Liquid
		Gas/Vapor
	Sensitivity	
	10 mg/l nickel (II)	

Directions and Comments
This test is dependable for nickel(II) with little interference. Merck test strips often use buffers and masking agents to reduce interference. Sample dilutions can be tested to eliminate some interference. Consult the test insert. Solid surfaces may be wipe tested after wetting the test pad with ammonia solution. More than 4000 mg/l copper(II) forms an orange color; >750 mg/l mercury(II) forms a yellow color; 220–750 mg/l mercury(II) forms a gray color; >50 mg/l cobalt(II) forms a yellow to brown color; a blue color can be formed from various molybdenum compounds.

References	Manufacturer
Merckoquant® Nickel Test 1.10006.0001, Merck KGaA, Darmstadt, Germany, April 2006.	Merck KGaA, Darmstadt, Germany

Merckoquant® Nitrate Tests 1.10020.0001, 1.10020.0002

Detects	Test Principle
• Nitrate • Nitrite	Nitrate is reduced to nitrite. In a buffered acid solution, nitrite reacts with an aromatic amine to

(*continued*)

Merckoquant® Nitrate Tests 1.10020.0001, 1.10020.0002 (*Continued*)

	form a diazonium salt, which in turn reacts with N-(1-naphthyl)-ethylenediamine to form a red-violet azo compound.	
	Test Phase	
		Solid
	✓	Liquid
		Gas/Vapor
	Sensitivity	
	10 mg/l nitrate; 2.3 mg/l NO_3-N.	

Directions and Comments

The test uses a modified Griess reaction (see Merckoquant® Nitrite Tests 1.10007.0001). The test strip contains two test pads; one is for nitrite and the other for nitrate. The test can only determine nitrate but the nitrite pad does not contain the reducing agent necessary to convert nitrate to nitrite. The test pads are wetted for 1 second and read at 1 minute. This test is only conditionally suited for seawater due to false or low results. To eliminate nitrite in order to determine only nitrate, add 5 drops of a 10% aqueous amidosulfonic acid solution to 5 ml of sample that has been pH adjusted to <10. Boil briefly, allow to cool, and repeat the test.

References	Manufacturer
Merckoquant® Nitrate Tests 1.10020.0001, 1.10020.0002, Merck KGaA, Darmstadt, Germany, November 2007.	Merck KGaA, Darmstadt, Germany

Merckoquant® Nitrite Tests 1.10007.0001, 1.10007.0002

Detects	Test Principle	
• Nitrite	Nitrite reacts with an aromatic amine in an acid buffered solution to form a diazonium salt, which then reacts with N-(1-naphthyl)ethylenediamine to form a red-violet color.	
	Test Phase	
		Solid
	✓	Liquid
		Gas/Vapor
	Sensitivity	
	2 mg/l nitrite; 0.6mg/l NO_2-N.	

Directions and Comments

The test uses a Griess reaction in pH 1 to 13 solution. A Griess reaction involves a primary aromatic amine in acid solution reacting with nitrous acid to form diazonium cations, which may react with the same or another primary aromatic amine to produce colored cations of an aminoazo compound. Chromate and permanganate interfere; nitrate does not.

References	Manufacturer
Merckoquant® Nitrite Tests 1.10007.0001, 1.10007.0002, Merck KGaA, Darmstadt, Germany, September 2007.	Merck KGaA, Darmstadt, Germany

Merckoquant® Nitrite Tests 1.10022.0001

Detects	Test Principle	
• Nitrite	Nitrite reacts with an aromatic amine to form an orange-red azo compound.	
	Test Phase	
		Solid
	✓	Liquid
		Gas/Vapor
	Sensitivity	
	0.1 g/l nitrite; 0.03 g/l NO_2-N.	

Directions and Comments	
Sample pH must be adjusted to 2 to 11 for quantitative measurement. The test strip is immersed for 1 second and read in 1 minute. This test uses a different method to detect higher concentration of nitrate than Merckoquant® Nitrite Tests 1.10007.0001.	

References	Manufacturer
Merckoquant® Nitrite Tests 1.10022.0001, Merck KGaA, Darmstadt, Germany, September 2007.	Merck KGaA, Darmstadt, Germany

Merckoquant® Peracetic Acid Test

Detects	Test Principle	
• Peracetic acid	Peracetic acid reacts with a phenol derivative to form a violet color.	
	Test Phase	
		Solid
	✓	Liquid
		Gas/Vapor
	Sensitivity	
	100 mg/l peracetic acid.	

Directions and Comments	
1000 mg/l nitrate and 1000 mg/l hydrogen peroxide do not interfere with this test. Strong oxidizing agents such as chlorine, bromine, iodine, and hypochlorite can produce false positive results.	

References	Manufacturer
Merckoquant® Peracetic Acid Test 1.10001.0001, Merck KGaA, Darmstadt, Germany, June 2007.	Merck KGaA, Darmstadt, Germany

Merckoquant® Peroxide Test 1.10337.0001

Detects	Test Principle
• Hydrogen peroxide • Hydroperoxide • Organic peroxide • Peroxide	A peroxidase enzyme transfers peroxide oxygen to an organic redox indicator, which then produces a yellow-brown color.

(continued)

Merckoquant® Peroxide Test 1.10337.0001 (*Continued*)		
	Test Phase	
		Solid
	✓	Liquid
		Gas/Vapor
	Sensitivity	
	100 mg/l hydrogen peroxide	

Directions and Comments

The peroxidase enzyme, probably horseradish peroxidase, is specific to reaction with organic peroxides, hydroperoxides, and hydrogen peroxide. Polymeric peroxides are not detected confidently with this method. The yellow-brown color is probably from iodine liberated from an iodide salt; other strong oxidizers present in the sample may interfere. The strip has a second reaction zone that will change color above 2000 mg/l hydrogen peroxide. If the second pad changes color, the first pad may indicate a false or low result. 500 mg/l nitrate and 100 mg/l hypochlorite do not interfere with this test. Consult the test insert.

References	Manufacturer
Merckoquant® Peroxide Test 1.10337.0001, Merck KGaA, Darmstadt, Germany, September 2006.	Merck KGaA, Darmstadt, Germany

Merckoquant® Peroxide Tests 1.10011.0001, 1.10011.0002, 1.10081.0001		
Detects	**Test Principle**	
• Hydrogen peroxide • Hydroperoxide • Organic peroxide • Peroxide	A peroxidase enzyme transfers peroxide oxygen to an organic redox indicator, which then produces a blue color.	
	Test Phase	
		Solid
	✓	Liquid
		Gas/Vapor
	Sensitivity	
	1 mg/l hydrogen peroxide	

Directions and Comments

The peroxidase enzyme, probably horseradish peroxidase, is specific to reaction with organic peroxides, hydroperoxides, and hydrogen peroxide. Polymeric peroxides are not detected confidently with this method. The blue color is probably from iodine liberated from an iodide salt, which then reacts with starch to form a blue indication; other strong oxidizers present in the sample may interfere. Merck test strips often use buffers and masking agents to reduce interference. 10 mg/l chromate and 2 mg/l permanganate do not interfere with this test. Sample dilutions can be tested to eliminate some interference. Consult the test insert.

References	Manufacturer
Merckoquant® Peroxide Tests 1.10011.0001, 1.10011.0002, 1.10081.0001, Merck KGaA, Darmstadt, Germany, July 2007.	Merck KGaA, Darmstadt, Germany

Merckoquant® Phosphate Test	
Detects	**Test Principle**
• Orthophosphate • Phosphate	Orthophosphate (PO_4^{3-}) reacts with molybdate in the presence of sulfuric acid to form molybdophosphoric acid, which is then reduced to phosphomolybdenum blue.

	Test Phase	
		Solid
	✓	Liquid
		Gas/Vapor

Sensitivity
10 mg/l orthophosphate

Directions and Comments
This test measures only orthophosphate. Other forms of phosphate must be decomposed by digestion before testing. The sample must be adjusted to pH 4 to 10. After wetting the test strip with the sample liquid, a 10% to 15% solution of sulfuric acid is added and the result is read 15 seconds later.

References	**Manufacturer**
Merckoquant® Phosphate Test 1.10428.0001, Merck KGaA, Darmstadt, Germany, September 2006. Merck Safety Data Sheet for 1.10428.0001, Merck KGaA, Darmstadt, Germany, October 14, 2004.	Merck KGaA, Darmstadt, Germany

Merckoquant® Potassium Test	
Detects	**Test Principle**
• Potassium	Potassium ions react with dipicrylamine to form an orange complex.

	Test Phase	
		Solid
	✓	Liquid
		Gas/Vapor

Sensitivity
250 mg/l potassium

Directions and Comments
The sample pH must be adjusted to 5 to 14. The test strip, which originally is red, is wetted with pH-adjusted sample and then immersed in the accompanying test solution and read in 1 minute. Merck test strips often use buffers and masking agents to reduce interference. Sample dilutions can be tested to eliminate some interference. Consult the test insert. This test can discern potassium when sodium is present at 10 times the potassium concentration.

References	**Manufacturer**
Merckoquant® Potassium Test 1.10042.0001, Merck KGaA, Darmstadt, Germany, July 2007.	Merck KGaA, Darmstadt, Germany

Merckoquant® Quaternary Ammonium Compounds		
Detects	**Test Principle**	
• Quat • Quaternary amine • Quaternary ammonium compounds	Certain quaternary ammonium compounds react with a yellow-green indicator to produce a turquoise-blue color.	
	Test Phase	
		Solid
	✓	Liquid
		Gas/Vapor
	Sensitivity	
	10 mg/l benzalkonium chloride	
Directions and Comments		
This test detects some quaternary ammonium compounds usually present in disinfection materials such as alkylbenzyldimethylammonium chloride (benzalkonium chloride), cetyltrimethylammonium bromide (CTAB), hexadecylpyridinium chloride, dodecyltrimethylammonium bromide (lauryltrimethylammonium bromide, LTAB), and octadecyltrimethylammonium chloride. A positive indication with other quaternary ammonium compounds is likely, but a complete list is not available. No interference is caused by protein concentrations below 1 g/l.		
References	**Manufacturer**	
Merckoquant® Quaternary Ammonium Compounds, 1.17920.0001, Merck KGaA, Darmstadt, Germany, June 2007.	Merck KGaA, Darmstadt, Germany	

Merckoquant® Sulfate Test		
Detects	**Test Principle**	
• Sulfate	Sulfate reacts with a red thorin-barium complex to form barium sulfate and release yellow thorin.	
	Test Phase	
		Solid
	✓	Liquid
		Gas/Vapor
	Sensitivity	
	200 mg/l sulfate	
Directions and Comments		
The sample solution pH must be adjusted to 4 to 8 and then the strip is immersed in the sample solution. The result is read after 2 minutes.		
References	**Manufacturer**	
Merckoquant® Sulfate Test 1.10019.0001, Merck KGaA, Darmstadt, Germany, October 2007.	Merck KGaA, Darmstadt, Germany	

Merckoquant® Sulfite Test

Detects	Test Principle	
• Sulfite	Sulfite reacts with potassium hexacyanoferrate, zinc sulfate, and sodium nitroprusside to form a red color.	
	Test Phase	
		Solid
	✓	Liquid
		Gas/Vapor
	Sensitivity	
	10 mg/l sulfite	

Directions and Comments

Sample pH must be adjusted to 8 to 10 for quantitative results. The test strip is immersed in the sample and read in 15 seconds. 1000 mg/l sulfate does not interfere, but some oxidizers interfere. A pH greater than 10 is acceptable for qualitative results. Adjusting the pH to an acidic value will produce hydrogen sulfide.

References	Manufacturer
Merckoquant® Sulfite Test 1.10013.0001, Merck KGaA, Darmstadt, Germany, July 2007.	Merck KGaA, Darmstadt, Germany

Merckoquant® Tin Test

Detects	Test Principle	
• Tin	Any tin present is reduced to tin(II) through reaction with thioglycolic acid and hydrochloric acid. Tin(II) reacts with toluene-3,4-dithiol to form a red color.	
	Test Phase	
		Solid
	✓	Liquid
		Gas/Vapor
	Sensitivity	
	10 mg/l tin	

Directions and Comments

The sample solution is adjusted to pH <10 to avoid reagent neutralization and tin is then reduced with thioglycolic acid and hydrochloric acid. The test strip containing toluene-3,4-dithiol is immersed in the treated sample for 5 minutes and then read. The test is designed for determination of small amounts of tin in food products. Lead, silver, copper, and mercury may interfere.

References	Manufacturer
Merckoquant® Tin Test 1.10028.0001, Merck KGaA, Darmstadt, Germany, January 2006. Merck Safety Data Sheet 1.10028.0001, Merck KGaA, Darmstadt, Germany, March 25, 2006.	Merck KGaA, Darmstadt, Germany

Merckoquant® Zinc Test	
Detects	**Test Principle**
• Cadmium • Copper • Lead • Mercury • Silver • Zinc	Zinc reacts with dithizone in basic solution to form a red color.

	Test Phase	
		Solid
✓		Liquid
		Gas/Vapor

Sensitivity
10 mg/l zinc

Directions and Comments
Strongly acidic samples must be adjusted to pH 4 to 7. Additional base (sodium hydroxide) is added to the sample and any precipitate is filtered. The test strip is immersed in the filtered sample for 1 second and results are read after 15 seconds. Dithizone will produce a red color with many metals; however, many of these metals are precipitated and filtered in this method. If the test strip is immersed in a sample adjusted to neutral pH without the step for precipitation, yellow-brown may indicate copper; red may indicate lead; orange may indicate mercury; violet may indicate silver; purple-red may indicate zinc. If the test strip is immersed in acidic sample (optimally pH 1 to 5) without the precipitation step, other metals can produce pink to brown colors including cadmium, cobalt, and lead. Further testing is necessary.

References	**Manufacturer**
Merckoquant® Zinc Test 1.10038.0001, Merck KGaA, Darmstadt, Germany, January 2007. Merck Safety Data Sheet 1.10038.0001, Merck KGaA, Darmstadt, Germany, August 22, 2007. Feigl, F., *Spot Tests in Inorganic Analysis,* Fifth Edition, Elsevier Publishing Company, New York, 1958.	Merck KGaA, Darmstadt, Germany

Methylene Blue Catalytic Reduction Test	
Detects	**Test Principle**
• Selenite	Complex selenium sulfide ions catalytically reduce methylene blue to a colorless compound.

	Test Phase	
		Solid
✓		Liquid
		Gas/Vapor

Sensitivity
0.08 µg/ml selenite

Directions and Comments
Selenite is converted to complex selenium sulfide ions by adding excess sodium sulfide. To perform the test, place a drop of sample, adjusted to alkaline, on a spot plate and a drop of water on a second spot plate to function as a blank. Add a drop of 0.2 M sodium sulfide (CAS 1313-84-4) to each followed by a drop of methylene blue (CAS 7220-79-3) to each. Selenite is indicated if faster decolorization occurs in the test spot compared to the blank. If both spots decolorize at the same rate, selenium is not present.

References	**Manufacturer**
Feigl, F., *Spot Tests in Inorganic Analysis,* Fifth Edition, Elsevier Publishing Company, New York, 1958.	

Methylene Blue Reagent		
Detects	**Test Principle**	
• Chromate • Ferricyanate • Ferrocyanate • Mercurous chloride • Perchlorate • Thiocyanate	Methylene blue forms colored complexes with several anions.	
	Test Phase	
	✓	Solid
	✓	Liquid
		Gas/Vapor
	Sensitivity	
	0.001 M perchlorate	

Directions and Comments

A 0.3% solution of methylene blue (3,7-bis(Dimethylamino)phenazathionium chloride, CAS 7220-79-3) in water is mixed with the sample and observed for the formation of colored precipitates. Strongly acidic samples must be diluted or neutralized since the complex is often soluble in low pH and will not be visible. For example, 70% perchloric acid will destroy the reagent, lesser amounts form soluble, invisible precipitate, and very dilute solutions are easily discerned at 0.1 to 0.01 M. A violet color indicates perchlorate. Other colors are possible from dilute acid or salts, such as green (thiocyanate); dark green (ferricyanide); dark blue (ferrocyanide); blue (zinc chloride, titanium tetranitrate, vanadate, tungstate); violet-red (mercurous chloride); violet-blue (molybdate); brown-red (chromate). Chrostovsky et al. use a spray of 0.05% methylene blue on filter paper impregnated with a solution of 1N zinc sulfate (CAS 7446-20-0) and 1N potassium nitrate (CAS 7757-79-1) for explosive residue. The Chrostovsky method states persulfate yields a result similar to perchlorate. See color image 2.36 and 2.37.

References	Manufacturer
F. J. Welcher, *Organic Analytical Reagents,* D. Van Nostrand Co., 1948, Vol IV, pp. 517-531. Wolsey, W., *A Simple Qualitative Detection Test for Perchlorate Contamination in Hoods,* http://www.orcbs.msu.edu/chemical/resources_links/contamhoods.htm, Office of Radiation, Chemical and Biological Safety, Michigan State University, East Lansing, Michigan, accessed January 6, 2006. Chrostovsky, J., Thurman, W. and Javorsky, J., *Arson Anal. Newsl.* 5, 14 (1981).	

Michler's Thioketone		
Detects	**Test Principle**	
• Bromine • Chlorine • Iodine • Lewisite • Nitrogen tetroxide • Nitrous acid	Michler's thioketone reacts with chlorine, bromine or iodine to produce a blue color.	
	Test Phase	
	✓	Solid
	✓	Liquid
		Gas/Vapor
	Sensitivity	
	0.2 µg/ml aqueous chlorine; 0.2 µg/ml aqueous bromine; 0.5 µg/ml aqueous iodine.	

(continued)

Michler's Thioketone (*Continued*)
Directions and Comments
Michler's thioketone (4,4′-Bis[dimethylamino]thiobenzophenone, CAS 1226-46-6), which is also present in the M256 Kit as the Lewisite crayon, will produce a blue color in the presence of chlorine, bromine, or iodine in aqueous solution or as a gas. The test may be performed in water as a spot test or with impregnated paper strips to detect gas or liquid solution. A similar result is observed from oxidizing cations, nitrous acid, and nitrogen tetroxide aqueous solutions. The three halogen gases can often be differentiated from these interfering liquids by detection in gas form with paper test strips. For example, a few drops of suspected bleach in a test tube combined with a few drops of acid will generate chlorine gas, which could be detected with a test strip suspended in the neck of the tube. The paper test is made by soaking and drying paper test strips with a 0.1% (w/v) solution of Michler's thioketone in benzene as directed by Feigl. Substitute an appropriate solvent to avoid carcinogenic benzene. Test strips "store well" if kept dry and protected from light.

References	Manufacturer
Feigl, F. and Anger, V., *Spot Tests in Inorganic Analysis*, Sixth Edition, Elsevier Publishing Company, Amsterdam, 1972.	

Molybdenum Metal Test	
Detects	**Test Principle**
• Molybdenum metal	Molybdenum metal reacts with concentrated nitric acid to form a brown and blue stain when blotted with filter paper.

Test Phase	
✓	Solid
✓	Liquid
	Gas/Vapor

Sensitivity
Not available

Directions and Comments
This test is used to detect molybdenum metal on metal surfaces. Place a drop of concentrated nitric acid (CAS 7697-37-2) on the metal surface for 10 seconds; bubbles differentiate molybdenum metal from stainless steel and nickel-chromium alloys. Gently blot and then press a filter paper on the drop and metal surface for another 10 seconds. A brown spot surrounded by a blue ring indicates molybdenum metal.

References	Manufacturer
Feigl, F. and Anger, V., *Spot Tests in Inorganic Analysis*, Sixth Edition, Elsevier Publishing Company, Amsterdam, 1972. *Chemist-Analyst*, 50, 77 (1961).	

Morin Reagent	
Detects	**Test Principle**
• Aluminum • Beryllium • Indium • Gallium • Thorium • Scandium	Aluminum salts and aluminum in neutral or slightly acidic solution react with morin (3,5,7,2′,4′-pentahydroxyflavanol) to form an intense green fluorescence in sunlight or UV light.

(*continued*)

Morin Reagent (*Continued*)

Test Phase	
	Solid
✓	Liquid
	Gas/Vapor
Sensitivity	
0.2 µg/ml aluminum	

Directions and Comments

Dissolve the sample in 2N sodium hydroxide; aluminum should dissolve easily in basic solution while most metals will not. Mix three drops of the sample solution in a black spot plate and add two drops of saturated alcoholic morin (3,5,7,2′,4′-pentahydroxyflavanol, CAS 654055-01-3) solution and then add two drops of 2N hydrochloric acid. A bright green fluorescent color under a UV light indicates aluminum. Other, less likely encountered, metals will react similarly and these include beryllium, indium, gallium, thorium, zirconium, hafnium, and scandium. The test is sensitive to pH, with the exception of zirconium. Aluminum and zirconium can be differentiated from the other metals forming a positive result by adding excess hydrochloric acid; other metals will dissolve but aluminum and zirconium will remain as a green fluorescence. Morin Reagent is the most characteristic reagent for Al but is easily influenced by the presence of iron. Alizarin Red S Reagent is the most sensitive to aluminum but also the most sensitive to iron. Eriochrome Cyanine R Reagent is sensitive and the best reagent to use in the presence of iron but is not practical for field use.

References	Manufacturer
Eegriwe, E., *Z. Anal. Chem.* 76, 440 (1929). Jungreis, E., *Spot Test Analysis: Clinical, Environmental, Forensic, and Geochemical Applications,* Second Edition, John Wiley & Sons, Inc., New York, 1997. Eegriwe, E., *Univ. Riga, Zeitschrift fuer Analytische Chemie* 76, 438–443 (1929). Feigl, F. and Anger, V., *Spot Tests in Inorganic Analysis,* Sixth Edition, Elsevier Publishing Company, Amsterdam, 1972.	

Naphthylenediamine Test

Detects	Test Principle	
• Nitrite • Nitrous acid • Selenite	Nitrites react in neutral solution with 1,8-naphthyldiamine to form an orange-red precipitate.	
	Test Phase	
		Solid
	✓	Liquid
		Gas/Vapor
	Sensitivity	
	0.1 µg/ml nitrous acid	

Directions and Comments

A drop of 0.1% 1,8-naphthyldiamine (1,8-diazanaphthalene, CAS 254-60-4) in 10% acetic acid (CAS 64-19-7) solution is added to an equal volume of sample solution. Slight warming that produces an orange-red color indicates nitrite. Selenite produces a brown precipitate. Other acids produce invisible precipitates.

References	Manufacturer
Feigl, F., *Spot Tests in Inorganic Analysis,* Fifth Edition, Elsevier Publishing Company, New York, 1958.	

Neothorin Reagent	
Detects	**Test Principle**
• Hafnium • Zirconium	Neothorin reacts with hafnium to produce a red-violet color.

		Test Phase
		Solid
✓		Liquid
		Gas/Vapor

Sensitivity

3 μg/ml hafnium or zirconium

Directions and Comments

Hafnium reacts almost identically to zirconium and is difficult to determine with field tests. One difference is the reaction with neothorin (2-[2-arsonophenylazo]chromotropic acid disodium salt, CAS 3547-38-4). Mix the neothorin as a 0.1% aqueous solution. Dissolve the sample in concentrated hydrochloric acid and add a drop of sample to a spot plate followed by a drop of test solution. Hafnium produces a red-violet solution; zirconium produces a blue-violet solution. Other cations do not interfere, but several anions will complex with hafnium or zirconium, such as citrate, fluoride, oxalate, phosphate, and sulfate. Neothorin produces colored complexes with other compounds and is used for spectrophotometric determination of aluminum, beryllium, calcium, thorium, uranium, vanadium(IV) fluoride, sulfate, palladium, ruthenium, gallium, and indium.

References	**Manufacturer**
Feigl, F. and Anger, V., *Spot Tests in Inorganic Analysis,* Sixth Edition, Elsevier Publishing Company, Amsterdam, 1972. *Anal. Abstracts*, 16 (1969).	

Nickel Dimethylglyoxime Equilibrium Solution	
Detects	**Test Principle**
• Carbonate • Nickel • Organic base • Oxide • Phosphate • Protein • Silicate	Nickel dimethylglyoxime equilibrium solution is clear or red depending on pH, which is influenced by certain ions.

		Test Phase
✓		Solid
✓		Liquid
		Gas/Vapor

Sensitivity

Variable but sensitive. Sensitivity is based on the precision of the reagent mixture.

Directions and Comments

This is a very sensitive test for materials that consume hydronium ions, especially basic materials that are barely soluble in water, such as oxides, carbonates, phosphates, some silicates, organic bases, proteins, and others that may not be detected by altering the color of conventional pH indicators. Nickel dimethylglyoxime equilibrium solution is clear or red depending on pH with equilibrium at pH 1.9. Nickel dimethylglyoxime equilibrium solution is made by dissolving 2.3 g nickel sulfate (CAS 10101-97-0) in 300 ml water and mixing with 2.8 g nickel dimethylglyoxime (CAS 13478-93-8) dissolved in 300 ml methanol. Allow the mixture to stand for 30 minutes and filter any remaining solids. The solution will keep several weeks if tightly capped. Add a drop of the nickel dimethylglyoxime equilibrium solution to a small amount of powdered solid or liquid test sample. The clear solution will produce red nickel dimethylglyoxime if the sample material consumes hydronium ions.

(continued)

Nickel Dimethylglyoxime Equilibrium Solution (*Continued*)

References	Manufacturer
Feigl, F., *Spot Tests in Inorganic Analysis,* Fifth Edition, Elsevier Publishing Company, New York, 1958.	

Nickel Hydroxide Test

Detects	Test Principle
• Bromite • Chlorite • Iodite • Persulfate • Persulfuric acid	Persulfate reacts with light green nickel(II) hydroxide to form black nickel(IV) oxyhydrate.

Test Phase	
	Solid
✓	Liquid
	Gas/Vapor

Sensitivity
2.5 µg/ml potassium persulfate

Directions and Comments

Mix equal volumes of 1 N sodium hydroxide and 1% nickel sulfate hexahydrate (CAS 10101-97-0). Add equal volumes of test solution and sample solution. A black or gray precipitate indicates persulfate. Hydrogen peroxide and other per-compounds such as perchlorate or periodate have no oxidizing affect on nickel hydroxide but peroxide will work to retard the test and must be removed with the addition of silver nitrate (CAS 7761-88-8). Halogenites act similarly to persulfate; however mixtures of halogenites and persulfate are incompatible. Some halogenites may require heating in a boiling water bath for a few minutes to cause the formation of black precipitate or spots. Halogenates have no effect on the test.

References	Manufacturer
Feigl, F., *Spot Tests in Inorganic Analysis,* Fifth Edition, Elsevier Publishing Company, New York, 1958. Feigl, F. and Anger, V., *Spot Tests In Inorganic Analysis*, Sixth Edition, Elsevier Publishing Company, Amsterdam, 1972.	

Ninhydrin Solution

Detects	Test Principle
• Adrenalin • Amino acid • Peptone • Polypeptide • Protein	Ninhydrin forms a purple complex with amino acids and other protein-like compounds when heated.

Test Phase	
✓	Solid
✓	Liquid
	Gas/Vapor

Sensitivity
Field tests have been used to detect proteins left by fingerprints. Laboratory method: 0.01 mg protein/ml.

(continued)

Ninhydrin Solution (*Continued*)
Directions and Comments
The test solution is 3% solution (w/v) of ninhydrin (CAS 485-47-2) in water. Another method uses acetone or a mixture of acetone and water to improve ninhydrin solubility. Mix the sample and test solution and heat gently. A blue color indicates protein, amino acid, peptides, or adrenalin. Alternately, a powder or liquid sample may be collected on a swab or paper strip, moistened with the test solution and then suspended in a test tube with a little water in the bottom. The test tube base is gently heated to boiling and the resulting steam will heat the sample and produce a purple color if protein is present.

References	Manufacturer
Welcher, F., *Chemical Solutions: Reagents Useful to the Chemist, Biologist, and Bacteriologist,* D. Van Nostrand Company, New York, 1942. *Chemical Abstracts 7*, 3765 (1913).	

Nitrate and Nitrite Test Strip		
Detects	**Test Principle**	
• Nitrate • Nitrite • Nitric acid • Nitrous acid	Nitrate reacts with a reducing agent to form nitrite. Nitrite is converted to nitrous acid, which in turn reacts with an aromatic amine. This reaction product couples with N-[1-napthyl] ethylenediamine to form a red-violet color.	
	Test Phase	
		Solid
	✓	Liquid
		Gas/Vapor
	Sensitivity	
	10 ppm nitrate; 1 ppm nitrite.	
Directions and Comments		
The test uses a modified Griess reaction (see Merckoquant® Nitrite Tests 1.10007.0001). The manufacturer knows of no interfering compounds. If necessary, a method to eliminate nitrite in order to determine only nitrate involves the addition of five drops of a 10% aqueous amidosulfonic acid (CAS 5329-14-6) solution to 5 ml of sample that has been pH adjusted to < 10. Boil briefly, allow it to cool, and repeat the test. See color image 2.38.		

References	Manufacturer
Kimel, H., Notes from correspondence, Precision Laboratories, Inc., Cottonwood, Arizona, USA, December 26, 2007.	Nitrite and Nitrate Test Strip (CZNIT600) is available from Precision Laboratories, Inc., Cottonwood, Arizona, USA (www.precisionlabs.co.uk). Nitrate/Nitrite Test Strip is available from Industrial Test Systems, Inc., Rock Hill, South Carolina, USA (www.sensafe.com).

Nitrate Test (Ammonia)	
Detects	**Test Principle**
• Nitrate	Nitrate reacts with aluminum and dilute sodium hydroxide to form ammonia. Ammonia is then detected by a suitable test.

(continued)

Nitrate Test (Ammonia) (*Continued*)

	Test Phase	
	✓	Solid
	✓	Liquid
		Gas/Vapor
	Sensitivity	
	Not available	

Directions and Comments

Place 1 ml of test solution or a few milligrams of solid sample with 1 ml of water in a test tube. Add some aluminum powder or some aluminum foil. Ammonia will be generated if nitrate is present; the ammonia is detected by a Nessler's reagent (CAS 7783-33-7 or see Ammonium Test [Gutzeit Scheme]). Ammonia may also be detected with a wet pH test strip; basic pH indicates ammonia gas generation but be careful not to contaminate the strip with sodium hydroxide.

References	Manufacturer
Engelder, C., Dunkelberger, T. and Schiller, W., *Semi-Micro Qualitative Analysis,* Wiley, New York, 1936.	

Nitrate Test (Gutzeit Scheme)

Detects	Test Principle	
• Nitrate	Nitrate reacts with 1,5-dihydro-oxyanthraquinone and sulfuric acid to produce a violet reaction product.	
	Test Phase	
		Solid
	✓	Liquid
		Gas/Vapor
	Sensitivity	
	Not available	

Directions and Comments

The sample is first reacted with a soluble carbonate compound such as sodium carbonate (CAS 497-19-8) and acetic acid (CAS 64-19-7) is added until no more carbon dioxide is produced. A drop of sulfuric acid solution of 1,5-dihydro-oxyanthraquinon (1,5-dihydroxyanthraquinone, CAS 117-12-4) is placed on filter paper. A drop of sample solution containing nitrate will produce a violet color.

References	Manufacturer
Gutzeit, G., *Helv. Chim. Acta,* 12, 829 (1929).	

Nitric Acid - Silver Nitrate Solution

Detects	Test Principle
• Azide • Bromide • Chloride • Cyanide	A solution of nitric acid and silver nitrate will form colored precipitates with chloride, bromide, iodide, cyanide, thiocyanide, ferricyanide, and azide.

(*continued*)

Nitric Acid - Silver Nitrate Solution (*Continued*)

Iodide Thiocyanide	Test Phase	
	✓	Solid
	✓	Liquid
		Gas/Vapor
	Sensitivity	
	Various, but designed as a spot test	

Directions and Comments

Mix a 1% solution of silver nitrate (CAS 7761-88-8) in nitric acid. Mix a drop of the test solution with the sample. The formation of white, yellow, or orange precipitate indicates the presence of chloride, bromide, iodide, cyanide, thiocyanide, ferricyanide, or azide. A black/brown precipitate might be due to the formation of silver sulfide. Sulfide ion can be confirmed by adding acid and testing with a lead acetate test strip.

References	Manufacturer
Feigl, F., *Spot Tests in Inorganic Analysis,* Fifth Edition, Elsevier Publishing Company, New York, 1958.	

Nitric Acid Reagent

Detects	Test Principle	
Heroin Illegal drug Mescaline Morphine Oxycodone Protein	Concentrated nitric acid reacts with various substances to produce colored reaction products.	
	Test Phase	
	✓	Solid
	✓	Liquid
		Gas/Vapor
	Sensitivity	
	1 µg mescaline HCl	

Directions and Comments

This is a presumptive test for illegal drugs. Concentrated nitric acid (CAS 7697-37-2) is added to the sample and any color change is noted. The sample must not contain too much water, which will dilute the nitric acid and prevent nitrification of certain functional groups. It is most helpful to compare any color formed in this test with results from other presumptive drug tests (Marquis Reagent, etc.) in a matrix to determine a pattern of unique results. Refer to the appendix for drug screen results. Many proteins will produce a yellow color.

References	Manufacturer
U.S. Department of Justice. Color Test Reagents/Kits for Preliminary Identification of Drugs of Abuse, NIJ Standard–0604.01, National Institute of Justice, National Law Enforcement and Corrections Technology Center, Rockville, MD, July 2000.	Several commercial test kits are available.

Nitrite Test (Gutzeit Scheme)

Detects	Test Principle
Nitrite	Nitrite reacts with naphthylamine-sulfanilic acid reagent and dilute sulfuric acid to produce a red-brown color.

(continued)

Nitrite Test (Gutzeit Scheme) (*Continued*)			
	Test Phase		
		Solid	
	✓	Liquid	
		Gas/Vapor	
	Sensitivity		
	Not available		

Directions and Comments

The test solution is formed by boiling 0.1 g 1-naphthylamine (CAS 134-32-7) in 20 ml of water. After filtering, add 50 ml of acetic acid (CAS 64-19-7). Separately dissolve 0.5 g of sulfanilic acid (CAS 121-57-3) in 150 ml of acetic acid and then mix the two solutions. The sample is first reacted with a soluble carbonate compound such as sodium carbonate (CAS 497-19-8) and then acetic acid (CAS 64-19-7) is added until no more carbon dioxide is produced. A drop of naphthylamine-sulfanilic acid reagent is placed on filter paper and treated with dilute sulfuric acid. A drop of the test solution will produce a red-brown color if nitrite is present.

References	**Manufacturer**
Gutzeit, G., *Helv. Chim. Acta*, 12, 829 (1929). Welcher, F., *Chemical Solutions: Reagents Useful to the Chemist, Biologist, and Bacteriologist*, D. Van Nostrand Company, Inc., New York, 1942.	

Nitrocellulose Test			
Detects	**Test Principle**		
• Nitrocellulose • Nitrous acid	The -ONO$_2$ group in nitrocellulose will fuse with benzoin and at 150°C and nitrous acid is then formed. Nitrous acid is detected with a suitable test.		
	Test Phase		
	✓	Solid	
	✓	Liquid	
		Gas/Vapor	
	Sensitivity		
	Not available		

Directions and Comments

A few milligrams of solid sample or two drops of a liquid sample are placed in a test tube followed by about 0.1 mg of benzoin (CAS 119-53-9). The tube is heated in a glycerol bath to 150°C to 160°C and nitrous acid is detected at the opening by a Griess reagent (see CZ NIT600 Nitrate & Nitrite Test Strip). A red color indicates nitrocellulose.

References	**Manufacturer**
Feigl, F. and Liebergott, E., *Chemist-Analyst* 52, 47 (1963). Jungreis, E., *Spot Test Analysis: Clinical, Environmental, Forensic, and Geochemical Applications*, Second Edition, John Wiley & Sons, Inc., New York, 1997.	

Nitrotoluene Test	
Detects	**Test Principle**
• Dinitrotoluene • DNT • Nitrotoluene • TNT • Trinitrotoluene • Explosive	Nitrotoluene reacts with N,N-dimethylformamide and di-n-butylamine to form a dark blue color.
	Test Phase
	✓ Solid
	✓ Liquid
	Gas/Vapor
	Sensitivity
	100 ppm TNT in soil
Directions and Comments	
The test solution (v/v) is 80% N,N-dimethylformamide (CAS 68-12-2) and 20% di-n-butylamine (CAS 111-92-2). The test is most sensitive if a drop of liquid sample or a few milligrams of solid sample are placed on a filter paper. Add a drop or two of the test solution. An immediate purple to black color is presumptive for a nitrotoluene compound. The filter paper can be used as a wipe to collect the sample. This presumptive test for nitrotoluene was developed for use at Los Alamos National Laboratory.	
References	**Manufacturer**
Baytos, J., *Field Spot-Test Kit for Explosives*, Los Alamos National Laboratory, Los Alamos, New Mexico, July 1991.	

Nitrophenylhydrazine Solution	
Detects	**Test Principle**
• Aldehyde • Carbonyl • Ketone	p-Nitrophenylhydrazine will react with ketones and aldehydes to form insoluble crystalline compounds.
	Test Phase
	Solid
	✓ Liquid
	Gas/Vapor
	Sensitivity
	Not available
Directions and Comments	
Mix 3 g of p-nitrophenylhydrazine (CAS 100-16-3) with 90 g of 40% acetic acid (CAS 64-19-7). Mix equal volumes of test solution and sample solution. Insoluble crystalline compounds indicate aldehyde or ketone. Caution: p-nitrophenylhydrazine must be stored with at least 30% water content or it is considered to be explosive.	
References	**Manufacturer**
Welcher, F., *Chemical Solutions: Reagents Useful to the Chemist, Biologist, and Bacteriologist*, D. Van Nostrand Company, Inc., New York, 1942. *J. Biol. Chem.* 4, 235 (1908). p-Nitrophenylhydrazine MSDS, Sciencelab.com, Inc., Houston, Texas, 2005.	

Nitron Solution

Detects	Test Principle		
• Nitrate	Nitrate in acetic acid reacts with nitron to form a voluminous white precipitate.		
	Test Phase		
	✓	Solid	
	✓	Liquid	
		Gas/Vapor	
	Sensitivity		
	3 ppm nitrate		

Directions and Comments

Just before use, mix a little nitron (1,4-Diphenyl-3-[phenylamino]-1H-1,2,4-triazolium, CAS 2218-94-2) with acetic acid (CAS 64-19-7). Add a drop of nitron to a spot plate and then add a drop of the sample solution. A voluminous white precipitate immediately forms if nitrate is present. Trace amounts of nitrate form white needle-like crystals after several hours. Sensitivity improves about 25% if the reaction is run at 0°C.

References	Manufacturer
Welcher, F., *Chemical Solutions: Reagents Useful to the Chemist, Biologist, and Bacteriologist*, D. Van Nostrand Company, Inc., New York, 1942. Engelder, C., Dunkelberger, T. and Schiller, W., *Semi-Micro Qualitative Analysis*, Wiley, New York, 1936. *Analyst* 32, 349 (1907).	

Nitroso-R-Salt Solution

Detects	Test Principle		
• Cobalt	Cobalt reacts with nitroso-R-salt in boiling acid to form a red color.		
	Test Phase		
		Solid	
	✓	Liquid	
		Gas/Vapor	
	Sensitivity		
	0.003 µg/ml cobalt with ion exchange resin, slightly reduced sensitivity on a spot plate		

Directions and Comments

Mix a 0.5% (w/v) aqueous solution of nitroso-R-salt (3-hydroxy-4-nitroso-2,7-naphthalenedisulfonic acid disodium salt, CAS 525-05-3). Dissolve 1 g of sodium acetate (CAS 127-09-3) into 2 ml of sample solution in a test tube and then add 2 ml of test solution. Slowly bring to a boil, slowly add 1 ml of concentrated nitric acid, and continue to gently boil for 1 minute. A red color indicates cobalt. The nitric acid treatment prevents interference by copper and nickel.

References	Manufacturer
Welcher, F., *Chemical Solutions: Reagents Useful to the Chemist, Biologist, and Bacteriologist,* D. Van Nostrand Company, Inc., New York, 1942. *J. Am. Chem. Soc.* 43, 746 (1921). Feigl, F. and Anger, V., *Spot Tests in Inorganic Analysis*, Sixth Edition, Elsevier Publishing Company, Amsterdam, 1972. *Anal. Abstracts,* 16 (1969).	

Offord's Reagent	
Detects	**Test Principle**
• Chlorate • Cobalt • Cupric • Ferric • Nickel	Chlorate reacts with ammonium thiocyanate upon heating to form a yellow color.

Test Phase	
✓	Solid
✓	Liquid
	Gas/Vapor

Sensitivity
0.5 µg/ml cobalt

Directions and Comments

Form a 3N aqueous solution of ammonium thiocyanate (CAS 1762-95-4) by mixing 2.28 g in 10 ml water. The test may also be performed as a solution in a test tube or spot plate. To make filter papers for later use, soak filter papers with the ammonium thiocyanate solution and then dry at less than 70°C. When ready to test, do not use any paper that has discolored. Heat the paper to 60°C for 10 minutes, wet with test solution, and heat at 100°C for 30 minutes. Do not allow the paper to contact metal at anytime, even when heating. A yellow color indicates chlorate. The test paper will turn green to blue if exposed to an acid solution of cobalt salt that also contains ethanol or acetone. Under identical treatment as just described for cobalt, large amounts of nickel salts produce a light blue color; ferric salts produce a red color; cupric salts produce a red-brown color.

References	**Manufacturer**
Welcher, F., *Chemical Solutions: Reagents Useful to the Chemist, Biologist, and Bacteriologist,* D. Van Nostrand Company, Inc., New York, 1942. *Ind. Eng. Chem., Anal. Ed.* 7, 93 (1935). Feigl, F. and Anger, V., *Spot Tests in Inorganic Analysis,* Sixth Edition, Elsevier Publishing Company, Amsterdam, 1972. *Anal. Abstracts,* 16 (1969).	

Orange IV Solution	
Detects	**Test Principle**
• Zinc	Orange IV solution reacts with zinc and acidified potassium ferricyanide to change a red solution to green.

Test Phase	
✓	Solid
✓	Liquid
	Gas/Vapor

Sensitivity
Not available

Directions and Comments

Dissolve 1 mg of orange IV (sodium 4-[(2E)-2-(2-hydroxy-4-oxo-1-cyclohexa-2,5-dienylidene)hydrazinyl] benzenesulfonate, CAS 547-57-9) in 10 ml of water. Place a drop of test solution in a test tube and add a drop of dilute sulfuric acid (1:24) and 5 drops of freshly prepared 2% (w/v) potassium ferricyanide solution (CAS 13746-66-2). The solution should be red. Add a drop of sample solution or a few milligrams of solid sample. A green color indicates zinc.

(continued)

Orange IV Solution (*Continued*)	
References	Manufacturer
Welcher, F., *Chemical Solutions: Reagents Useful to the Chemist, Biologist, and Bacteriologist,* D. Van Nostrand Company, Inc., New York, 1942. Engelder, C., Dunkelberger, T. and Schiller, W., *Semi-Micro Qualitative Analysis,* Wiley, New York, 1936.	

Palladium Chloride Paper		
Detects	**Test Principle**	
• Amalgam • Mercury vapor	Mercury vapor reacts with palladium chloride to form black palladium.	
	Test Phase	
	✓	Solid
	✓	Liquid
	✓	Gas/Vapor
	Sensitivity	
	Variable, but can detect vapor from mercury amalgams.	

Directions and Comments

Filter paper is moistened with a 1% solution of palladium chloride (CAS 7647-10-1) and dried. The paper will be brown with palladium chloride and the test works better with dry paper. The paper is held over a sample that is slowly heated to red hot. A black area of palladium deposition forms over the sample. Briefly holding the paper over ammonia vapor (strong ammonium hydroxide) will clear the original brown color due to colorless palladium ammonia chloride formation and a gray to black spot of palladium metal remains to indicate mercury. The test paper may be used in air; if the paper is held over elemental mercury at 25°C to 30°C, a distinct gray spot will form within 10 minutes and will gradually darken to black. Interference will be caused by other material that can reduce palladium chloride; however, this is a good screening test for mercury in soil and other materials. See color images 2.39 and 2.40.

References	Manufacturer
Feigl, F., *Spot Tests in Inorganic Analysis,* Fifth Edition, Elsevier Publishing Company, New York, 1958.	

Palladous Chloride Reagent		
Detects	**Test Principle**	
• Hydroiodic acid • Iodide	Palladous chloride reacts with iodide to form a brown-black precipitate that is insoluble in hydrochloric acid but soluble in concentrated solutions of sodium chloride or ammonium hydroxide.	
	Test Phase	
		Solid
	✓	Liquid
		Gas/Vapor
	Sensitivity	
	1 μg/ml iodide	

(*continued*)

Palladous Chloride Reagent (*Continued*)

Directions and Comments

Mix a 1% aqueous solution of palladous chloride (palladium(II) chloride, CAS 7647-10-1). Dissolve the sample into an aqueous solution and place a drop on a filter paper. Add a drop of test solution. A brown-black spot indicates iodide. The test may also be performed in a test tube and then confirmed by testing for solubility. The brown-black precipitate should remain in concentrated hydrochloric acid but become soluble in concentrated solution of sodium chloride or ammonium hydroxide.

References	Manufacturer
Feigl, F. and Anger, V., *Spot Tests in Inorganic Analysis,* Sixth Edition, Elsevier Publishing Company, Amsterdam, 1972. *Anal. Abstracts,* 16 (1969).	

Permanganate Test

Detects	Test Principle		
• Permanganate	Purple permanganate on paper is oxidized by air to form a brown color.		
	Test Phase		
			Solid
		✓	Liquid
			Gas/Vapor
	Sensitivity		
	Not available		

Directions and Comments

Permanganates are generally purple-black solids that form a deep violet color in water. Permanganates can be screened with potassium iodide-test strips, which will turn a deep blue to black; high concentrations may immediately bleach the black color to white. A drop of neutral or slightly acidic permanganate solution placed on thick filter paper will form insoluble brown manganese dioxide within 3 minutes of exposure to air. Interfering colors from other materials may be washed away with water to leave the insoluble brown manganese dioxide.

References	Manufacturer
General Information Bulletin, 74-8, 4, Federal Bureau of Investigation, Washington, DC, 1974. Jungreis, E., *Spot Test Analysis: Clinical, Environmental, Forensic, and Geochemical Applications,* Second Edition, John Wiley & Sons, Inc., New York, 1997.	

Peroxide Test Strip (PER 400)

Detects	Test Principle
• Hydrogen peroxide • Hydroperoxide • Organic peroxide • Perborate • Peroxide	Peroxidase enzyme transfers oxygen from the peroxide to produce a brown reaction product, which may be iodine formed from iodide.

(continued)

Peroxide Test Strip (PER 400) (*Continued*)

	Test Phase	
	✓	Solid
	✓	Liquid
		Gas/Vapor
	Sensitivity	
	100 ppm hydrogen peroxide	

Directions and Comments

This strip is specific for peroxide using a peroxidase enzyme mediated reaction mechanism. The enzyme method can detect peroxides and hydroperoxides, but sometimes not polymeric peroxides that might form in simple ethers. If the test strip is used in an anhydrous organic solution, it must be briefly dipped in water to detect peroxide.

References	Manufacturer
Kimel, H., Notes from correspondence, Precision Laboratories, Inc., Cottonwood, Arizona, USA, December 26, 2007.	Precision Laboratories, Inc., Cottonwood, Arizona, USA

Peroxide Test Strip (10%)

Detects	Test Principle	
• Hydrogen peroxide	Peroxidase enzyme transfers oxygen from the peroxide to produce a brown reaction product, which may be iodine formed from iodide.	
	Test Phase	
		Solid
	✓	Liquid
		Gas/Vapor
	Sensitivity	
	0.5% hydrogen peroxide	

Directions and Comments

This test is similar to more sensitive peroxide test strips (see Peroxide Test Strip [PER 400]) but is used for detection of high-concentration peroxide test solutions in applications such as airport screening tests. This strip is specific for peroxide using a peroxidase (CAS 9003-99-0) mediated reaction mechanism. If the test strip is used in an anhydrous organic solution, it must be briefly dipped in water to detect peroxide. The manufacturer designed this test for higher concentration hydrogen peroxide detection. Use a more sensitive test when testing for hydroperoxides, organic peroxides, or polymeric peroxides.

References	Manufacturer
Kimel, H., Notes from correspondence, Precision Laboratories, Inc., Cottonwood, Arizona, USA, December 26, 2007.	Precision Laboratories, Inc., Cottonwood, Arizona, USA

pH Test

Detects	Test Principle
• pH • Hydroxide • Carbonate	Various pH indicators indicate the presence of certain materials when preformed in relatively small amounts of water.

(*continued*)

pH Test (*Continued*)		
• Sulfide • Ammonium	**Test Phase**	
		Solid
	✓	Liquid
		Gas/Vapor
	Sensitivity	
	Various	

Directions and Comments

A very small amount of solid is wetted just enough with water to react a pH test strip upon contact. The test also works with aqueous solutions and suspensions. pH of less than 4.3 indicates free acid, acid salts, or hydrolyzed salts of weak bases; pH of 3 or less indicates strong acid. pH of greater than 8.1 indicates alkali or alkaline earth hydroxides, carbonates, sulfides, or hydrolyzed salts of strong bases; pH 8.1 to 10.5 indicates ammonia or alkali salts of very weak bases; pH greater than 12 indicate free alkali or alkaline earth hydroxides. Other results are possible due to specific formulation, dilution, or buffers used in some products. A pH test strip moistened with neutral water may be used to test air for corrosive vapor, although the accuracy is degraded. See color image 2.41.

References	Manufacturer
Feigl, F., *Spot Tests in Inorganic Analysis,* Fifth Edition, Elsevier Publishing Company, New York, 1958.	

Phenylanthranilic Acid Test		
Detects	**Test Principle**	
• Chlorate • Chloride • Perchlorate	Chlorate forms an orange or red color when mixed with N-phenylanthranilic acid dissolved in concentrated sulfuric acid.	
	Test Phase	
	✓	Solid
	✓	Liquid
		Gas/Vapor
	Sensitivity	
	Not quantified but described as "sensitive"	

Directions and Comments

This sensitive test is used to determine chlorate in the presence of nitrate, perchlorate, or chloride in pyrotechnic mixtures. Dissolve 100 mg of N-phenylanthranilic acid (CAS 91-40-7) in 0.5 ml of concentrated sulfuric acid and agitate for a few minutes. Dissolve 500 mg of sample in 2 ml of water and agitate well to dissolve any chlorate. Remove any insoluble material by filtering or decanting after centrifuging for the most sensitive result. Add the phenylanthranilic acid solution dropwise. An orange or red color indicates presence of chlorate. Perchlorate or chloride in the absence of chlorate forms a white precipitate.

References	Manufacturer
Visser, W., *Practical Pyrotechniques,* http://www. wfvisser.dds.nl/EN/analysis_EN.html, Wissel Naer, Nederlands, January 15, 2008.	

Phenylenediamine Reagent	
Detects	**Test Principle**
• Aldehyde • Carbonyl group • Ketone • Ozone	m-Phenylenediamine reacts with a carbonyl group (C=O) in ketones and aldehydes to form a brilliant green color.

Test Phase	
✓	Solid
✓	Liquid
	Gas/Vapor

Sensitivity
Not available

Directions and Comments

Mix a 1% solution (w/v) of m-phenylenediamine (CAS 108-45-2) in water or alcohol. Mix the test solution with the sample. Carbonyl group is indicated by the appearance of a brilliant green fluorescence after a few minutes that will peak in about 2 hours. Serial dilutions may be necessary if the target ketone or aldehyde is in a mixture. Mineral acid may enhance the older test. Ozone may be detected by addition of 5% sodium hydroxide to the m-phenylenediamine; a red color develops. Nitrous acid and hydrogen peroxide do not interfere in the ozone test.

References	Manufacturer
Welcher, F., *Chemical Solutions: Reagents Useful to the Chemist, Biologist, and Bacteriologist,* D. Van Nostrand Company, Inc., New York, 1942. *Zeitschr. Anal. Chem.* 36, 371 (1987). *J. Amer. Chem. Soc.,* 63, 240–242 (1941).	

Polynitroaromatic Screening Test	
Detects	**Test Principle**
• Explosive • Nitrite • Polynitroaromatic	Polynitroaromatics react with tetraalkyl ammonium hydroxide or phosphonium hydroxide in dimethyl sulfoxide and water to liberate nitrite. Nitrite is detected with a suitable test.

Test Phase	
✓	Solid
✓	Liquid
	Gas/Vapor

Sensitivity
Not available

Directions and Comments

A patented test kit uses 2.5% to 20% tetraalkyl ammonium hydroxide, such as tetramethyl ammonium hydroxide (CAS 75-59-2) or a tetraalkyl phosphonium hydroxide such as tetra-N-butylphosphonium hydroxide (CAS 14518-69-5) in 60% dimethyl sulfoxide (CAS 67-68-5) and water to liberate nitrite. Nitrite is detected with a Griess reagent or other suitable test (see Merckoquant® Nitrate Tests or Nitrate & Nitrite Test Strip - CZ NIT600).

References	Manufacturer
Glattstein, B., *Pat. Specif.* (Aust) AU 602, 734 (1987). Jungreis, E., *Spot Test Analysis: Clinical, Environmental, Forensic, and Geochemical Applications,* Second Edition, John Wiley & Sons, Inc., New York, 1997.	

Potassium Bismuth Iodide Reagent		
Detects	**Test Principle**	
• Cesium • Thallium	Cesium ions react with potassium bismuth iodide to form a bright red precipitate.	
	Test Phase	
	✓	Solid
	✓	Liquid
		Gas/Vapor
	Sensitivity	
	1 µg/ml cesium	

Directions and Comments

Mix a saturated solution of 5 g of potassium iodide (CAS 7681-11-0) in water. Boil the solution and dissolve 1 g of bismuth oxide (CAS 1304-76-3) and then slowly add 25 ml of acetic acid while mixing. The test is performed in a test tube or on paper with solid or liquid sample. A blank may be useful to compare the original pale yellow color of the test solution to a brighter yellow or orange result, which indicates cesium. Metals that precipitate iodide interfere and may be removed from the sample by adding excess iodide, filtering, and testing the filtrate for cesium. Among the alkali and univalent metals, thallium is the only one to persist in the filtrate and will produce a brown result.

References	Manufacturer
Feigl, F. and Anger, V., *Spot Tests in Inorganic Analysis*, Sixth Edition, Elsevier Publishing Company, Amsterdam, 1972.	

Potassium Chlorate Reagent		
Detects	**Test Principle**	
• Osmium • Ruthenium	Osmium or ruthenium in neutral or very weakly acid solution induces chlorate to oxidize other compounds while chlorate is reduced to chloride.	
	Test Phase	
		Solid
	✓	Liquid
		Gas/Vapor
	Sensitivity	
	0.005 µg/ml osmium tetroxide	

Directions and Comments

Dissolve the sample in a neutral water solution. Mix a 1% aqueous solution (w/v) of potassium chlorate (CAS 3811-04-9). Place a drop in a test tube. Add a drop of concentrated sulfuric acid to 50 ml of water and add one drop of this solution to the drop of potassium chlorate (note: chlorate becomes increasingly oxidizing in increasingly acidic solution). Place a drop of the test solution on a potassium iodide-starch test strip and then add a drop of the sample solution. A blue-black color indicates osmium or ruthenium. This test is specific for osmium or ruthenium only if other oxidizing materials or darkly colored compounds are absent.

References	Manufacturer
Feigl, F. and Anger, V., *Spot Tests in Inorganic Analysis*, Sixth Edition, Elsevier Publishing Company, Amsterdam, 1972.	

Potassium Chromate Reagent	
Detects	**Test Principle**
• Silver • Silver salt	Potassium chromate reacts with silver salts to produce red silver chromate.

Test Phase	
✓	Solid
✓	Liquid
	Gas/Vapor

Sensitivity
2 µg/ml silver ion

Directions and Comments

Dissolve the sample in acetic acid (CAS 64-19-7). Mix a 2% solution (w/v) of potassium chromate (CAS 7789-00-6). Add a drop of test solution to a drop of liquid sample or a few milligrams of solid sample. A red color indicates silver salts.

References	Manufacturer
Feigl, F. and Anger, V., *Spot Tests in Inorganic Analysis,* Sixth Edition, Elsevier Publishing Company, Amsterdam, 1972.	

Potassium Ethyl Xanthate Reagent	
Detects	**Test Principle**
• Molybdate	Potassium ethyl xanthate reacts with molybdate in mineral acid to produce a deep red-blue color.

Test Phase	
	Solid
✓	Liquid
	Gas/Vapor

Sensitivity
0.04 µg/ml molybdate

Directions and Comments

Place a grain of potassium ethyl xanthate (ethylxanthic acid potassium salt, CAS 140-89-6) in a spot plate or test tube. Dissolve the sample in neutral or slightly acidic water solution and add a drop to the spot plate. Add two drops of 2N hydrochloric acid. A pink to red to violet to black color indicates molybdate, depending on concentration. The complex is soluble in organic liquids and can be separated by shaking with hexane. Interference is caused by the presence of anions that form stable complexes with molybdenum, such as fluorides, oxalates, tartrates, etc.

References	Manufacturer
Feigl, F. and Anger, V., *Spot Tests in Inorganic Analysis,* Sixth Edition, Elsevier Publishing Company, Amsterdam, 1972.	

Potassium Iodide Reagent (Gold)	
Detects	**Test Principle**
• Gold(III) • Palladium • Platinum(IV) salts	Palladium, platinum(IV), or gold(III) salts react with potassium iodide to produce a color that can be cleared with sodium sulfide.

Test Phase	
	Solid
✓	Liquid
	Gas/Vapor

Sensitivity
0.5 µg/ml platinum by drop method

Directions and Comments
Mix a 5% aqueous solution (w/v) of potassium iodide (CAS 7681-11-0). Adjust the sample to as weakly acidic as possible. Mix a drop of the sample solution with a drop of test solution. A brown color indicates platinum(IV); a blue color indicates gold(III); a red-brown color indicates palladium. Confirm by adding a drop of 10% sodium sulfide (CAS 1313-82-2) solution; the color should clear. Alternatively and with less sensitivity, add a drop of the pH-adjusted sample solution to a potassium iodide-starch strip. The colored results as indicated above should clear with the addition of a drop of sodium sulfide solution. Strong acids and oxidizers interfere. See color images 2.42 through 2.44.

References	**Manufacturer**
Feigl, F. and Anger, V., *Spot Tests in Inorganic Analysis*, Sixth Edition, Elsevier Publishing Company, Amsterdam, 1972. Houghton, R., unpublished studies, St. Johns, Michigan, 2008.	

Potassium Iodide Reagent (Thallium)	
Detects	**Test Principle**
• Lead • Mercury • Silver • Thallium(I)	Potassium iodide reacts with thallium(I), or mercury, silver or lead, to form a bright yellow precipitate. Only the thallium(I) precipitate remains upon reaction with sodium thiosulfate.

Test Phase	
	Solid
✓	Liquid
	Gas/Vapor

Sensitivity
0.6 µg/ml thallium(I)

Directions and Comments
Mix a 10% solution of potassium iodide (CAS 7681-11-0). Dissolve the sample in a weakly acid solution and place a drop on a spot plate. Add a drop of the test solution. A yellow precipitate indicates thallium(I) or salts of mercury, silver, or lead. Mix a 2% solution of sodium thiosulfate (sodium thiosulfate pentahydrate, CAS 10102-17-7) and add two drops to the test; if the yellow precipitate remains, thallium(I) is indicated. If the yellow precipitate clears, salts of mercury, silver, or lead are indicated.

References	**Manufacturer**
Feigl, F. and Anger, V., *Spot Tests in Inorganic Analysis*, Sixth Edition, Elsevier Publishing Company, Amsterdam, 1972.	

Potassium Iodide Starch Test Strip (JD260)

Detects	Test Principle
• Chlorine • Hydrogen peroxide • Iodine • Oxidizer • Ozone	A strong oxidizer liberates free iodine from potassium iodide. Iodine in the presence of a starch compound in the paper strip forms a blue to black color.

Test Phase

✓	Solid
✓	Liquid
✓	Gas/Vapor

Sensitivity

10 ppm chlorine; 10 ppm iodine; 5 ppm hydrogen peroxide; 6 ppm ozone.

Directions and Comments

The oxidizer must be able to displace iodide from potassium. Many oxidizers such as nitrate, perchlorate, and many organic peroxides are not able to displace iodide from potassium and are not detected. Concentrated strong oxidizers (e.g., >400 ppm chlorine) can produce a colored result that is instantly bleached white. As the liquid wicks up the strip, look for a blue line at the leading edge of the liquid. A test strip moistened with water can detect these strong oxidizers in air or by touching the surface of a solid sample. See color images 2.45 and 2.46.

References	Manufacturer
Kimel, H., Notes from correspondence, Precision Laboratories, Inc., Cottonwood, Arizona, USA, December 26, 2007.	Precision Laboratories, Inc., Cottonwood, Arizona, USA

Potassium Permanganate Solution

Detects	Test Principle
• Alkyne • Alkene • Thiocyanate	Purple permanganate solution is cleared by alkene and alkyne structures.

Test Phase

✓	Solid
✓	Liquid
	Gas/Vapor

Sensitivity

Not available

Directions and Comments

Dilute a solution of potassium permanganate (CAS 7722-64-7) in a test tube to a transparent light purple color. Add 1 ml of the dilute solution to a test tube and then add one drop of liquid sample or 1 mg of solid sample. The purple color will clear when permanganate participates in an addition reaction with alkene and alkyne carbon bonds. Alkanes will not clear the purple color. Potassium permanganate can be used to oxidize other compounds for detection. For example, thiocyanates are undetectable by most cyanide tests. Permanganate can be reacted with thiocyanates to form cyanide, which can then be detected with a suitable cyanide test.

References	Manufacturer
Shriner, R., Fuson, R. and Curtin, D., *The Systematic Identification of Organic Compounds, A Laboratory Manual,* Fifth Edition, John Wiley and Sons, Inc., New York, 1964. *J. Chem. Educ.,* 37, 205 (1960). Feigl, F. and Anger, V., *Spot Tests in Inorganic Analysis,* Sixth Edition, Elsevier Publishing Company, Amsterdam, 1972.	

Povidone Test for Starch	
Detects	**Test Principle**
• Starch • Polysaccharide	Iodine will interact with the coil structure of a polysaccharide and produce a black color.
	Test Phase
	✓ Solid
	✓ Liquid
	Gas/Vapor
	Sensitivity
	2 mg/ml corn starch in water; a visible speck as dry corn starch.

Directions and Comments

Povidone has about 10 times the available iodine and has a 5-year shelf life compared to dilute Lugol's solution. Povidone is a complex of 1-vinyl-2-pyrrolidinone polymers and iodine (CAS 25655-41-8). At full strength the brown color of povidone is so strong that the color change to black is difficult to see. Povidone may be applied full strength or diluted 10:1 or 20:1. Testing on a white background such as absorbent paper or cotton will help wick the excess brown liquid away and expose the blackened starch. Povidone (Betadine®) is a good choice for long shelf life. Iodine will stain starches due to its interaction with the coil structure of the polysaccharide (starch) but will not detect simple sugars such as glucose or fructose. See color image 2.47.

References	**Manufacturer**
Houghton, R., *Emergency Characterization of Unknown Materials*, CRC Press, 2007.	Povidone is available from several retailers.

Procke-Uzel's Reagent	
Detects	**Test Principle**
• Lithium	Potassium periodate and ferric chloride react in basic solution with lithium to form a yellow precipitate or turbidity.
	Test Phase
	Solid
	✓ Liquid
	Gas/Vapor
	Sensitivity
	25 μg/ml lithium

Directions and Comments

Dissolve 0.2 g potassium periodate (CAS 7790-21-8) in 1 ml of 2N potassium hydroxide and then dilute to 5 ml with water. Add 0.3 ml 10% solution (w/v) ferric chloride hexahydrate (CAS 10025-77-1) and then dilute to 10 ml with 2N potassium hydroxide. Solid samples should be dissolved by acid as necessary. Neutralize the sample solution if necessary, place a drop on a spot plate, and add a drop of test solution. A yellow precipitate or turbidity indicates potassium. Sodium interferes if present in higher concentrations.

References	**Manufacturer**
Welcher, F., *Chemical Solutions: Reagents Useful to the Chemist, Biologist, and Bacteriologist*, D. Van Nostrand Company, Inc., New York, 1942. *Chemical Abstracts* 32, 5329 (1938).	

Prussian Blue Test	
Detects	**Test Principle**
• Cyanide • Ferrocyanide	Free cyanide anion in aqueous solution will react with a mixture of ferric and ferrous salts to produce blue complexes (one of which is Prussian blue or ferric ferrocyanide) when acidified.

	Test Phase	
	✓	Solid
	✓	Liquid
		Gas/Vapor

Sensitivity
0.07 µg/ml potassium ferrocyanide

Directions and Comments

Any soluble ferric salt such as ferric chloride (CAS 10025-77-1) and any soluble ferrous salt such as ferrous sulfate (CAS 7782-63-0) may be used. A small amount of aqueous ferrous solution is added to the sample. The solution is acidified with hydrochloric or other suitable acid and a small amount of aqueous ferric solution is added. The reaction is very sensitive and good results occur if reagent concentration does not exceed cyanide concentration. Excess acid will produce hydrogen cyanide gas; work with small amounts and adequate ventilation. Heating may improve the solubility of free cyanide from strongly bound cyanide salts. The test may be modified for ferrocyanides by omitting the ferrous solution. An alternate method is to acidify the sample in a test tube and heat to boiling for 30 seconds while holding a paper strip moistened with sodium hydroxide solution over the opening. The basic liquid will capture hydrocyanic acid. Add a drop of ferrous sulfate solution, a drop of concentrated hydrochloric acid, and a drop of ferric chloride solution. If cyanide is present, Prussian blue will form on the paper.

References	**Manufacturer**
Engelder, C., Dunkelberger, T. and Schiller, W., *Semi-Micro Qualitative Analysis*, Second Edition, John Wiley & Sons, Inc., New York, May 1946.	

Pyridine Reagent	
Detects	**Test Principle**
• Auric • Gold	Gold(III) ions react with pyridine in hydrobromic acid to form an orange to maroon colored crystal.

	Test Phase	
		Solid
	✓	Liquid
		Gas/Vapor

Sensitivity
100 ppm as auric ion

Directions and Comments

Mix 40% hydrobromic acid (CAS 10035-10-6) with pyridine (CAS 110-86-1) in a proportion of 9:1 by volume. Add the sample solution or the soluble solid to 1 ml of test solution. An orange to maroon color indicates gold(III) ions. Metallic gold may be detected if first dissolved by aqua regia.

(continued)

Pyridine Reagent (*Continued*)	
References	**Manufacturer**
Welcher, F., *Chemical Solutions: Reagents Useful to the Chemist, Biologist, and Bacteriologist*, D. Van Nostrand Company, Inc., New York, 1942. Short Microscopic Determination of the Ore Minerals, Bull. 825 U.S. Geol. Survey, Pt. 4 Microscopical Methods (1931).	

Pyridylazo Resorcinol (PAR) Reagent		
Detects	**Test Principle**	
• Niobium • Uranium(VI)	Pyridylazo resorcinol (PAR) reacts with many cations in neutral to mildly acidic solution to form a red color. The red color can be cleared by EDTA with the exception of niobium and uranium(VI) ions.	
	Test Phase	
		Solid
	✓	Liquid
		Gas/Vapor
	Sensitivity	
	0.1 µg/ml niobium	
Directions and Comments		
Mix a 0.03% solution of PAR (4-[2-Pyridylazo]resorcinol monosodium salt hydrate, CAS 16593-81-0). Dissolve the sample in water, optimally with a little tartrate buffer (pH 5.8-6.4) and add a drop to a spot plate or test tube. Add a drop of 0.75% aqueous solution of EDTA (Ethylenediaminetetraacetic acid, CAS 60-00-4). Add a drop of PAR test solution. Add two drops of sodium acetate-acetic acid buffer solution, pH 5.8 (CAS 126-96-5). An orange to red color that develops within two minutes indicates niobium. Uranium(VI) interferes.		
References	**Manufacturer**	
Feigl, F. and Anger, V., *Spot Tests in Inorganic Analysis*, Sixth Edition, Elsevier Publishing Company, Amsterdam, 1972.		

Quinalizarin Reagent (Aluminum)		
Detects	**Test Principle**	
• Aluminum • Aluminum salt • Beryllium	Quinalizarin reacts with aluminum to form a red to red-violet spot that persists in acetic acid.	
	Test Phase	
		Solid
	✓	Liquid
		Gas/Vapor
	Sensitivity	
	0.005 µg of aluminum in 0.01 ml water when performed on filter paper	

(continued)

Quinalizarin Reagent (Aluminum) (*Continued*)

Directions and Comments

A few milligrams or a drop of the sample are placed in a spot plate followed by a drop of 0.02% quinalizarin (1,2,5,8-tetrahydroxyanthraquinone, CAS 81-61-8) in ethanol (w/v). A drop of ammonium hydroxide will turn the quinalizarin a violet color. Allow time for the caustic to dissolve aluminum metal; aluminum salts will react immediately. Add acetic acid until the violet of unreacted quinalizarin becomes yellow-brown. Aluminum will remain as a red to red-violet color. Beryllium, which reacts similarly in basic solution, will not remain as a red color in acidic solution. Performing the test on filter paper will improve visibility; be sure any fleck of red has been contacted by acetic acid. If the test is performed on paper, exposure to fumes of concentrated ammonium hydroxide and glacial acetic acid are sufficient to cause reaction. Compare to Quinalizarin Reagent (Magnesium).

References	Manufacturer
Feigl, F. and Anger, V., *Spot Tests in Inorganic Analysis*, Sixth Edition, Elsevier Publishing Company, Amsterdam, 1972.	

Quinalizarin Reagent (Magnesium)

Detects	Test Principle
• Beryllium salt • Cerium salt • Lanthanum salt • Magnesium • Magnesium salt • Neodymium salt • Praseodymium salt • Thorium salt • Zirconium salt	Magnesium salts react with alkaline solutions of quinalizarin to form a light blue color.

Test Phase	
✓	Solid
✓	Liquid
	Gas/Vapor

Sensitivity

0.25 µg/ml magnesium in water; 10 µg beryllium detected in 1000 µg/ml magnesium.

Directions and Comments

Magnesium metal must first be dissolved in 2N hydrochloric acid; magnesium salts are tested directly. A few milligrams or a drop of the sample are placed in a spot plate followed by a drop of 0.02% quinalizarin (1,2,5,8-tetrahydroxyanthraquinone, CAS 81-61-8) in ethanol (w/v). The acidic quinalizarin is orange. Add 2N sodium hydroxide until the color changes to violet. Magnesium is indicated by a light blue precipitate or color that intensifies on standing. Using a second test on water may be helpful to compare the original violet to a slight appearance of blue; higher concentrations are obvious. Alkali earth metals and aluminum in moderate amounts do not interfere. Large amounts of ammonium salts and phosphate interfere. Less likely to be encountered cerium, beryllium, zirconium, thorium, lanthanum, praseodymium, and neodymium react in the same way as magnesium. Beryllium can be differentiated from magnesium by the resistance of the magnesium-quinalizarin complex in sodium hydroxide solution to bromine water; beryllium-quinalizarin complex in sodium hydroxide solution is destroyed by bromine water. Conversely, the magnesium-quinalizarin complex is destroyed in ammonium hydroxide solution; the beryllium-quinalizarin complex remains in ammonium hydroxide solution. Compare to Quinalizarin Reagent (Aluminum).

References	Manufacturer
Hahn, F., Wolf, H., and Jaeger, G., *Ber. Dtsch. Chem. Ges.* 57, 1394 (1924). Jungreis, E., *Spot Test Analysis: Clinical, Environmental, Forensic, and Geochemical Applications,* Second Edition, John Wiley & Sons, Inc., New York, 1997. Feigl, F. and Anger, V., *Spot Tests in Inorganic Analysis*, Sixth Edition, Elsevier Publishing Company, Amsterdam, 1972. *Z. Anal. Chem.*, 94, 274 (1933). *Z. Anal. Chem.*, 73, 57 (1928).	

Quinizarin Reagent

Detects	Test Principle
• Gallium(III) • Indium(III) • Thallium(III)	Fluorescent species form when thallium (III), gallium(III), or indium(III) ions interact with quinizarin in aqueous or nonaqueous solution.

Test Phase	
✓	Solid
✓	Liquid
	Gas/Vapor

Sensitivity
0.0007 mg/ml indium; 0.008 mg/ml gallium; 1 mg/ml thallium.

Directions and Comments

Mix 50 mg of quinizarin (1,4-dihydroxyanthraquinone, CAS 81-64-1) in 1 ml of concentrated ammonium hydroxide. Neutralize the sample solution as necessary and then add 1 ml of saturated ammonium chloride (CAS 12125-02-9) solution followed by 8 drops of the test solution. An immediate precipitate indicates indium, gallium, or thallium. Aluminum, thallium, beryllium, and zirconium interfere but can be masked with sodium fluoride (CAS 7681-49-4).

References	Manufacturer
Welcher, F., *Chemical Solutions: Reagents Useful to the Chemist, Biologist, and Bacteriologist*, D. Van Nostrand Company, Inc., New York, 1942. *Chemical Abstracts* 29, 1742 (1935). Bridgeman, R. et al., US Patent 7301025, Xerox Corp., Stamford, Connecticut, November 27, 2007.	

Raikow's Reagent

Detects	Test Principle
• Mercaptan • Sulfur, Organic • Thiol	Vanillin and phloroglucinol dried on paper form a red color when exposed to combustion products of organic material containing sulfur.

Test Phase	
✓	Solid
✓	Liquid
✓	Gas/Vapor

Sensitivity
Not available

Directions and Comments

Dissolve 10 mg of vanillin (CAS 121-33-5) and 10 mg of phloroglucinol (CAS 6099-90-7) in 1 ml of ether (CAS 60-29-7). Burn about 0.5 g of the material in a test tube by heating with a torch. Capture a portion of the smoke in the test tube and cover. Soak filter papers with the test solution and place the paper in the combustion byproducts. A red color indicates the material contained sulfur. Keep unused ether away from ignition sources.

References	Manufacturer
Welcher, F., *Chemical Solutions: Reagents Useful to the Chemist, Biologist, and Bacteriologist*, D. Van Nostrand Company, Inc., New York, 1942. *Zeitschr. Anal. Chem.*, 726 (1906); 701 (1910).	

RDX Test (J-Acid)	
Detects	**Test Principle**
• C-4 • Cyclonite • Cyclotrimethylenetrinitramine • Formaldehyde • Hexogen • RDX	RDX (cyclotrimethylenetrinitramine) reacts with sulfuric acid to form formaldehyde, which then reacts with J-acid to form a yellow, UV-fluorescent xanthilium dyestuff.

Test Phase

	Solid
✓	Liquid
	Gas/Vapor

Sensitivity

Not available

Directions and Comments

One drop of 0.1% J-acid solution (7-amino-4-hydroxy-2-naphthalenesulfonic acid, CAS 87-02-5) in concentrated sulfuric acid is placed in a spot plate or on an acid-resistant medium, such as glass filter paper (concentrated sulfuric acid will destroy cellulose paper). One drop of sample dissolved in acetone is added to the test solution. The appearance of yellow fluorescence under UV light indicates RDX.

References	**Manufacturer**
Jungreis, E., *Spot Test Analysis: Clinical, Environmental, Forensic, and Geochemical Applications,* Second Edition, John Wiley & Sons, Inc., New York, 1997. Sawicki, E., Stanley, T. and Pfaff, J., *Chemist-Analyst* 51, 5 (1962).	

RDX Test (Pyrolytic Oxidation)	
Detects	**Test Principle**
• C-4 • Cyclonite • Cyclotrimethylenetrinitramine • Hexogen • RDX • T-4	Cyclotrimethylenetrinitramine (RDX) heated to 180°C in the presence of manganese dioxide forms formaldehyde and nitrous acid. Formaldehyde and nitrous acid are detected with suitable tests.

Test Phase

✓	Solid
✓	Liquid
	Gas/Vapor

Sensitivity

Not available

Directions and Comments

Colorless RDX will dissolve in acetone; it will not dissolve in water. Place 1 to 2 mg of the RDX sample in a test tube and add a few grains of manganese dioxide (CAS 1313-13-9). Cover the tube with a moistened test strip for nitrite (Griess test) and heat in a water bath to 180°C; a red stain indicates nitrous acid. Repeat, but cover the tube with a moistened test strip for formaldehyde (Nessler's reagent) and heat in a water bath to 180°C; a brown or black color indicates formaldehyde. Appropriate colorimetric air-monitoring tubes may also be used.

(continued)

RDX Test (Pyrolytic Oxidation) (*Continued*)	
References	**Manufacturer**
Jungreis, E., *Spot Test Analysis: Clinical, Environmental, Forensic, and Geochemical Applications*, Second Edition, John Wiley & Sons, Inc., New York, 1997.	

RDX Test (Thymol)		
Detects	**Test Principle**	
<ul><li>C-4</li><li>Cyclonite</li><li>Cyclotrimethylenetrinitramine</li><li>Hexogen</li><li>HMX</li><li>RDX</li></ul>	RDX (cyclotrimethylenetrinitramine) and thymol react in the presence of sulfuric acid to produce a red color.	
	Test Phase	
	✓	Solid
	✓	Liquid
		Gas/Vapor
	Sensitivity	
	Not available, but considered specific for RDX	

Directions and Comments

Place a few milligrams of sample in a test tube followed by approximately 200 mg of thymol (CAS 89-83-8) and about six drops of concentrated sulfuric acid. Heat the mixture for 5 minutes in a boiling water bath and then add 5 ml ethanol. The formation of a rich blue solution indicates RDX. HMX produces a blue-green tint to the solution. Sugars and aldehydes interfere and produce a brown color. Repeating the test but heating to 150°C will yield the same rich blue color for RDX but HMX will produce an olive color. The test is considered specific for RDX.

References	**Manufacturer**
Jungreis, E., *Spot Test Analysis: Clinical, Environmental, Forensic, and Geochemical Applications,* Second Edition, John Wiley & Sons, Inc., New York, 1997. Amas, S., and Yallop, H., *J. Forensic Sci. Soc.* 6, 185 (1966). Amas, S., and Yallop, H., *Analyst* (London) 94, 828 (1969).	

Reducing Substance Test		
Detects	**Test Principle**	
<ul><li>Alcohol</li><li>Aldehyde</li><li>Carbohydrate</li><li>Dimethysulfoxide</li><li>DMSO</li><li>Reducers</li></ul>	Easily oxidized materials react with orange potassium dichromate/nitric acid solution to form a blue-violet color. Heating produces a green color.	
	Test Phase	
	✓	Solid
	✓	Liquid
		Gas/Vapor
	Sensitivity	
	Not available. Use of a blank is suggested for low concentration samples.	

(*continued*)

Reducing Substance Test (*Continued*)

Directions and Comments

Mix a 0.5% solution (w/v) of potassium dichromate (CAS 7778-50-9) in strong nitric acid (specific gravity 1.33 or about 50% to 60% nitric acid). Add a drop of the orange solution to a drop or a few milligrams of sample. A blue-violet color indicates easily oxidized substances such as alcohols, glycols, aldehydes, carbohydrates, etc. Additional heating will convert the blue-violet color to green. Sometimes the heat of the reaction will produce the green color from the blue-violet without supplemental heating.

References	Manufacturer
Welcher, F., *Chemical Solutions: Reagents Useful to the Chemist, Biologist, and Bacteriologist,* D. Van Nostrand Company, Inc., New York, 1942. *Chemical Abstracts* 6, 204 (1912).	

Resorcinol Reagent

Detects	Test Principle
• Zinc	Resorcinol reacts with ammoniacal zinc to form a blue color over a few minutes.

	Test Phase	
		Solid
✓		Liquid
		Gas/Vapor

Sensitivity
2 µg/ml zinc

Directions and Comments

Resorcinol forms colors with many compounds, but only zinc is known to form a blue color over several minutes. Add a few milligrams of resorcinol (CAS 108-46-3) to 1 ml of alcohol, and then add a few drops to the test solution followed by 1 ml of 6N ammonium hydroxide (CAS 1336-21-6). A blue color that develops over a few minutes indicates zinc.

References	Manufacturer
Feigl, F. and Anger, V., *Spot Tests in Inorganic Analysis,* Sixth Edition, Elsevier Publishing Company, Amsterdam, 1972. *Ann. Chim. Applicata,* 19, 383 (1929). Yoe, J., *A Laboratory Manual of Qualitative Analysis,* John Wiley and Sons, Inc., New York, 1938, p. 176.	

Rhodamine B Reagent (Antimony)

Detects	Test Principle
• Antimony • Bismuth chlorides • Gold • Mercury chlorides • Molybdate salt • Thallium • Tungstate salt	Rhodamine B reacts with pentavalent antimony in strong hydrochloric acid solution to form violet or blue precipitates.

(*continued*)

Rhodamine B Reagent (Antimony) (*Continued*)

	Test Phase	
		Solid
	✓	Liquid
		Gas/Vapor
	Sensitivity	
	0.2 µg/ml antimony	

Directions and Comments

Pentavalent antimony is detected directly; trivalent antimony can be reacted to pentavalent antimony with sodium nitrite (CAS 7632-00-0). Mix 10 mg rhodamine B (CAS 81-88-9) in 100 ml water and add a few drops to the sample. Mix a few milligrams of the sample with a drop of concentrated hydrochloric acid and oxidize if necessary with a few milligrams of sodium nitrite to produce a red color. Add 10 drops of the test solution and mix. A color change from red to bright red to violet indicates antimony. This test is used as a screen for post-blast residue. Gold(III), thallium(III), basic bismuth chlorides, and mercury chlorides also form violet precipitates in concentrated hydrochloric acid, as do molybdates and tungstates. Organic antimony compounds are first ignited to ash to produce antimony pentoxide and tetroxide; add alkali iodide and then test directly with the test solution. Organic arsenic does not interfere. Organic bismuth and organic gold compounds react as organic antimony.

References	Manufacturer
Jungreis, E., *Spot Test Analysis: Clinical, Environmental, Forensic, and Geochemical Applications,* Second Edition, John Wiley & Sons, Inc., New York, 1997. Feigl, F. and Anger, V., *Spot Tests in Inorganic Analysis*, Sixth Edition, Elsevier Publishing Company, Amsterdam, 1972.	

Rhodamine B Reagent (Gallium)

Detects	Test Principle	
• Antimony • Gallium • Gold • Thallium(III)	Rhodamine B reacts with the gallium, gold, antimony, or thallium(III) in hydrochloric acid to form a fluorescent orange-red in benzene.	
	Test Phase	
		Solid
	✓	Liquid
		Gas/Vapor
	Sensitivity	
	0.5 µg/ml gallium	

Directions and Comments

Mix 0.2% solution (w/v) of rhodamine B (CAS 81-88-9) in water and then add an equal volume of concentrated hydrochloric acid. Dissolve the sample in a drop of concentrated hydrochloric acid and then add three drops of the test solution. Add three drops of benzene and shake. Gallium will cause a pink to red color in the benzene layer that will fluoresce orange-red under UV light. Gold reacts similarly as gallium. Thallium reacts as gallium, but the benzene layer fluoresces yellow. Antimony will produce a color change from red to bright red-violet without the benzene step; benzene extraction is needed only for antimony(III) iodide. Gold(III), thallium(III), basic bismuth chlorides, and mercury chlorides also form violet precipitates in concentrated hydrochloric acid, as do molybdates and tungstates.

(*continued*)

Rhodamine B Reagent (Gallium) (*Continued*)	
References	**Manufacturer**
Feigl, F. and Anger, V., *Spot Tests in Inorganic Analysis*, Sixth Edition, Elsevier Publishing Company, Amsterdam, 1972. *Anal. Abstracts,* 16 (1969).	

Rubeanic Acid (Cobalt)			
Detects	**Test Principle**		
• Cobalt • Copper • Nickel	Rubeanic acid in alcohol solution in the presence of ammonia reacts with cobalt to form a brown color.		
	Test Phase		
			Solid
		✓	Liquid
			Gas/Vapor
	Sensitivity		
	0.03 µg/ml cobalt		

Directions and Comments

The test solution is a 1% alcoholic solution of rubeanic acid (dithiooxamide, CAS 79-40-3). Place a drop of the sample solution on a filter paper and hold over ammonia fumes. Dissolve the solid sample in acid, adjust the pH to slightly acidic to neutral, and then add a drop of the test solution to the filter paper. A brown color indicates cobalt; blue indicates nickel; black indicates copper. Compare to Rubeanic Acid (Ruthenium).

References	**Manufacturer**
Feigl, F. and Anger, V., *Spot Tests in Inorganic Analysis*, Sixth Edition, Elsevier Publishing Company, Amsterdam, 1972.	

Rubeanic Acid (Ruthenium)			
Detects	**Test Principle**		
• Palladium salt • Platinum salt • Ruthenium salt	Rubeanic acid in strong mineral acid reacts with ruthenium salts to form a deep blue color.		
	Test Phase		
			Solid
		✓	Liquid
			Gas/Vapor
	Sensitivity		
	0.2 µg/ml ruthenium		

Directions and Comments

Mix a 0.2% solution (w/v) of rubeanic acid (dithiooxamide, CAS 79-40-3) in glacial acetic acid (CAS 64-19-7). Dissolve the sample in concentrated hydrochloric acid, add one or two drops of test solution, and warm gently. A blue color indicates ruthenium. A red color indicates palladium or platinum. Osmium does not react.

References	**Manufacturer**
Feigl, F. and Anger, V., *Spot Tests in Inorganic Analysis*, Sixth Edition, Elsevier Publishing Company, Amsterdam, 1972.	

Rubeanic Acid Reagent

Detects	Test Principle
<ul><li>Copper</li><li>Ferric</li><li>Mercurous</li><li>Palladium</li><li>Platinum</li><li>Silver</li></ul>	Rubeanic acid produces various colors with metal ions.

Test Phase	
	Solid
✓	Liquid
	Gas/Vapor

Sensitivity
9 µg/ml copper

Directions and Comments

Mix a 1% solution (w/v) of rubeanic acid (dithiooxamide, CAS 79-40-3). A drop of sample solution is placed on a dry filter paper followed by a drop of test solution. The following colors indicate the associated ion: green is copper; faint orange is ferric; yellow turning to green-black is silver; brown turning to black is mercurous; tan is bismuth; rose is platinum; brown is palladium. To mask all of these ions but copper, perform the following test. A drop of 20% malonic acid (CAS 141-82-2) solution is placed on a filter paper followed by a drop of the neutral sample solution. Next, a drop of 10% ethylenediamine (CAS 107-15-3) solution is added and finally a drop of the rubeanic acid test solution. A green color indicates copper. Compare to Copper Metal Test.

References	Manufacturer
Feigl, F., *Spot Tests in Inorganic Analysis,* Fifth Edition, Elsevier Publishing Company, New York, 1958.	

Safranin O Solution

Detects	Test Principle
<ul><li>Hypohalogenous acid</li><li>Hypochloric acid</li><li>Hypobromous acid</li><li>Hypoiodous acid</li><li>Hypochlorite</li><li>Hypobromite</li><li>Hypoiodite</li></ul>	Hypohalogenites react with red safranin O dye to form a blue-violet color in basic pH.

Test Phase	
	Solid
✓	Liquid
	Gas/Vapor

Sensitivity
0.5 µg/ml hypochlorite or hypobromite; 2.5 µg/ml hypoiodite.

Directions and Comments

A drop of saturated safranin O dye (CAS 477-73-6) solution is placed on a filter paper and dried in air. The hypohalogenous acid solution is combined with sodium hydroxide solution until the pH is basic. A drop of test solution is applied to the dried paper and a violet color forms if hypohalogenite is present. If a brown stain forms, wash the paper with water to remove the water-soluble brown compound. The violet-colored compound will remain if present.

References	Manufacturer
Feigl, F., *Spot Tests in Inorganic Analysis,* Fifth Edition, Elsevier Publishing Company, New York, 1958.	

Safranin T Solution	
Detects	**Test Principle**
• Nitrite • Nitrate	Safranin T reacts with nitrite to form a blue color in dilute sulfuric acid. Nitrate may be reduced to nitrite with magnesium powder and detected as nitrite.

	Test Phase	
		Solid
	✓	Liquid
		Gas/Vapor

Sensitivity
0.02 mg/5 ml nitrite

Directions and Comments
Dissolve 0.3 g of safranin T in 100 ml of water or dilute safranin T solution (CAS 477-73-6). Add five drops of test solution to 5 ml of sample solution and acidify with dilute sulfuric acid. A blue color indicates nitrite. Add magnesium powder (CAS 7439-95-4) or other suitable reducer to convert any nitrate to nitrite. An increase in blue color indicates nitrite.

References	**Manufacturer**
Welcher, F., *Chemical Solutions: Reagents Useful to the Chemist, Biologist, and Bacteriologist,* D. Van Nostrand Company, Inc., New York, 1942. *Chemical Abstracts* 21, 873 (1927).	

Schiff's Reagent (Aldehyde)	
Detects	**Test Principle**
• Aldehyde • Ketone	Schiff's reagent forms a violet red color in the presence of the carbonyl group of aldehydes or ketones.

	Test Phase	
	✓	Solid
	✓	Liquid
		Gas/Vapor

Sensitivity
Not available, but used in tissue staining

Directions and Comments
Schiff's reagent for aldehydes (Sigma-Aldrich 84655, no CAS number) is made from p-rosaniline (CAS 632-99-5), a pink dye, treated with sulphurous acid. The sulphurous acid clears the color of the dye, but the colorless Schiff's reagent is unstable. Aldehydes cause a shift in the sulfurous acid to produce a violet-purple color that is different from the original pink of p-rosaniline. Add a drop of colorless test solution to a drop of liquid sample or a few milligrams of solid sample in a spot plate. An intense red-violet to purple-violet color indicates ketone or aldehyde. Some ketones and unsaturated compounds are able to react with sulfurous acid to reverse the reaction that created the colorless Schiff's reagent and a pink color is formed when p-rosaniline is reformed. This pink color does not indicate aldehydes and is different from the purple-violet caused by aldehydes.

(continued)

Schiff's Reagent (Aldehyde) (*Continued*)

References	Manufacturer
Welcher, F., *Chemical Solutions: Reagents Useful to the Chemist, Biologist, and Bacteriologist*, D. Van Nostrand Company, Inc., New York, 1942. *Sigma-Aldrich Material Safety Data Sheet for Schiff's Reagent for Aldehydes*, Version 1.8, www.sigmaaldrich.com, April 11, 2006. Shriner, R., Fuson, R. and Curtin, D., *The Systematic Identification of Organic Compounds, A Laboratory Manual*, Fifth Edition, John Wiley and Sons, Inc., New York, 1964. *J. Chem. Educ.*, 37, 205 (1960).	Sigma-Aldrich Corp., St. Louis, Missouri, USA (www.sigmaaldrich.com)

Sensidyne Deluxe Haz Mat III Tube 131 (Br_2)

Detects	Test Principle		
• Bromine	The B layer produces a yellow color from a pH indicator detecting an acid. C layer produces a yellow color from bromine reacting with o-toluidine. E layer produces a brown color from bromine reacting with lead acetate to produce a brown reaction product.		
	Test Phase		
			Solid
			Liquid
		✓	Gas/Vapor
	Sensitivity		
	Not available		

Directions and Comments

Bromine produces yellow in the B and C layers, and brown in the E layer. This result is not in the ID chart but was provided by the manufacturer. Follow with Tube 114 (bromine). This tube is used as a screen for unknown material. Many compounds react by class or individually. Consult the test instructions for more detail.

References	Manufacturer
Sensidyne, Inc., *Sensidyne Gas Detector Tube Handbook*, Clearwater, Florida, undated. Roberson, R., Correspondence and notes from conversation, Sensidyne, Inc., Clearwater, Florida, February 2007.	Sensidyne, Inc., Clearwater, Florida, USA (www.sensidyne.com)

Sensidyne Deluxe Haz Mat III Tube 131 (Cl_2)

Detects	Test Principle		
• Chlorine	The B layer produces a white color when chlorine reacts to bleach a pH indicator. Layer C produces a yellow-orange color when chlorine reacts with o-toluidine.		
	Test Phase		
			Solid
			Liquid
		✓	Gas/Vapor

Sensidyne Deluxe Haz Mat III Tube 131 (Cl₂) (*Continued*)	
	Sensitivity
	20 ppm chlorine

Directions and Comments

Chlorine produces white discoloration in the B layer and yellow in layer E. This tube is used as a screen for unknown material. Many compounds react by class or individually. Consult the test instructions for more detail.

References	**Manufacturer**
Sensidyne, Inc., *Sensidyne Gas Detector Tube Handbook,* Clearwater, Florida, undated.	Sensidyne, Inc., Clearwater, Florida, USA (www.sensidyne.com)

Sensidyne Deluxe Haz Mat III Tube 131 (H₂S)		
Detects	**Test Principle**	
• Hydrogen sulfide	The D layer produces a brown layer when hydrogen sulfide reacts with lead acetate.	
	Test Phase	
		Solid
		Liquid
	✓	Gas/Vapor
	Sensitivity	
	10 ppm hydrogen sulfide	

Directions and Comments

Hydrogen sulfide produces brown in the D layer. This tube is used as a screen for unknown material. Many compounds react by class or individually. Consult the test instructions for more detail.

References	**Manufacturer**
Sensidyne, Inc., *Sensidyne Gas Detector Tube Handbook,* Clearwater, Florida, undated.	Sensidyne, Inc., Clearwater, Florida, USA (www.sensidyne.com)

Sensidyne Deluxe Haz Mat III Tube 131 (HBr)		
Detects	**Test Principle**	
• Hydrogen bromide	The B layer is expected to produce a yellow (as acid) or pink (as halogen) color.	
	Test Phase	
		Solid
		Liquid
	✓	Gas/Vapor
	Sensitivity	
	Not available	

Directions and Comments

The test is expected to produce yellow (as acid) or pink (as halogen) in the B layer. This result is not in the ID chart and is extrapolated. There is no hydrogen bromide tube to provide confirmation testing. This tube is used as a screen for unknown material. Many compounds react by class or individually. Consult the test instructions for more detail.

References	**Manufacturer**
Sensidyne, Inc., *Sensidyne Gas Detector Tube Handbook,* Clearwater, Florida, undated.	Sensidyne, Inc., Clearwater, Florida, USA (www.sensidyne.com)

Sensidyne Deluxe Haz Mat III Tube 131 (HCl)	
Detects	**Test Principle**
• Hydrogen chloride	Layer B contains a pH indicator that produces a pink color when exposed to halogen and yellow when exposed to acid.

	Test Phase	
		Solid
		Liquid
	✓	Gas/Vapor

Sensitivity
20 ppm hydrogen chloride

Directions and Comments
The B layer produces a pink color. Compare this test to Tube 173SA (HCl), which uses another pH indicator. This tube is used as a screen for unknown material. Many compounds react by class or individually. Consult the test instructions for more detail.

References	**Manufacturer**
Sensidyne, Inc., *Sensidyne Gas Detector Tube Handbook,* Clearwater, Florida, undated.	Sensidyne, Inc., Clearwater, Florida, USA (www.sensidyne.com)

Sensidyne Deluxe Haz Mat III Tube 131 (HCN)	
Detects	**Test Principle**
• Hydrogen cyanide	Hydrogen cyanide will produce a pale yellow layer as an acid in the B layer. Hydrogen cyanide will produce a white color in the E layer when it reacts with potassium disulfide palladate(II).

	Test Phase	
		Solid
		Liquid
	✓	Gas/Vapor

Sensitivity
Not available

Directions and Comments
Hydrogen cyanide is usually not indicated in the Sensidyne Deluxe Haz Mat III screen. Some concentrations of hydrogen cyanide produce a pale yellow discoloration in the A layer and a white discoloration in the E layer. This result is not in the ID chart but was provided by the manufacturer. Follow either result with Tube 112SB (HCN) as a confirmation test if hydrogen cyanide is suspected. This tube is used as a screen for unknown material. Many compounds react by class or individually. Consult the test instructions for more detail.

References	**Manufacturer**
Sensidyne, Inc., *Sensidyne Gas Detector Tube Handbook,* Clearwater, Florida, undated. Roberson, R., Correspondence and notes from conversation, Sensidyne, Inc., Clearwater, Florida, February 2007.	Sensidyne, Inc., Clearwater, Florida, USA (www.sensidyne.com)

Sensidyne Deluxe Haz Mat III Tube 131 (NH₃)		
Detects	**Test Principle**	
• Amine • Ammonia • Hydrazine	Layer A produces a yellow color when ammonia reacts with a pH indicator.	
	Test Phase	
		Solid
		Liquid
	✓	Gas/Vapor
	Sensitivity	
	5 ppm ammonia	
Directions and Comments		
Ammonia produces a yellow color in the A layer. Any basic gas or vapor such as an amine or hydrazine will produce this result. This tube is used as a screen for unknown material. Many compounds react by class or individually. Consult the test instructions for more detail.		
References	**Manufacturer**	
Sensidyne, Inc., *Sensidyne Gas Detector Tube Handbook,* Clearwater, Florida, undated.	Sensidyne, Inc., Clearwater, Florida, USA (www.sensidyne.com)	

Sensidyne Deluxe Haz Mat III Tube 131 (PH₃)		
Detects	**Test Principle**	
• Arsine • Phosphine	The E layer produces a gray-blue to black layer when phosphine or arsine reacts with potassium disulfide palladate II to form a reaction product.	
	Test Phase	
		Solid
		Liquid
	✓	Gas/Vapor
	Sensitivity	
	2 ppm phosphine	
Directions and Comments		
Phosphine and arsine produce gray to blue in the E layer. Follow with Tube 121U (arsine) for other possibilities. This tube is used as a screen for unknown material. Many compounds react by class or individually. Consult the test instructions for more detail.		
References	**Manufacturer**	
Sensidyne, Inc., *Sensidyne Gas Detector Tube Handbook,* Clearwater, Florida, undated.	Sensidyne, Inc., Clearwater, Florida, USA (www.sensidyne.com)	

Sensidyne Deluxe Haz Mat III Tube 131 (SO₂)	
Detects	**Test Principle**
• Sulfur dioxide • Acetic acid	Sulfur dioxide produces a yellow color from a pH indicator in Layer B.

(continued)

Sensidyne Deluxe Haz Mat III Tube 131 (SO₂) (*Continued*)

	Test Phase	
		Solid
		Liquid
	✓	Gas/Vapor
	Sensitivity	
	10 ppm sulfur dioxide	

Directions and Comments

Sulfur dioxide produces a yellow color in the B layer. Acetic acid may react similarly, but the result is not dependable. Follow with Tube 103SA (SO₂) as a confirmation test. This tube is used as a screen for unknown material. Many compounds react by class or individually. Consult the test instructions for more detail.

References	Manufacturer
Sensidyne, Inc., *Sensidyne Gas Detector Tube Handbook,* Clearwater, Florida, undated.	Sensidyne, Inc., Clearwater, Florida, USA (www.sensidyne.com)

Sensidyne Deluxe Haz Mat III Tube 131 (SO₃)

Detects	Test Principle	
• Sulfur trioxide	Sulfur trioxide is expected to produce a yellow color in layer B when it reacts with a pH indicator.	
	Test Phase	
		Solid
		Liquid
	✓	Gas/Vapor
	Sensitivity	
	Not available	

Directions and Comments

Sulfur trioxide is expected to produce a yellow color in the B layer. This result is not in the ID chart and is extrapolated. Sulfur trioxide will immediately form sulfuric acid in water. A Sensidyne sulfuric acid tube is available for confirmation testing but it is not reliable due to interference. This tube is used as a screen for unknown material. Many compounds react by class or individually. Consult the test instructions for more detail.

References	Manufacturer
Sensidyne, Inc., *Sensidyne Gas Detector Tube Handbook,* Clearwater, Florida, undated.	Sensidyne, Inc., Clearwater, Florida, USA (www.sensidyne.com)

Sensidyne Tube #101S

Detects	Test Principle
• Acetylene • Ammonia • Butadiene • Carbon disulfide • Chlorine • Hydrogen	Ammonium molybdate reacts with acetylene and palladium sulfate to produce molybdenum blue. The original pale yellow layer changes to a brown-blue color.

(continued)

Sensidyne Tube #101S (*Continued*)

	Test Phase	
		Solid
		Liquid
✓		Gas/Vapor
	Sensitivity	
	10 ppm acetylene	

Directions and Comments

The test is not specific to acetylene and many compounds will cause production of molybdenum blue. Results from this test should be viewed as suggestive. Hydrogen produces a blue color throughout the entire layer. Butadiene, hydrogen cyanide, and ammonia fade the original pale yellow to a white color. Chlorine, carbon disulfide, nitrogen dioxide, and benzene produce a yellow-orange to yellow-brown color. Consult the user manual for more detail.

References	Manufacturer
Sensidyne, Inc., *Sensidyne Gas Detector Tube Handbook,* Clearwater, Florida, undated.	Sensidyne, Inc., Clearwater, Florida, USA (www.sensidyne.com)

Sensidyne Tube #102SC

Detects	Test Principle
• Acetone • Acrolein • Aldehyde • Ketone	Acetone reacts with hydroxylamine hydrochloride to produce hydrogen chloride. Hydrogen chloride neutralizes sodium hydroxide, causing a pH indicator to change from yellow to a pink color.

	Test Phase	
		Solid
		Liquid
✓		Gas/Vapor
	Sensitivity	
	10 ppm acetone	

Directions and Comments

Aldehydes and ketones are indicated by a pink color. When this tube is used as part of the Sensidyne Deluxe Haz Mat III, aldehydes and ketones are indicated as the only remaining compounds that can liberate hydrogen chloride from hydroxyl amine hydrochloride. Using the test outside the Sensidyne qualitative analysis scheme can produce false positives from other compounds.

References	Manufacturer
Sensidyne, Inc., *Sensidyne Gas Detector Tube Handbook,* Clearwater, Florida, undated.	Sensidyne, Inc., Clearwater, Florida, USA (www.sensidyne.com)

Sensidyne Tube #102SD

Detects	Test Principle
• Acetone • Alcohol • Alkane • Aromatic	Chromium(VI) is reduced to chromium(III) in the presence of acetone and sulfuric acid. The yellow layer changes to a dark brown color to indicate acetone.

(*continued*)

Sensidyne Tube #102SD (*Continued*)		
• Ester • Ketone	**Test Phase**	
		Solid
		Liquid
	✓	Gas/Vapor
	Sensitivity	
	20 ppm acetone	
Directions and Comments		
Chromium(VI) can be reduced by many hydrocarbons. Alkanes smaller than C_3 are not detected.		
References	**Manufacturer**	
Sensidyne, Inc., *Sensidyne Gas Detector Tube Handbook,* Clearwater, Florida, undated.	Sensidyne, Inc., Clearwater, Florida, USA (www.sensidyne.com)	

Sensidyne Tube #105SB		
Detects	**Test Principle**	
• Amine • Ammonia • Hydrazine	Ammonia and other basic vapors neutralize phosphoric acid, which causes a pH indicator to change from pale purple to a pale yellow color.	
	Test Phase	
		Solid
		Liquid
	✓	Gas/Vapor
	Sensitivity	
	5 ppm ammonia	
Directions and Comments		
Any basic gas or vapor will cause a positive result. The test cannot differentiate ammonia from amines, hydrazine, or other basic gases or vapors.		
References	**Manufacturer**	
Sensidyne, Inc., *Sensidyne Gas Detector Tube Handbook,* Clearwater, Florida, undated.	Sensidyne, Inc., Clearwater, Florida, USA (www.sensidyne.com)	

Sensidyne Tube #105SH		
Detects	**Test Principle**	
• Ammonia	Ammonia reacts with cobalt chloride and water to form a complex salt. The original pink layer becomes a blue to brownish green color.	
	Test Phase	
		Solid
		Liquid
	✓	Gas/Vapor
	Sensitivity	
	100 ppm ammonia	

(*continued*)

Sensidyne Tube #105SH (*Continued*)	
Directions and Comments	
This test is specific to ammonia. Amines, hydrazine, and other basic gases and vapors do not interfere.	
References	**Manufacturer**
Sensidyne, Inc., *Sensidyne Gas Detector Tube Handbook,* Clearwater, Florida, undated.	Sensidyne, Inc., Clearwater, Florida, USA (www.sensidyne.com)

Sensidyne Tube #107SA		
Detects	**Test Principle**	
• Acetylene • Diethyl ether	Diethyl ether reacts with chromium(VI) and sulfuric acid to produce green chromium(III). Some hydrocarbons can be oxidized by chromium (VI) to a black color. A green to brown color is a positive result.	
	Test Phase	
		Solid
		Liquid
	✓	Gas/Vapor
	Sensitivity	
	10 ppm diethyl ether	
Directions and Comments		
Many organic vapors and gases, but not halohydrocarbons, will also be indicated. Acetylene will turn the entire layer brown.		
References	**Manufacturer**	
Sensidyne, Inc., *Sensidyne Gas Detector Tube Handbook,* Clearwater, Florida, undated.	Sensidyne, Inc., Clearwater, Florida, USA (www.sensidyne.com)	

Sensidyne Tube #109SB		
Detects	**Test Principle**	
• Bromine • Chlorine • Chlorine dioxide • Nitrogen dioxide • Nitrogen trichloride	Chlorine oxidizes o-toluidine to form orthoquinone. The original white layer changes to a pale orange color.	
	Test Phase	
		Solid
		Liquid
	✓	Gas/Vapor
	Sensitivity	
	0.06 ppm chlorine	
Directions and Comments		
Chlorine produces a pale orange color. Bromine, chlorine dioxide, nitrogen dioxide, and nitrogen trichloride form a pale yellow color. These pale yellow tints may be concentration dependent and should be considered suggestive and not conclusive as a differentiation test.		
References	**Manufacturer**	
Sensidyne, Inc., *Sensidyne Gas Detector Tube Handbook,* Clearwater, Florida, undated.	Sensidyne, Inc., Clearwater, Florida, USA (www.sensidyne.com)	

Sensidyne Tube #111U	
Detects	**Test Principle**
• Acetate • Alcohol • Ester • Ethyl acetate • Hydrocarbons (larger aliphatic) • Ketone	Ethyl acetate reacts with chromium(VI) and sulfuric acid to produce chromium(III). The yellow tube layer yellow changes to a brown color.

	Test Phase	
		Solid
		Liquid
✓		Gas/Vapor

Sensitivity
5 ppm ethyl acetate

Directions and Comments
These vapors and gases are indicated in the Sensidyne qualitative analysis scheme as the only remaining compounds that can reduce chromium(VI). Alcohols, esters, acetates, ketones, and kerosene produce a brown color. Paraffin hydrocarbons and halogenated hydrocarbons can produce a pale brown color throughout the indicator layer. Compare to Sensidyne Tube #186 Organic Gas Checker, which uses the same reaction and can qualitatively detect hydrocarbons (except methane and ethane), alcohols, esters, ketones, aromatics, and hydrogen sulfide.

References	**Manufacturer**
Sensidyne, Inc., *Sensidyne Gas Detector Tube Handbook,* Clearwater, Florida, undated.	Sensidyne, Inc., Clearwater, Florida, USA (www.sensidyne.com)

Sensidyne Tube #112SA	
Detects	**Test Principle**
• Acetone • Carbon disulfide • Dicyanide • Hydrogen cyanide • Hydrogen sulfide • Sulfur dioxide	Hydrogen cyanide reacts with sodium picrate to form a brown-red color.

	Test Phase	
		Solid
		Liquid
✓		Gas/Vapor

Sensitivity
10 ppm hydrogen cyanide

Directions and Comments
This test offers an alternative reaction principle to detect several compounds. Other compounds are likely to produce similar results.

References	**Manufacturer**
Sensidyne, Inc., *Sensidyne Gas Detector Tube Handbook,* Clearwater, Florida, undated.	Sensidyne, Inc., Clearwater, Florida, USA (www.sensidyne.com)

Sensidyne Tube #112SB	
Detects	**Test Principle**
• Hydrogen cyanide • Hydrogen sulfide • Sulfur dioxide	Hydrogen cyanide reacts with mercuric chloride to produce hydrogen chloride. Hydrogen chloride alters a pH indicator that changes from a yellow to a red color.

(continued)

Sensidyne Tube #112SB (Continued)		
	Test Phase	
		Solid
		Liquid
	✓	Gas/Vapor
	Sensitivity	
	0.2 ppm hydrogen cyanide	
Directions and Comments		
Sulfur dioxide and hydrogen sulfide produce similar indications. Many other substances react with mercuric chloride.		
References	**Manufacturer**	
Sensidyne, Inc., *Sensidyne Gas Detector Tube Handbook,* Clearwater, Florida, undated.	Sensidyne, Inc., Clearwater, Florida, USA (www.sensidyne.com)	

Sensidyne Tube #114		
Detects	**Test Principle**	
• Bromine • Nitrogen dioxide • Chlorine • Chlorine dioxide	o-Toluidine is oxidized by bromine to form orthoquinone. The original white tube layer changes to an orange color.	
	Test Phase	
		Solid
		Liquid
	✓	Gas/Vapor
	Sensitivity	
	0.1 ppm bromine	
Directions and Comments		
This test can be used to detect gas or vapor in air with higher oxidation potential than bromine.		
References	**Manufacturer**	
Sensidyne, Inc., *Sensidyne Gas Detector Tube Handbook*, Clearwater, Florida, undated.	Sensidyne, Inc., Clearwater, Florida, USA (www.sensidyne.com)	

Sensidyne Tube #118SB		
Detects	**Test Principle**	
• Benzene	Iodine pentoxide is reduced by benzene in the presence of sulfuric acid. Iodine causes the formation of a green-brown color.	
	Test Phase	
		Solid
		Liquid
	✓	Gas/Vapor
	Sensitivity	
	0.02 ppm benzene	

(continued)

Sensidyne Tube #118SB (*Continued*)	
Directions and Comments	
Other aromatic hydrocarbons are indicated by a stain in the tube filter, by no color formation, or by a yellow-brown stain. Hexane will turn the entire layer a dark gray color.	
References	**Manufacturer**
Sensidyne, Inc., *Sensidyne Gas Detector Tube Handbook,* Clearwater, Florida, undated.	Sensidyne, Inc., Clearwater, Florida, USA (www.sensidyne.com)

Sensidyne Tube #120SB			
Detects	**Test Principle**		
• Hydrogen sulfide	Hydrogen sulfide reacts with white lead acetate to form brown lead sulfide.		
	Test Phase		
			Solid
			Liquid
	✓		Gas/Vapor
	Sensitivity		
	0.3 ppm hydrogen sulfide		
Directions and Comments			
Hydrogen sulfide produces a dark brown color. The Sensidyne qualitative analysis scheme uses this test to differentiate hydrogen sulfide from arsine and other less toxic gases or vapors.			
References	**Manufacturer**		
Sensidyne, Inc., *Sensidyne Gas Detector Tube Handbook,* Clearwater, Florida, undated.	Sensidyne, Inc., Clearwater, Florida, USA (www.sensidyne.com)		

Sensidyne Tube #121U			
Detects	**Test Principle**		
• Arsine • Diborane • Hydrogen cyanide • Hydrogen selenide • Mercaptan • Phosphine	Arsine reacts with mercury chloride to produce hydrogen chloride. Hydrogen chloride causes a pH indicator to change from yellow to a pink color.		
	Test Phase		
			Solid
			Liquid
	✓		Gas/Vapor
	Sensitivity		
	0.02 ppm arsine		
Directions and Comments			
A pink color is formed by arsine, phosphine, mercaptans, hydrogen selenide, and hydrogen sulfide. Hydrogen cyanide and sulfur dioxide form a red color throughout the tube layer. Arsine in the presence of sulfur dioxide can still be indicated as a purple tint within the red color. Diborane is expected to produce a result similar to arsine in sulfur dioxide. Acetylene and carbon monoxide do not react here when this test is used as part of the Sensidyne qualitative analysis scheme.			
References	**Manufacturer**		
Sensidyne, Inc., *Sensidyne Gas Detector Tube Handbook,* Clearwater, Florida, undated.	Sensidyne, Inc., Clearwater, Florida, USA (www.sensidyne.com)		

Sensidyne Tube #126SA	
Detects	**Test Principle**
• Carbon dioxide • Hydrogen cyanide • Hydrogen sulfide • Sulfur dioxide	Carbon dioxide reacts with sodium hydroxide in the tube, which causes a pH indicator to change from pale blue to a pale pink color.

	Test Phase	
		Solid
		Liquid
✓		Gas/Vapor

Sensitivity
100 ppm carbon dioxide

Directions and Comments

Hydrogen sulfide, hydrogen cyanide, and sulfur dioxide are indicated as carbon dioxide. Other compounds may interfere. Consult the package insert for detailed information.

References	**Manufacturer**
Sensidyne, Inc., *Sensidyne Gas Detector Tube Handbook,* Clearwater, Florida, undated.	Sensidyne, Inc., Clearwater, Florida, USA (www.sensidyne.com)

Sensidyne Tube #126UH	
Detects	**Test Principle**
• Carbon dioxide	Carbon dioxide reacts with hydrazine to produce an acidic compound. A pH indicator changes from white to a purple color.

	Test Phase	
		Solid
		Liquid
✓		Gas/Vapor

Sensitivity
1% carbon dioxide

Directions and Comments

No interference is cause by sulfur dioxide, ammonia, hydrogen cyanide, chlorine, hydrogen chloride, nitrogen dioxide, or hydrogen sulfide. Large amounts of acid gases will cause a false positive.

References	**Manufacturer**
Sensidyne, Inc., *Sensidyne Gas Detector Tube Handbook,* Clearwater, Florida, undated.	Sensidyne, Inc., Clearwater, Florida, USA (www.sensidyne.com)

Sensidyne Tube #128SC	
Detects	**Test Principle**
• Acrylonitrile • Hydrogen cyanide • Nitrile	Acrylonitrile is oxidized by chromate and sulfuric acid to produce hydrogen cyanide, which then reacts with mercuric chloride to produce hydrogen chloride. A pH indicator changes from yellow to a pink color.

(continued)

Sensidyne Tube #128SC (*Continued*)

	Test Phase	
		Solid
		Liquid
	✓	Gas/Vapor
	Sensitivity	
	0.5 ppm acrylonitrile	

Directions and Comments
Most nitriles should be detected; however, a complete list is not available. Hydrogen cyanide will be indicated directly. Consult the package insert for detailed information.

References	Manufacturer
Sensidyne, Inc., *Sensidyne Gas Detector Tube Handbook*, Clearwater, Florida, undated.	Sensidyne, Inc., Clearwater, Florida, USA (www.sensidyne.com)

Sensidyne Tube #133A

Detects	Test Principle	
• Acetaldehyde • Acetone • Acrolein • Methyl ethyl ketone • Methyl isobutyl ketone • Sulfur dioxide	Acetaldehyde reacts with hydroxylamine hydrochloride and produces hydrogen chloride. Hydrogen chloride is indicated by a pH indicator.	
	Test Phase	
		Solid
		Liquid
	✓	Gas/Vapor
	Sensitivity	
	5 ppm acetaldehyde	

Directions and Comments
Other aldehydes and ketones such as acetone, acrolein, methyl ethyl ketone, and methyl isobutyl ketone may react similarly. Sulfur dioxide produces a similar result.

References	Manufacturer
Sensidyne, Inc., *Sensidyne Gas Detector Tube Handbook,* Clearwater, Florida, undated.	Sensidyne, Inc., Clearwater, Florida, USA (www. sensidyne.com)

Sensidyne Tube #134SB

Detects	Test Principle	
• Dichloroethylene • Halohydrocarbon • Tetrachloroethylene • Trichloroethylene • Vinyl chloride	Trichloroethylene reacts with lead oxide and sulfuric acid to produce hydrogen chloride. Hydrogen chloride reacts with a pH indicator to form a blue-purple color.	
	Test Phase	
		Solid
		Liquid
	✓	Gas/Vapor

(*continued*)

Sensidyne Tube #134SB (*Continued*)	
	Sensitivity
	0.1 ppm trichloroethylene
Directions and Comments	
Trichloroethylene, tetrachloroethylene, dichloroethylene, and vinyl chloride are specifically indicated by a purple color. Halogenated hydrocarbons are expected to produce a similar result.	
References	**Manufacturer**
Sensidyne, Inc., *Sensidyne Gas Detector Tube Handbook*, Clearwater, Florida, undated.	Sensidyne, Inc., Clearwater, Florida, USA (www.sensidyne.com)

Sensidyne Tube #137U			
Detects	**Test Principle**		
• Alcohol • Hydrogen	Hydrogen reacts with atmospheric oxygen within the tube to produce water vapor. Water vapor reacts with magnesium perchlorate to cause a pH indicator to change from yellow to a green color.		
	Test Phase		
			Solid
			Liquid
		✓	Gas/Vapor
	Sensitivity		
	0.03% (300 ppm) hydrogen		
Directions and Comments			
Ethanol and other alcohols react similarly. Other materials can produce a false positive if they can cause a condensation reaction in the presence of magnesium perchlorate.			
References	**Manufacturer**		
Sensidyne, Inc., *Sensidyne Gas Detector Tube Handbook,* Clearwater, Florida, undated.	Sensidyne, Inc., Clearwater, Florida, USA (www.sensidyne.com)		

Sensidyne Tube #141SB			
Detects	**Test Principle**		
• Acid • Carbon disulfide • Hydrogen sulfide • Sulfur dioxide	Carbon disulfide is oxidized by chromate to produce sulfur dioxide, which then neutralizes sodium hydroxide. A pH indicator changes from pink to a yellow color.		
	Test Phase		
			Solid
			Liquid
		✓	Gas/Vapor
	Sensitivity		
	0.3 ppm carbon disulfide		

Sensidyne Tube #141SB (*Continued*)	
Directions and Comments	

Acid gases capable of passing through the reaction layers can neutralize the sodium hydroxide and produce a false positive result. Hydrogen sulfide is detected similar to carbon disulfide. Chlorine will bleach the original pink to a pale pink.

References	**Manufacturer**
Sensidyne, Inc., *Sensidyne Gas Detector Tube Handbook,* Clearwater, Florida, undated.	Sensidyne, Inc., Clearwater, Florida, USA (www.sensidyne.com)

Sensidyne Tube #142S		
Detects	**Test Principle**	
• Hydrogen sulfide • Mercury • Nitrogen dioxide	Mercury reacts with cupric iodide to produce cupric mercury iodide, which is indicated as a pale orange color.	
	Test Phase	
		Solid
		Liquid
	✓	Gas/Vapor
	Sensitivity	
	0.02 mg/m2 mercury	

Directions and Comments	

A pale orange color is characteristic of mercury. Nitrogen dioxide and hydrogen sulfide are differentiated from mercury by formation of a brown color.

References	**Manufacturer**
Sensidyne, Inc., *Sensidyne Gas Detector Tube Handbook,* Clearwater, Florida, undated.	Sensidyne, Inc., Clearwater, Florida, USA (www.sensidyne.com)

Sensidyne Tube #147S		
Detects	**Test Principle**	
• Carbon tetrachloride • Chlorohydrocarbon • Chloropicrin • Phosgene	Carbon tetrachloride reacts with iodine pentoxide and sulfuric acid to form phosgene. Phosgene reacts with 4-[p-nitrobenzyl]-pyridine and benzylaniline to produce a dye that changes from white to a red color.	
	Test Phase	
		Solid
		Liquid
	✓	Gas/Vapor
	Sensitivity	
	0.2 ppm carbon tetrachloride	

Directions and Comments	

This is a reliable test for low levels of chlorohydrocarbons. Detection is not affected by halogens or organic fluorine, bromine, or iodine. The test is suitable only for very low concentrations of chlorohydrocarbon in air. Concentrations slightly above the maximum detection limit of 60 ppm will cause the indicator layer to change to dark blue. Very high concentrations may cause the indicating color to be completely obscured.

(*continued*)

Sensidyne Tube #147S (*Continued*)	
References	**Manufacturer**
Sensidyne, Inc., *Sensidyne Gas Detector Tube Handbook,* Clearwater, Florida, undated.	Sensidyne, Inc., Clearwater, Florida, USA (www.sensidyne.com)

Sensidyne Tube #156S			
Detects	**Test Principle**		
• Acid • Chlorine • Hydrogen chloride • Hydrogen fluoride	Hydrogen fluoride reacts with a pH indicator to change green-yellow to a pink color.		
	Test Phase		
			Solid
			Liquid
		✓	Gas/Vapor
	Sensitivity		
	0.05 ppm hydrogen fluoride		
Directions and Comments			
All acid gases are expected to produce a pink result.			
References	**Manufacturer**		
Sensidyne, Inc., *Sensidyne Gas Detector Tube Handbook,* Clearwater, Florida, undated.	Sensidyne, Inc., Clearwater, Florida, USA (www.sensidyne.com)		

Sensidyne Tube #157SB			
Detects	**Test Principle**		
• Halogen • Halohydrocarbon • Methyl bromide	Methyl bromide reacts with iodine pentoxide, chromate, and sulfuric acid to produce bromine. Bromine reacts with o-toluidine to form a yellow reaction product.		
	Test Phase		
			Solid
			Liquid
		✓	Gas/Vapor
	Sensitivity		
	1 ppm methyl bromide		
Directions and Comments			
This test is not specific to methyl bromide. All halohydrocarbons and halogens produce a yellow indication.			
References	**Manufacturer**		
Sensidyne, Inc., *Sensidyne Gas Detector Tube Handbook,* Clearwater, Florida, undated.	Sensidyne, Inc., Clearwater, Florida, USA (www.sensidyne.com)		

Sensidyne Test #165SA	
Detects	**Test Principle**
• Acetylene • Ethyl mercaptan • Hydrogen sulfide • Methyl mercaptan	Ethyl mercaptan reacts with palladium sulfate to change the original white layer to a yellow color.

	Test Phase	
		Solid
		Liquid
✓		Gas/Vapor

Sensitivity
0.2 ppm ethyl mercaptan

Directions and Comments
The test is designed to detect ethyl mercaptan; other mercaptans are expected to react, but a complete list is not available. Carbon monoxide produces a gray color. Hydrogen sulfide and acetylene produce a brown color. Methyl mercaptan produces a red-yellow color.

References	**Manufacturer**
Sensidyne, Inc., *Sensidyne Gas Detector Tube Handbook,* Clearwater, Florida, undated.	Sensidyne, Inc., Clearwater, Florida, USA (www.sensidyne.com)

Sensidyne Tube #165SB	
Detects	**Test Principle**
• Arsine • Ethyl mercaptan • Hydrogen cyanide • Hydrogen selenide • Mercaptan • Phosphine	Mercaptan reacts with mercuric chloride to produce hydrogen chloride. A pH indicator changes from yellow to a pink color.

	Test Phase	
		Solid
		Liquid
✓		Gas/Vapor

Sensitivity
1 ppm ethyl mercaptan

Directions and Comments
This test is not specific for ethyl mercaptan but is designed to quantitatively measure ethyl mercaptan. Other mercaptans will also be detected. Hydrogen sulfide, phosphine, arsine, hydrogen selenide, and hydrogen cyanide will be indicated with a similar result.

References	**Manufacturer**
Sensidyne, Inc., *Sensidyne Gas Detector Tube Handbook,* Clearwater, Florida, undated.	Sensidyne, Inc., Clearwater, Florida, USA (www.sensidyne.com)

Sensidyne Tube #167S	
Detects	**Test Principle**
• Arsine • Hydrogen selenide • Hydrogen sulfide	Hydrogen selenide reacts with gold(III) chloride to form a dark brown color.

(continued)

Sensidyne Tube #167S (*Continued*)		
• Mercury • Nickel carbonyl • Sulfur dioxide	**Test Phase**	
		Solid
		Liquid
	✓	Gas/Vapor
	Sensitivity	
	0.5 ppm hydrogen selenide	

Directions and Comments

The test produces a colloid gold that is brown in color. Hydrogen selenide, mercury vapor, and hydrogen sulfide also produce a brown color. Sulfur dioxide produces a pale blue color.

References	Manufacturer
Sensidyne, Inc., *Sensidyne Gas Detector Tube Handbook,* Clearwater, Florida, undated.	Sensidyne, Inc., Clearwater, Florida, USA (www.sensidyne.com)

Sensidyne Tube #171SB		
Detects	**Test Principle**	
• Acetaldehyde • Formaldehyde • Monomer • Styrene	Formaldehyde reacts with an aromatic compound and sulfuric acid in a pretreatment tube where a condensation reaction produces a polymer. The polymer is detected in a second, indicating tube where a brown-orange color is formed.	
	Test Phase	
		Solid
		Liquid
	✓	Gas/Vapor
	Sensitivity	
	0.5 ppm formaldehyde	

Directions and Comments

This test is designed to detect formaldehyde and may be useful in detecting several monomers, but a complete list is not available from the manufacturer. Styrene and acetaldehyde produce a similar indication. Trichloroethylene, ethyl acetate, and ethyl ether increased quantitative results when formaldehyde was present, but did not give false positive indications if formaldehyde was absent.

References	Manufacturer
Sensidyne, Inc., *Sensidyne Gas Detector Tube Handbook,* Clearwater, Florida, undated.	Sensidyne, Inc., Clearwater, Florida, USA (www.sensidyne.com)

Sensidyne Tube #174A	
Detects	**Test Principle**
• Chlorine • Nitrogen dioxide • Nitrogen monoxide • Nitrogen oxide	Nitrogen oxide is oxidized by chromium(VI) and sulfuric acid to produce nitrogen dioxide. Nitrogen dioxide reacts with o-toluidine to form a yellow-orange color.

(*continued*)

Sensidyne Tube #174A (Continued)

		Test Phase	
		Solid	
		Liquid	
	✓	Gas/Vapor	
		Sensitivity	
		1 ppm nitrogen oxide; 0.5 ppm nitrogen dioxide.	

Directions and Comments
This test can differentiate nitrogen oxide from nitrogen dioxide. Nitrogen dioxide is indicated directly by o-toluidine. Nitrogen oxide is converted to nitrogen dioxide and then indicated. By using two reaction layers in series with the oxidizing layer in this test, chlorine will produce a similar color but a different pattern than nitrogen oxide. Consult the package insert for more detailed information.

References	**Manufacturer**
Sensidyne, Inc., *Sensidyne Gas Detector Tube Handbook,* Clearwater, Florida, undated.	Sensidyne, Inc., Clearwater, Florida, USA (www.sensidyne.com)

Sensidyne Tube #181S

Detects	**Test Principle**	
• Aniline • Toluidine	Aniline reacts with p-dimethylamino-benzaldehyde to produce yellow azomethine.	
	Test Phase	
		Solid
		Liquid
	✓	Gas/Vapor
	Sensitivity	
	0.05 ppm aniline	

Directions and Comments
Toluidine produces a result similar to aniline. Ammonia does not interfere.

References	**Manufacturer**
Sensidyne, Inc., *Sensidyne Gas Detector Tube Handbook,* Clearwater, Florida, undated.	Sensidyne, Inc., Clearwater, Florida, USA (www.sensidyne.com)

Sensidyne Tube #183U

Detects	**Test Principle**	
• Amine • Ammonia • Cresol • Phenol	Cresol is oxidized by cerium and a cresol polymer is formed. The original pale yellow layer turns pale brown, probably through a chromium(VI) oxidation reaction.	
	Test Phase	
		Solid
		Liquid
	✓	Gas/Vapor

(continued)

Sensidyne Tube #183U (*Continued*)	
	Sensitivity
	0.3 ppm cresol
Directions and Comments	
Cresol and phenols produce a pale brown color. Aromatic amines produce a blue color. Aliphatic amines and ammonia cause the original pale yellow to change to a white color.	
References	**Manufacturer**
Sensidyne, Inc., *Sensidyne Gas Detector Tube Handbook,* Clearwater, Florida, undated.	Sensidyne, Inc., Clearwater, Florida, USA (www.sensidyne.com)

Sensidyne Tube #186		
Detects	**Test Principle**	
• Alcohol • Aromatic • Hydrocarbon • Hydrogen sulfide • Ketone • Organic gas	Hydrocarbons react with an orange layer of chromium(VI) to produce green chromium(III). Organic material may also be oxidized to a black color. Positive result is indicated by a green to black color.	
	Test Phase	
		Solid
		Liquid
	✓	Gas/Vapor
	Sensitivity	
	5 ppm benzene; 1500 ppm propane; 2500 ppm acetylene; 600 ppm acetone.	
Directions and Comments		
This test detects many organic materials and some inorganic materials. It detects organic material (except methane and ethane) such as, alcohols, aldehydes, esters, ketones, aromatics, and others as well as hydrogen sulfide and other strong reducers.		
References	**Manufacturer**	
Sensidyne, Inc., *Sensidyne Gas Detector Tube Handbook,* Clearwater, Florida, undated.	Sensidyne, Inc., Clearwater, Florida, USA (www.sensidyne.com)	

Sensidyne Tube #190U		
Detects	**Test Principle**	
• Alcohol • Butanol • Toluene	Chromium(VI) is reduced by alcohol in the presence of sulfuric acid changing the yellow layer to a pale blue color. The test probably uses a pH indicator sensitive to the reduced amount of sulfuric acid in the layer.	
	Test Phase	
		Solid
		Liquid
	✓	Gas/Vapor

Sensidyne Tube #190U (*Continued*)	
	Sensitivity
	2 ppm 1-butanol
Directions and Comments	
The test is designed to detect a wide range of alcohols. Toluene will turn the entire layer light blue.	
References	**Manufacturer**
Sensidyne, Inc., *Sensidyne Gas Detector Tube Handbook,* Clearwater, Florida, undated.	Sensidyne, Inc., Clearwater, Florida, USA (www.sensidyne.com)

Sensidyne Tube #214S			
Detects	**Test Principle**		
• Alcohol • Alkane • Aromatic • Ester • Halohydrocarbon • o-Dichlorobenzene	o-Dichlorobenzene is oxidized by iodine pentoxide and sulfuric acid to form iodine, which causes a yellow color.		
	Test Phase		
		Solid	
		Liquid	
	✓	Gas/Vapor	
	Sensitivity		
	2 ppm o-dichlorobenzene		
Directions and Comments			
This test detects compounds that can reduce iodine pentoxide. Alcohols produce a similar stain. Alkanes (>C3), halohydrocarbons, esters, and aromatics produce a dark brown color throughout the indicating layer. Compare to Sensidyne Tube #215S.			
References	**Manufacturer**		
Sensidyne, Inc., *Sensidyne Gas Detector Tube Handbook,* Clearwater, Florida, undated.	Sensidyne, Inc., Clearwater, Florida, USA (www.sensidyne.com)		

Sensidyne Tube #215S			
Detects	**Test Principle**		
• p-Dichlorobenzene	p-Dichlorobenzene is oxidized by lead oxide and sulfuric acid to form hydrogen chloride. Hydrogen chloride changes a pH indicator from white to a blue-purple color.		
	Test Phase		
		Solid	
		Liquid	
	✓	Gas/Vapor	
	Sensitivity		
	6 ppm p-dichlorobenzene		
Directions and Comments			
Benzene and toluene produce higher results. Hexane produces a weaker green-blue stain. Ethanol does not interfere. Compare to Sensidyne Tube 214S.			

(*continued*)

Sensidyne Tube #215S (*Continued*)	
References	**Manufacturer**
Sensidyne, Inc., *Sensidyne Gas Detector Tube Handbook,* Clearwater, Florida, undated.	Sensidyne, Inc., Clearwater, Florida, USA (www.sensidyne.com)

Sensidyne Tube #216S		
Detects	**Test Principle**	
• Acetic acid • Chlorine • Hydrogen chloride • Nitrogen dioxide • Sulfur dioxide	Acetic acid reacts with sodium hydroxide which causes a pH indicator to change color from pale pink to a yellow color.	
	Test Phase	
		Solid
		Liquid
	✓	Gas/Vapor
	Sensitivity	
	2 ppm acetic acid	
Directions and Comments		
This test will detect any compound in air capable of neutralizing sodium hydroxide. Sulfur dioxide and chlorine produce a yellow color similar to acetic acid. Hydrogen chloride produces a pink color. Nitrogen dioxide produces an "unclear stain" according to the manufacturer.		
References	**Manufacturer**	
Sensidyne, Inc., *Sensidyne Gas Detector Tube Handbook,* Clearwater, Florida, undated.	Sensidyne, Inc., Clearwater, Florida, USA (www.sensidyne.com)	

Sensidyne Tube #232SA		
Detects	**Test Principle**	
• Aldehyde • Ethylene glycol • Ethylene oxide • Hydrogen sulfide • Sulfur dioxide	Ethylene glycol is oxidized and formic acid is produced. Formic acid reacts with a pH indicator to form a pink color.	
	Test Phase	
		Solid
		Liquid
	✓	Gas/Vapor
	Sensitivity	
	5 mg/m3 ethylene glycol	
Directions and Comments		
Aldehydes, sulfur dioxide, and ethylene oxide produce a yellow color. Hydrogen sulfide produces an orange-pink color.		
References	**Manufacturer**	
Sensidyne, Inc., *Sensidyne Gas Detector Tube Handbook,* Clearwater, Florida, undated.	Sensidyne, Inc., Clearwater, Florida, USA (www.sensidyne.com)	

Sensidyne Tube #235S (*Continued*)	
Detects	**Test Principle**
• Dichloroethane • Halogen • Halohydrocarbon • Nitrogen oxide	Dichloroethane reacts with chromate and sulfuric acid to form chlorine in a pretreatment tube. Chlorine reacts with 3,3′-dimethylnaphthidine to form a purple nitroso-compound in the second section of the tube.

Test Phase	
	Solid
	Liquid
✓	Gas/Vapor

Sensitivity
3 ppm 1,1-dichloroethane

Directions and Comments

Nitrogen oxides, halogens, and halohydrocarbons produce purple colors. Alcohol, hexane, and toluene do not interfere. The tube detects 1,1- and 1,2-dichloroethane with varying sensitivity (compare to Tube #230S).

References	**Manufacturer**
Sensidyne, Inc., *Sensidyne Gas Detector Tube Handbook,* Clearwater, Florida, undated.	Sensidyne, Inc., Clearwater, Florida, USA (www.sensidyne.com)

Sensidyne Tube #239S	
Detects	**Test Principle**
• Carbon disulfide • Carbonyl sulfide • Hydrogen sulfide • Sulfur dioxide	Carbonyl sulfide is oxidized by chromate and sulfur dioxide is produced. Sulfur dioxide neutralizes sodium hydroxide and a pH indicator changes pink to a yellow color.

Test Phase	
	Solid
	Liquid
✓	Gas/Vapor

Sensitivity
2 ppm carbonyl sulfide

Directions and Comments

The test will detect any sulfur compound that can be oxidized by chromate.

References	**Manufacturer**
Sensidyne, Inc., *Sensidyne Gas Detector Tube Handbook,* Clearwater, Florida, undated.	Sensidyne, Inc., Clearwater, Florida, USA (www.sensidyne.com)

Sensidyne Tube #247S	
Detects	**Test Principle**
• Hydrogen peroxide	Hydrogen peroxide reacts with titanium sulfate to produce a yellow complex.

(*continued*)

Sensidyne Tube #247S (*Continued*)		
	Test Phase	
		Solid
		Liquid
	✓	Gas/Vapor
	Sensitivity	
	0.2 ppm hydrogen peroxide	

Directions and Comments	
This test is nearly as specific to hydrogen peroxide as the peroxidase method used in test strips. Other strong oxidizers such as chlorine, ozone, and nitrogen dioxide do not interfere. Aldehydes do not interfere.	
References	**Manufacturer**
Sensidyne, Inc., *Sensidyne Gas Detector Tube Handbook,* Clearwater, Florida, undated.	Sensidyne, Inc., Clearwater, Florida, USA (www.sensidyne.com)

Sensidyne Tube #282S		
Detects	**Test Principle**	
• Hydrogen sulfide • Mercaptan	Hydrogen sulfide reacts with lead acetate in a treatment tube to produce a dark brown color; mercaptans are unaffected. In the second tube mercaptans react with mercuric chloride to produce hydrogen chloride. Hydrogen chloride changes a pH indicator from yellow to a pink color.	
	Test Phase	
		Solid
		Liquid
	✓	Gas/Vapor
	Sensitivity	
	0.2 ppm hydrogen sulfide and/or mercaptan	

Directions and Comments	
The test is able to differentiate hydrogen sulfide from mercaptans while providing simultaneous quantitative measurement. Other gases or vapors that react with mercuric chloride will produce similar results.	
References	**Manufacturer**
Sensidyne, Inc., *Sensidyne Gas Detector Tube Handbook,* Clearwater, Florida, undated.	Sensidyne, Inc., Clearwater, Florida, USA (www.sensidyne.com)

Silver Ferrocyanide Paper	
Detects	**Test Principle**
• Bromide • Chloride • Iodide	Chloride, bromide, or iodide in solution reacts with silver ferrocyanide in the presence of a ferric salt to form Prussian blue.

(*continued*)

Silver Ferrocyanide Paper (*Continued*)

	Test Phase	
	✓	Solid
	✓	Liquid
		Gas/Vapor
	Sensitivity	
	Various	

Directions and Comments

Photographic silver nitrate paper (preferred) is soaked for 10 minutes in 1% potassium ferrocyanide (CAS 14459-95-1) solution and then washed with water and soaked in dilute silver nitrate (CAS 7761-88-8) solution and washed with water again. The paper is then soaked in 10% ferric sulfate (CAS 15244-10-7) solution and dried. This test may also be performed on a filter paper strip. The test strip may be used with aqueous solutions or on solid material once wetted with deionized water. The appearance of Prussian blue indicates chloride, bromide, or iodide.

References	Manufacturer
Feigl, F., *Spot Tests in Inorganic Analysis,* Fifth Edition, Elsevier Publishing Company, New York, 1958.	

Silver Nitrate Paper

Detects	Test Principle	
• Arsenic • Arsenite • Chromate • Phosphorus	Silver nitrate forms various colored precipitates with certain anions.	
	Test Phase	
	✓	Solid
	✓	Liquid
		Gas/Vapor
	Sensitivity	
	Not available	

Directions and Comments

Mix a solution of silver nitrate (CAS 7761-88-8) or purchase a solution with a concentration of 2% to 5%. Soak filter paper in the solution and dry. Apply a drop of sample solution to a paper or moisten a paper with deionized water and touch it to a grain of solid sample. Orange-red indicates chromate; yellow indicates arsenite or arsenic; black indicates phosphorus. Strips are most effective when prepared and used immediately; exposure to light will turn the strip black.

References	Manufacturer
Welcher, F., *Chemical Solutions: Reagents Useful to the Chemist, Biologist, and Bacteriologist*, D. Van Nostrand Company, Inc., New York, 1942.	

Silver Nitrate Reagent

Detects	Test Principle
• Chloride • Reactive organic chlorine	Silver nitrate forms a white precipitate with chloride.

(*continued*)

Silver Nitrate Reagent (*Continued*)

	Test Phase	
	✓	Solid
	✓	Liquid
		Gas/Vapor
	Sensitivity	
	Not available	

Directions and Comments

Mix a solution of silver nitrate (CAS 7761-88-8) or purchase a solution with a concentration of 2% to 5%. Add a drop of liquid solution or a few milligrams of solid sample to 1 ml or less of the test solution. The immediate production of a thick, opaque white indicates chloride. A slow production of cloudy white color that forms evenly throughout the solution over a few minutes can indicate the presence of an organic compound with a reactive chloride. See color images 2.48 and 2.49.

References	Manufacturer
Petrucci, R. and Wismer, R., *General Chemistry with Qualitative Analysis,* Second Edition, Macmillan Publishing Company, New York, 1987.	

Simon's Reagent

Detects	Test Principle
• 3,4-methylenedioxy-N-methamphetamine • Amphetamine • d-Methamphetamine HCl • Ecstasy • MDMA HCl • Methylphenidate HCl	Simon's reagent (nitroprusside and acetaldehyde) reacts with amphetamine. The solution is then reacted with sodium carbonate to form blue to violet colored reaction products in basic solution.

	Test Phase	
	✓	Solid
	✓	Liquid
		Gas/Vapor
	Sensitivity	
	10 µg d-methamphetamine HCl; 300 µg methylphenidate HCl.	

Directions and Comments

This test is a presumptive field test for illegal drugs. Solution A is 1 g of sodium nitroprusside (CAS 13755-38-9) in 50 ml of water and 2 ml of acetaldehyde (CAS 75-07-0) mixed well. Solution B is 2% solution (w/v) of sodium carbonate (CAS 6132-02-1). The sample is mixed with one volume of Solution A and then two volumes of Solution B. It is most helpful to compare any color formed in this test with results from other presumptive drug tests (Marquis Reagent, etc.) in a matrix to determine a pattern of unique results. Refer to the appendix for drug screen results.

References	Manufacturer
U.S. Department of Justice. Color Test Reagents/ Kits for Preliminary Identification of Drugs of Abuse, NIJ Standard–0604.01, National Institute of Justice, National Law Enforcement and Corrections Technology Center, Rockville, MD, July 2000.	Several commercial test kits are available.

Sodium Arsenite Reagent	
Detects	**Test Principle**
• Lithium	Lithium forms a white or rose colored precipitate with sodium arsenite upon heating.
	Test Phase
			Solid
		✓	Liquid
			Gas/Vapor
	Sensitivity
	Not available
Directions and Comments	
Dissolve 0.5 g sodium arsenite (CAS 7784-46-5) in 10 ml of methanol or ethanol. Mix 5 ml of test solution with 0.5 ml of sample solution and gently heat. A white or rose colored precipitate indicates lithium.	
References	**Manufacturer**
Welcher, F., *Chemical Solutions: Reagents Useful to the Chemist, Biologist, and Bacteriologist,* D. Van Nostrand Company, Inc., New York, 1942. *Chemical Abstracts* 26, 4269 (1932).	

Sodium Azide - Iodine Solution	
Detects	**Test Principle**
• Sulfide • Thiocyanate • Thiocyanuric acid • Thiosulfate • Thiosulfuric acid	Sulfide, thiocyanate, or thiosulfate reacts with sodium azide and iodine in a catalytic reaction to produce nitrogen gas.
	Test Phase
		✓	Solid
		✓	Liquid
			Gas/Vapor
	Sensitivity
	0.3 µg/ml sodium sulfide
Directions and Comments	
In this sensitive test, even minute amounts of sulfide are able to catalyze the rapid evolution of nitrogen gas. The test indicates both organic and inorganic forms of sulfide including thiocyanate and thiosulfate, but not other forms of sulfur such as elemental sulfur, sulfate, sulfite, etc. The test may be performed on both soluble and insoluble sulfides. The test solution is a mixture of 3 g sodium azide (CAS 26628-22-8) in 100 ml of a 0.1 N iodine (CAS 7553-56-2) solution. The mixture is stable. A soluble sample is placed in the solution and the production of nitrogen gas bubbles indicates sulfide. An insoluble sample will produce nitrogen bubbles from the sample surface with less intensity. For a more sensitive test, place the insoluble sample and test solution in a capillary tube in order to detect nitrogen bubble production before dissipation from the tube.	
References	**Manufacturer**
Feigl, F., *Z. Anal. Chem.,* 74, 369 (1928). Feigl, F., *Spot Tests in Inorganic Analysis,* Fifth Edition, Elsevier Publishing Company, New York, 1958.	

Sodium Nitroprusside Paper

Detects	Test Principle
• Hydrogen sulfide • Mercaptan • Sulfur • Sulfur dioxide	Sodium nitroprusside and sodium carbonate dried onto filter paper will form a violet color in the presence of sulfur compounds.

Test Phase	
✓	Solid
✓	Liquid
✓	Gas/Vapor

Sensitivity
1 ppm sulfur dioxide; 0.2 ppm hydrogen sulfide.

Directions and Comments

This is a diverse test for sulfur compounds in air. Mix an aqueous solution (w/v) of 4% sodium nitroprusside dihydrate (CAS 13755-38-9) and 2% sodium carbonate (CAS 497-19-8). Wet filter paper and expose to the sample gas. A violet color indicates a sulfur-containing gas such as sulfur dioxide or hydrogen sulfide. Strips may be dried and stored. Strips may be exposed to liquid samples directly or moistened with water and used on solid material.

References	Manufacturer
Welcher, F., *Chemical Solutions: Reagents Useful to the Chemist, Biologist, and Bacteriologist*, D. Van Nostrand Company, Inc., New York, 1942. *Chemical Abstracts* 35, 1353 (1941).	

Sodium Rhodizonate Solution (Barium)

Detects	Test Principle
• Barium • Strontium	Rhodizonate react with barium or strontium in acid solution to form a bright red color.

Test Phase	
✓	Solid
✓	Liquid
	Gas/Vapor

Sensitivity
0.25 µg/ml barium

Directions and Comments

This solution is unstable and must be mixed the day of use. Mix a 5% solution (w/v) of rhodizonic acid disodium salt (CAS 523-21-7) and add a few drops to the sample solution. Barium and strontium will form a brown-red color. To differentiate barium from strontium, add a drop or two of dilute hydrochloric acid. Barium will change to red with the addition of hydrochloric acid, but strontium will dissolve. Dimethylamine hydrochloride (CAS 506-59-2) can be added to differentiate barium from strontium; barium will cause the solution to change to bright red and strontium will cause the solution to change to violet-blue. Bivalent heavy metals interfere. Compare to Sodium Rhodizonate Solution (Calcium).

(continued)

Sodium Rhodizonate Solution (Barium) (*Continued*)

References	Manufacturer
Welcher, F., *Chemical Solutions: Reagents Useful to the Chemist, Biologist, and Bacteriologist*, D. Van Nostrand Company, Inc., New York, 1942. *Chemical Abstracts* 23, 4644 (1929); 19, 1108 (1925). Feigl, F. and Anger, V., *Spot Tests in Inorganic Analysis,* Sixth Edition, Elsevier Publishing Company, Amsterdam, 1972.	

Sodium Rhodizonate Solution (Calcium)

Detects	Test Principle	
• Calcium salt	Sodium rhodizonate reacts with calcium ions in alkaline solution to form a violet color.	
	Test Phase	
		Solid
	✓	Liquid
		Gas/Vapor
	Sensitivity	
	1 µg/ml calcium	

Directions and Comments

The test detects soluble calcium salts while insoluble salts, such as carbonate, fluoride, oxalate, and phosphate do not react. This solution is unstable and must be mixed the day of use. Mix a 0.2% solution (w/v) of rhodizonic acid disodium salt (CAS 523-21-7). Dissolve the sample in neutral or slightly acid solution in a test tube. Place a drop of sample solution on a filter paper followed by a drop of test solution and then a drop of 0.5N sodium hydroxide. A violet color indicates calcium ions. Barium and strontium interfere but can be differentiated if reacted with sulfuric acid and neutralized with sodium hydroxide before the test is performed again. Barium sulfate or strontium sulfate are not detected in alkali solution while calcium sulfate is detected. Compare to Sodium Rhodizonate Solution (Barium).

References	Manufacturer
Feigl, F. and Anger, V., *Spot Tests in Inorganic Analysis,* Sixth Edition, Elsevier Publishing Company, Amsterdam, 1972.	

Sodium Sulfide Reagent

Detects	Test Principle	
• Polysulfides • Sulfur (elemental)	Sulfur reacts with sodium sulfide solution and acetone to form a green-blue color.	
	Test Phase	
	✓	Solid
	✓	Liquid
		Gas/Vapor
	Sensitivity	
	6 µg/ml elemental sulfur	

(continued)

Sodium Sulfide Reagent (*Continued*)

Directions and Comments

Mix a 2% aqueous solution (w/v) of sodium sulfide (CAS 1313-84-4). Dissolve the solid sample in water and add alcohol as needed to dissolve. Place a drop of sample solution in a test tube. Add a drop of test solution followed immediately by 0.5 ml of acetone (CAS 67-64-1) and mix. A green-blue color indicates elemental sulfur or polysulfides.

References	Manufacturer
Feigl, F. and Anger, V., *Spot Tests in Inorganic Analysis*, Sixth Edition, Elsevier Publishing Company, Amsterdam, 1972.	

Sodium Thiosulfate Reagent

Detects	Test Principle		
• Cobalt	Cobalt(II) reacts with sodium thiosulfate to form a blue cobalt thiosulfate.		
	Test Phase		
			Solid
		✓	Liquid
			Gas/Vapor
	Sensitivity		
	8 µg/ml cobalt		

Directions and Comments

This test must be performed in neutral pH. Place 0.5 mg of sodium thiosulfate (CAS 10102-17-7) on a spot plate or in a test tube. Add a drop of neutral sample in aqueous solution. A blue color that appears immediately or within a few minutes indicates cobalt. A drop of alcohol may increase the rate of the reaction. Compare to Merckoquant® Cobalt Test. Iron(III) salts produce a violet color that is transient. Zinc, cadmium, copper, nickel, and palladium do not interfere.

References	Manufacturer
Feigl, F. and Anger, V., *Spot Tests in Inorganic Analysis*, Sixth Edition, Elsevier Publishing Company, Amsterdam, 1972. *Chem. Abstracts*, 42, 6698, 6699a (1948).	

Stannous Chloride Reagent (Molybdate)

Detects	Test Principle		
• Molybdate • Tungstate	Stannous chloride reacts with molybdates and tungstates to form blue compounds.		
	Test Phase		
		✓	Solid
		✓	Liquid
			Gas/Vapor
	Sensitivity		
	5 µg/ml tungstate		

(*continued*)

Stannous Chloride Reagent (Molybdate) (*Continued*)

Directions and Comments

Mix a 25% solution of stannous chloride (tin[II] chloride, CAS 10025-69-1) in concentrated hydrochloric acid. Place 1 mg of solid sample or two drops of liquid sample in a test tube and carefully add three to five drops of the test solution. A blue color indicates molybdate or tungstate. Use Hydrogen Peroxide Reagent to differentiate molybdate from tungstate if necessary.

References	Manufacturer
Feigl, F. and Anger, V., *Spot Tests in Inorganic Analysis*, Sixth Edition, Elsevier Publishing Company, Amsterdam, 1972.	

Stannous Chloride Reagent (Rhodium)

Detects	Test Principle	
• Rhodium salt	Rhodium salts react with tin chloride in the presence of ammonium chloride and potassium iodide to produce a cherry red color.	
	Test Phase	
		Solid
	✓	Liquid
		Gas/Vapor
	Sensitivity	
	"μg amounts" of rhodium salts	

Directions and Comments

Mix a saturated solution of ammonium chloride (CAS 12125-02-9) in 2 ml of water. Separately, mix a saturated solution of potassium iodide (CAS 7681-11-0) in 1 ml of water. Separately, mix a 20% solution of tin(II) chloride (CAS 10025-69-1) in water. Combine all three solutions and place a drop in a spot plate or test tube. Add a drop of the sample solution. A cherry red color indicates rhodium salt. Gold, palladium, and platinum interfere.

References	Manufacturer
Feigl, F. and Anger, V., *Spot Tests in Inorganic Analysis*, Sixth Edition, Elsevier Publishing Company, Amsterdam, 1972.	

Stannous Chloride Reagent (Technetium)

Detects	Test Principle	
• Technetium	Technetium, after reduction by tin(II) chloride, reacts with dimethylglyoxime to form a violet to green color that becomes bright green.	
	Test Phase	
		Solid
	✓	Liquid
		Gas/Vapor
	Sensitivity	
	0.04 μg/ml technetium	

(*continued*)

Stannous Chloride Reagent (Technetium) (*Continued*)

Directions and Comments

Mix a saturated solution of dimethylglyoxime (CAS 95-45-4) in ethanol and place 1 ml in a test tube and then add 1 ml of the sample solution. Separately, mix a 20% solution of tin(II) chloride (CAS 10025-69-1) in 10 N hydrochloric acid. Place 0.5 ml of the tin solution into the tube containing the dimethylglyoxime and sample. A violet to green color that appears immediately and then slowly becomes bright green indicates technetium.

References	Manufacturer
Feigl, F. and Anger, V., *Spot Tests in Inorganic Analysis*, Sixth Edition, Elsevier Publishing Company, Amsterdam, 1972. Jasmin, F. et al., *Talanta*, 2, 93 (1959).	

Sucrose Test	
Detects	**Test Principle**
• Sucrose • Sugar	Sucrose reacts with 1-napthol and sulfuric acid to produce a purple color.
	Test Phase
	✓ Solid
	✓ Liquid
	Gas/Vapor
	Sensitivity
	Not available, but has been used for post-blast detection of sucrose

Directions and Comments

A few milligrams of solid or two drops of liquid sample are placed in a spot plate. One drop of 15% 1-naphthol (CAS 90-15-3) in ethanol and two drops of concentrated sulfuric acid are added. A purple color indicates sucrose.

References	Manufacturer
General Information Bulletin, 74-8, 4, Federal Bureau of Investigation, Washington, DC, 1974. Jungreis, E., *Spot Test Analysis: Clinical, Environmental, Forensic, and Geochemical Applications*, Second Edition, John Wiley & Sons, Inc., New York, 1997.	

Sulfanilic Acid Reagent	
Detects	**Test Principle**
• Bromate • Bromic acid	Sulfanilic acid reacts with bromate in strong nitric acid to form a violet color.
	Test Phase
	✓ Solid
	✓ Liquid
	Gas/Vapor
	Sensitivity
	0.5 µg/ml bromate

(*continued*)

Sulfanilic Acid Reagent (*Continued*)

Directions and Comments

Mix a saturated aqueous solution of sulfanilic acid (CAS 121-57-3). Place a few milligrams of solid sample or a drop of liquid sample in a test tube and carefully add two drops of 6M nitric acid (CAS 7697-37-2). Add a drop of the sulfanilic acid solution. Bromic acid or bromates are indicated by the development of a violet color over a few minutes that will then slowly turn brown. Chlorates and iodates do not interfere. A substitute for sulfanilic acid is any aromatic amine containing a strong acid group in the para position.

References	Manufacturer
Feigl, F. and Anger, V., *Spot Tests in Inorganic Analysis,* Sixth Edition, Elsevier Publishing Company, Amsterdam, 1972.	

Sulfate Test (Gutzeit Scheme)

Detects	Test Principle	
• Sulfate	A red spot formed by sodium rhodizonate, barium chloride, and hydrochloric acid will disappear when reacted with sulfate.	
	Test Phase	
		Solid
	✓	Liquid
		Gas/Vapor
	Sensitivity	
	Not available	

Directions and Comments

The sample is first reacted with a soluble carbonate compound such as sodium carbonate (CAS 497-19-8) and acetic acid (CAS 64-19-7) is added until no more carbon dioxide is produced. The test uses one drop of the liquid from the carbonate solution. Mix the test solution as a drop each of 1% sodium rhodizonate (CAS 523-21-7), 1% barium chloride (CAS 10361-37-2), and 3N hydrochloric acid. A red spot is formed on paper from a freshly prepared solution of the first three reagents. Add a drop of sample solution to the red spot. Sulfate will clear the red color.

References	Manufacturer
Gutzeit, G., *Helv. Chim. Acta,* 12, 829 (1929).	

Sulfide Test (Gutzeit Scheme)

Detects	Test Principle	
• Sulfide	Sulfide reacts with acid to produce hydrogen sulfide, which produces an orange color with antimony chloride.	
	Test Phase	
	✓	Solid
	✓	Liquid
		Gas/Vapor
	Sensitivity	
	Not available	

(continued)

Sulfide Test (Gutzeit Scheme) (*Continued*)
Directions and Comments
This spot test will produce toxic hydrogen sulfide gas. Place a small amount of sample in a test tube and add one or two drops of acetic acid (CAS 64-19-7). Place a paper strip moistened with a 5% solution of antimony chloride (CAS 10025-91-9) over the test tube or suspend it in the neck of the tube. Alternatively, use a moistened piece of lead acetate paper or the appropriate colorimetric air monitoring tube to detect hydrogen sulfide if antimony chloride is not available.

References	Manufacturer
Gutzeit, G., *Helv. Chim. Acta*, 12, 829 (1929).	

Sulfite Test	
Detects	**Test Principle**
• Sulfite • Sulfurous acid	Sulfite reacts with mineral acids to form sulfur dioxide, which may be detected with an appropriate sulfur dioxide gas test.

	Test Phase	
	✓	Solid
	✓	Liquid
		Gas/Vapor

Sensitivity
3.5 µg/ml sulfur dioxide when nitroprusside indicator is used

Directions and Comments
The sample is mixed with a mineral acid (e.g., 2N hydrochloric acid) to generate sulfur dioxide. Feigl held moist zinc nitroprusside (CAS 14709-62-7) paper over the reaction to detect sulfur dioxide, which produced a salmon-pink color. Alternatively, an appropriate sulfur dioxide gas detector can also be used.

References	Manufacturer
Feigl, F., *Spot Tests in Inorganic Analysis,* Fifth Edition, Elsevier Publishing Company, New York, 1958.	

Sulfur Reagent	
Detects	**Test Principle**
• Hydrocarbon • Hydrogen (organic) • Organic compound	Nonvolatile organic material containing hydrogen reacts with molten sulfur to produce hydrogen sulfide.

	Test Phase	
	✓	Solid
	✓	Liquid
		Gas/Vapor

Sensitivity
0.1 µg/ml as phenol; 0.2 µg/ml as urea; 0.1 µg/ml as aniline.

(*continued*)

Sulfur Reagent (*Continued*)

Directions and Comments

Hydrogen from any nonvolatile organic compound when heated with molten sulfur produces hydrogen sulfide gas (toxic). Place 0.1 g of solid sample or a drop of nonvolatile liquid sample in a test tube and then add 0.1 g of sulfur (CAS 7704-34-9). Cover the opening of the test tube with a piece of lead acetate paper or a hydrogen sulfide test strip. Heat the base of the test tube, ideally to 250°C, in a glycerol bath. Alternately, use a low torch flame sparingly. Sulfur melts at 113°C. The appearance of a brown or black color on the test strip indicates hydrogen from organic material.

References	Manufacturer
Feigl, F. and Anger, V., *Spot Tests in Inorganic Analysis,* Sixth Edition, Elsevier Publishing Company, Amsterdam, 1972. *Anal. Abstracts,* 16 (1969).	

Sulfuric Acid Test

Detects	Test Principle	
• Sulfuric acid	Free sulfuric acid oxidizes methylenedisalicylic acid to an intensely red compound upon heating.	
	Test Phase	
	✓	Solid
	✓	Liquid
		Gas/Vapor
	Sensitivity	
	2.5 μg/ml sulfuric acid	

Directions and Comments

The sample is mixed with water as needed and a few milligrams of methylenedisalicylic acid (CAS 122-25-8) are added. The mixture is heated to 150°C for 3 minutes. Formation of a red color indicates sulfuric acid. Phosphoric acid does not produce a red color but decreases sensitivity if sulfuric acid is present.

References	Manufacturer
Feigl, F., *Spot Tests in Inorganic Analysis,* Fifth Edition, Elsevier Publishing Company, New York, 1958.	

TATB Test

Detects	Test Principle	
• Explosive • TATB • Triaminotrinitrobenzene	A caustic solution of DMSO is able to dissolve TATB and then produce a red color.	
	Test Phase	
	✓	Solid
	✓	Liquid
		Gas/Vapor
	Sensitivity	
	Not available. Qualitative result only.	

(*continued*)

TATB Test (*Continued*)

Directions and Comments

TATB is not soluble in many other explosive compound tests and thus is not detected in those tests. The TATB test solution is 5 g of potassium hydroxide (CAS 1310-58-3), 5 ml of water, and 90 ml of DMSO (dimethylsulfoxide, CAS 67-68-5) shaken until mixed completely to form 100 ml of solution. The test is most sensitive when performed on filter paper. Place a drop of liquid sample or a few milligrams of solid sample on the paper. Add two drops of test solution. The development of a red color in a few minutes is presumptive for TATB. Shine a UV lamp (optimal wavelength is 254 nm) on the paper after a few minutes to cause a sharper contrast with any slight, positive red color. TATB as a PBX (polymer bonded explosive) produces a darker red color under UV light. This presumptive test for TATB was developed for use at Los Alamos National Laboratory. The paper filter may be used as a wipe to collect the sample.

References	Manufacturer
Baytos, J., *Field Spot-Test Kit for Explosives,* Los Alamos National Laboratory, Los Alamos, New Mexico, July 1991.	

Thiocyanate Conversion Test

Detects	Test Principle	
• Thiocyanate	Thiocyanate is oxidized by hydrogen peroxide or permanganate solution to form hydrogen cyanide. Free cyanide will be retained in solution if the pH is slightly basic or released as a gas if acidic.	
	Test Phase	
	✓	Solid
	✓	Liquid
		Gas/Vapor
	Sensitivity	
	Sensitive, but pH and reagent dependent	

Directions and Comments

Place 1 ml of sample solution in a test tube and add a few drops of hydrogen peroxide (CAS 7722-84-1) or potassium permanganate solution (CAS 7722-64-7), which will convert thiocyanate to hydrogen cyanide in solution (hydrocyanic acid). Use an appropriate cyanide test such as ITS Cyanide ReagentStrip™. Alternatively, add acid to the sample to release toxic hydrogen cyanide gas, which can be detected in the neck of the test tube with colorimetric air detector tube, such as Sensidyne Tube #112SB.

References	Manufacturer
Feigl, F., *Spot Tests in Inorganic Analysis,* Fifth Edition, Elsevier Publishing Company, New York, 1958.	

Thiocyanate Solution Test (Cobalt)

Detects	Test Principle	
• Cobalt	Thiocyanate reacts with cobalt to form a blue color.	
• Copper(II)	**Test Phase**	
• Iron(III)	✓	Solid
	✓	Liquid
		Gas/Vapor

(*continued*)

Thiocyanate Solution Test (Cobalt) (*Continued*)	
	Sensitivity
	0.03 μg/ml

Directions and Comments

Mix a 5% solution of cobalt chloride (CAS 7791-13-1) in water. Place a few milligrams of solid sample or a drop of liquid sample in a test tube and add a drop of test solution. Cobalt thiocyanate forms a blue colored complex (cobalt thiocyanate is used as a reagent in other tests). Ferric thiocyanate yields red; copper(II) thiocyanate produces red-brown. Addition of acetone to the aqueous solution may aid solubility as needed.

References	Manufacturer
Feigl, F., *Spot Tests in Inorganic Analysis,* Fifth Edition, Elsevier Publishing Company, New York, 1958.	

Trinitrophenylmethylnitramine Test		
Detects	**Test Principle**	
• 1,3,5-Trinitro-1,3,5-triazine • 2,4,6-Trinitrophenylmethylnitramine • Cyclonite • RDX • Tetryl	2,4,6-Trinitrophenylmethylnitramine forms nitrous acid when heated to 150°C. The nitrous acid is detected by a test strip.	
	Test Phase	
	✓	Solid
	✓	Liquid
		Gas/Vapor
	Sensitivity	
	Not available	

Directions and Comments

A few milligrams of suspected 2,4,6-trinitrophenylmethylnitramine (tetryl) is placed in a test tube and is heated in a glycerol (CAS 56-81-5) bath to 150°C. Nitrous acid formed from tetryl is detected at the opening of the tube with a Griess reagent and the appearance of a red color. Other explosives might contain nitrate and cause a confusing result if using a nitrate/nitrite test strip. Nitrate and nitrite can be differentiated by using a test strip that utilizes two pads to determine nitrate and nitrite, such as the Nitrate and Nitrite Test Strip [CZ NIT600]. Consult the package insert. This test has been used to differentiate tetryl (2,4,6-trinitrophenylmethylnitramine) from RDX or cyclonite (1,3,5-trinitro-1,3,5-triazine) because RDX does not yield nitrous acid until it ignites. Therefore, RDX can be detected through careful ignition of a few milligrams and the subsequent detection of nitrite if tetryl has already been ruled out.

References	Manufacturer
Feigl, F. and Hagenauer-Castro, D., *Chemist-Analyst* 51, 16 (1962). Jungreis, E., *Spot Test Analysis: Clinical, Environmental, Forensic, and Geochemical Applications,* Second Edition, John Wiley & Sons, Inc., New York, 1997.	

Trinitrotoluene Test	
Detects	**Test Principle**
• 2,4,6-Trinitrotoluene • 2,4-Dinitrotoluene • DNT	TNT and DNT react with tetramethylammonium hydroxide to form unique colors.

(*continued*)

Trinitrotoluene Test (*Continued*)		
• TNT • Explosive	**Test Phase**	
	✓	Solid
	✓	Liquid
		Gas/Vapor
	Sensitivity	
	2 µg/ml 2,4,6-trinitrotoluene	

Directions and Comments	
A drop of liquid sample or 5 to 10 mg of solid sample is placed on a spot plate. One drop of alcohol-acetone (1:1) is added to the sample followed by one drop of 25% aqueous solution of tetramethylammonium hydroxide (CAS 75-59-2). 2,4-Dinitrotoluene (DNT) forms a blue color; 2,4,6-trinitrotoluene (TNT) forms a red color. The test is selective to DNT and TNT through a wide range of explosive materials. The presence of nitroglycerine adds a yellow color. A mixture of DNT and nitroglycerin may yield a green color.	
References	**Manufacturer**
Amas, A. and Yallop, H., *Analyst* (London) 91, 336 (1966). Jungreis, E., *Spot Test Analysis: Clinical, Environmental, Forensic, and Geochemical Applications,* Second Edition, John Wiley & Sons, Inc., New York, 1997.	

Triphenylmethylarsonium Reagent		
Detects	**Test Principle**	
• Antimony • Barium • Lead	Antimony reacts with triphenylmethylarsonium iodide to form an orange precipitate.	
	Test Phase	
		Solid
	✓	Liquid
		Gas/Vapor
	Sensitivity	

Directions and Comments	
The test solution is a 10% alcoholic solution of triphenylmethylarsonium iodide (CAS 1499-33-8). Mix a drop of the sample with a drop of the test solution; an orange precipitate forms in 30 to 120 seconds if antimony is present. The test may be performed as a swab; see Firearm Discharge Residue Test.	
References	**Manufacturer**
Harrison, H. and Gilroy, R., *J. Forensic Sci.* 4, 184 (1959). Jungreis, E., *Spot Test Analysis: Clinical, Environmental, Forensic, and Geochemical Applications,* Second Edition, John Wiley & Sons, Inc., New York, 1997.	

Universal pH Test Strip	
Detects	**Test Principle**
• Acid • Base	A single pad contains several pH sensitive dyes that produce unique colors through pH intervals of 0 to 14.

(continued)

Universal pH Test Strip (*Continued*)

Caustic / Corrosive	Test Phase	
• Caustic	✓	Solid
• Corrosive	✓	Liquid
		Gas/Vapor
	Sensitivity	
	+/− 1 pH interval	

Directions and Comments

The test strip pad uses a nonbleed system and does not disperse dye into the sample. The sample must contain water for the test to detect pH. The strip is simply wetted with the sample and the resulting color matched to a chart to determine pH. A pH test strip moistened with deionized water will detect acidic or basic water soluble gases and vapors in air.

References	Manufacturer
Kimel, H., Notes from correspondence, Precision Laboratories, Inc., Cottonwood, Arizona, USA, December 26, 2007.	Precision Laboratories, Inc., Cottonwood, Arizona, USA

Unsaturated Hydrocarbon Test

Detects	Test Principle	
• Aliphatic amine	Bromine reacts with unsaturated alkyl bonds, thus decolorizing the light brown solution.	
• Alkene	**Test Phase**	
• Alkyne	✓	Solid
• Enol	✓	Liquid
• Methyl ethyl ketone		Gas/Vapor
• Phenol	**Sensitivity**	
• Unsaturated hydrocarbon	Not available	

Directions and Comments

Mix a 2% to 3% aqueous solution (v/v) of bromine (CAS 7726-95-6). Dissolve 0.1 g of solid sample or two drops of liquid sample in carbon tetrachloride (CAS 56-23-5, carcinogenic). Add the test solution by the drop with agitation to the sample solution until the bromine color persists. If hydrogen bromide gas evolves, a substitution reaction is occurring and does not prove the sample is unsaturated. This result suggests enols, many phenols, aliphatic amines, and particularly methyl ethyl ketone; simple esters do not react. If the bromine color does not persist initially and hydrogen bromide gas does not evolve, an addition reaction is occurring and the sample is unsaturated. Aromatic amines are a special case and not readily detected in the field. Compare to Bromine Water.

References	Manufacturer
Welcher, F., *Chemical Solutions: Reagents Useful to the Chemist, Biologist, and Bacteriologist,* D. Van Nostrand Company, Inc., New York, 1942. Shriner, R., Fuson, R. and Curtin, D., *The Systematic Identification of Organic Compounds, A Laboratory Manual,* Fifth Edition, John Wiley and Sons, Inc., New York, 1964. *J. Chem. Educ.,* 37, 205 (1960).	

Van Eck's Reagent

Detects	Test Principle		
• Chromic Acid • Chromium(VI)	a-Naphthylamine and tartaric acid react with chromic acid to form a blue color.		
	Test Phase		
	✓	Solid	
	✓	Liquid	
		Gas/Vapor	
	Sensitivity		
	0.001 mg/ml chromic acid		

Directions and Comments

Mix an aqueous solution (w/v) consisting of 0.5% a-naphthylamine (CAS 134-32-7) and 50% tartaric acid (CAS 87-69-4). The test solution is colored blue by chromic acid and chromic salts in acid solution. Undetected are chromous cation and chromate anion. The test may be performed on a spot plate, filter paper, or in a test tube.

References	Manufacturer
Welcher, F., *Chemical Solutions: Reagents Useful to the Chemist, Biologist, and Bacteriologist,* D. Van Nostrand Company, Inc., New York, 1942.	

Villiers-Fayolle's Reagent

Detects	Test Principle		
• Chlorine	o-Toluidine reacts with aniline, acetic acid, and free chlorine to produce a blue color.		
	Test Phase		
		Solid	
	✓	Liquid	
		Gas/Vapor	
	Sensitivity		
	Not available		

Directions and Comments

Measure 2 ml of water into a test tube and add o-toluidine (CAS 95-53-4) to form a saturated solution. Measure 10 ml of water into a small beaker and add aniline (CAS 62-53-3) to form a saturated solution, decant the aqueous solution, and add 3 ml of acetic acid (CAS 64-19-7). Decant the 2 ml solution of o-toluidine and mix with the aniline solution to form the test solution. Mix a drop of sample solution and a drop of test solution. A blue color forms in the presence of free chlorine.

References	Manufacturer
Welcher, F., *Chemical Solutions: Reagents Useful to the Chemist, Biologist, and Bacteriologist,* D. Van Nostrand Company, Inc., New York, 1942. *Comptes rendus hebdomadaires des seances de l'académie des sciences,* 118, 1413 (1894).	

Water Test	
Detects	**Test Principle**
• Water	Anhydrous citric acid and sodium bicarbonate react with water to produce effervescence of carbon dioxide.

Test Phase	
✓	Solid
✓	Liquid
	Gas/Vapor

Sensitivity
Approximately 10% water

Directions and Comments

Use equal volumes of citric acid (CAS 77-92-9) and sodium bicarbonate (CAS 144-55-8) powders in the base of a test tube. Add enough of the liquid sample to cover the powders and watch for formation of bubbles. Aggressive effervescence indicates relatively high concentration of water. Formation of small bubbles at the surface of the liquid should not be confused with air displaced from the powder; watch for the bubbles to slowly form. An alternative is to use Alka-Seltzer, which will mildly effervesce at about 10% water; at less than 10% water the reaction is difficult to recognize. This mild effervescence appears as a fine trail of closely spaced bubbles and is different from the appearance of trapped air clinging to portions of the crushed tablet. These small bubbles may soon cease in low concentrations of water. Effervescence increases as the water concentration increases, and the solution will aggressively effervesce at about 25% or more water. Reagents should be kept sealed and dry as it will absorb moisture from the air and lessen the strength of the effervescing reaction. See color images 2.50 through 2.54.

References	**Manufacturer**
Houghton, R., *Emergency Characterization of Unknown Materials,* CRC Press, 2007.	Bayer HealthCare LLC, Morristown, New Jersey, USA

Watersafe® Lead Test	
Detects	**Test Principle**
• Lead(II)	The test uses lateral flow immunochromatography and monoclonal antibodies that are specifically attracted to lead(II).

Test Phase	
	Solid
✓	Liquid
	Gas/Vapor

Sensitivity
15 ppb lead

Directions and Comments

The test is specific to lead(II). The test is a patent-pending lateral flow immunochromatography test that uses proprietary monoclonal antibody. The test gives a positive/negative result with the transition point calibrated to approximately 15 ppb in drinking water, the EPA action level. The test is not quantitative, but there is no upper limit and the test is not susceptible to hook effect. The operating temperature range is 50°F to 86°F (10°C to 30°C) since it uses biological material. The test is not affected by pH as found in drinking water, but the test will be ineffective in extreme pH. A premeasured sample is placed in a container and mixed with a buffer that has been sprayed on the sidewall of the container. The lateral flow strip is placed in the liquid

(continued)

Directions and Comments (*Continued*)	
and left undisturbed for 10 minutes. The test contains a control to ensure response. The test can be used to detect lead(I) and organolead if the sample is first converted to lead(II). See color images 2.55 and 2.56.	
References	**Manufacturer**
Geisberg, M., *Personal communication,* Silver Lake Research Corporation, Monrovia, California, June 30, 2008.	Silver Lake Research Corporation, Monrovia, California, USA (www.silverlakeresearch.com). Industrial Test Systems, Inc., Rock Hill, South Carolina, USA (www.sensafe.com).

Zirconium-Alizarin Reagent		
Detects	**Test Principle**	
• Fluoride • Hydrofluoric acid	Fluoride added to a red complex of zirconium-alizarin in hydrochloric acid will decompose the complex and produce a yellow color.	
	Test Phase	
	✓	Solid
	✓	Liquid
		Gas/Vapor
	Sensitivity	
	1 µg/ml fluorine	

Directions and Comments

When testing for larger amounts of fluoride the test may be performed on paper or directly in the violet solution in a test tube or spot plate. To form the reagent, mix a 0.05 g of zirconium nitrate (CAS 13826-66-9, zirconium chloride, CAS 10026-11-6 may be substituted) in 50 ml of water and 10 ml of concentrated hydrochloric acid. Separately, mix 0.05 g of sodium alizarin sulfonate (CAS 130-22-3) in 50 ml of water and then combine the two solutions to form the test solution. Mix a drop of sample solution with a drop of violet test solution; a yellow color indicates fluoride. If very low concentration of fluoride is suspected, add more sample solution to the violet test solution. The test must be performed in acidic conditions (hydrochloric acid). To make test strips, simply soak the paper in a 5% zirconium nitrate solution, drain, and then dip it in a 2% aqueous solution of sodium alizarin sulfonate. The paper will take on the red color. Rinse the paper with water until the rinse water is clear and then dry the paper, which will be more or less red. A drop of the aqueous sample is placed on the red paper; yellowing indicates fluoride (check both sides of the paper). Fluorine in complex ions, such as boron fluoride and silicofluoride, react similarly. Very large amounts of sulfates, thiosulfates, phosphates, arsenates, or oxalates can interfere. Insoluble fluorides in solid form may be tested directly with a drop or two of the test solution or may be made soluble by the addition of borax and hydrochloric acid. Alternately, and with reduced sensitivity, mix a solution of 0.5% zirconium chloride (CAS 10026-11-6) and 0.5% sodium alizarin sulfonate with some hydrochloric acid and test the sample directly. See color images 2.57 through 2.60.

References	**Manufacturer**
Feigl, F., *Spot Tests in Inorganic Analysis,* Fifth Edition, Elsevier Publishing Company, New York, 1958. Feigl, F. and Anger, V., *Spot Tests in Inorganic Analysis,* Sixth Edition, Elsevier Publishing Company, Amsterdam, 1972. *Anal. Abstracts,* 16 (1969).	

Zwikker Reagent

Detects	Test Principle
• Barbiturate • Pentobarbital • Phenobarbital • Secobarbital • Tea • Tobacco	Zwikker Reagent uses a two-step process to complex reduced copper(II) with barbiturates in the presence of pyridine to form light purple reaction products.

Test Phase	
✓	Solid
✓	Liquid
	Gas/Vapor

Sensitivity
1000 µg phenobarbital

Directions and Comments

This test is a presumptive field test for illegal drugs. Zwikker Reagent consists of two solutions. Solution A is a 0.5% solution of copper sulfate (CAS 7758-99-8) in water (w/v). Solution B is 5 ml of pyridine (CAS 110-86-1) in 95 ml of chloroform (CAS 67-66-3). The sample is mixed in a spot plate or test tube with one volume of Solution A followed by an equal volume of Solution B. Light purple indicates barbiturates. Other colors are possible. It is most helpful to compare any color formed in this test with results from other presumptive drug tests (Marquis Reagent, etc.) in a matrix to determine a pattern of unique results. Refer to the appendix for drug screen results. Proteins may form purple compounds when mixed with Solution A; see Biuret Protein Test.

References	Manufacturer
U.S. Department of Justice. Color Test Reagents/Kits for Preliminary Identification of Drugs of Abuse, NIJ Standard–0604.01, National Institute of Justice, National Law Enforcement and Corrections Technology Center, Rockville, MD, July 2000.	Several commercial test kits are available.

Zwikker's Reagent

Detects	Test Principle
• Chromate • Perchlorate • Permanganate • Persulfate • Thiosulfate	An aqueous solution of cupric sulfate and pyridine yields characteristic crystals with certain ions.

Test Phase	
	Solid
	Liquid
	Gas/Vapor

Sensitivity
Not available

Directions and Comments

Mix 4 ml of a 10% solution (w/v) of cupric sulfate pentahydrate (CAS 7758-99-8), 1 ml of pyridine (CAS 110-86-1), and 5 ml of water. The test solution gives characteristic crystals with perchlorate, persulfate, thiosulfate chromate, permanganate, and other anions.

References	Manufacturer
Welcher, F., *Chemical Solutions: Reagents Useful to the Chemist, Biologist, and Bacteriologist,* D. Van Nostrand Company, Inc., New York, 1942. *Chemical Abstracts* 34, 5785 (1940).	

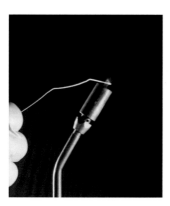

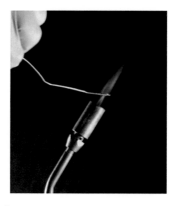

Color image 1.2 Arsenic produces a blue color with areas of lighter blue to lavender. (Chapter 1 – Heating by Torch.)

Color image 1.1 A non-reactive metal wire (in this case a paperclip) is heated in the torch flame until any interfering flame color becomes absent. Skin perspiration contains sodium, which produces a yellow-orange flame color that is very difficult to remove. (Chapter 1 – Heating by Torch.)

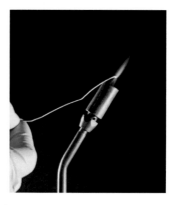

Color image 1.4 Boron produces a strong green color. (Chapter 1 – Heating by Torch.)

Color image 1.3 Lead produces a blue-gray color. Compare to arsenic. (Chapter 1 – Heating by Torch.)

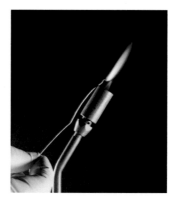

Color image 1.5 Copper produces a green color that is similar to boron. (Chapter 1 – Heating by Torch.)

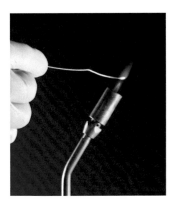

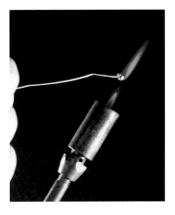

Color image 1.7 Calcium produces an orange color that is redder and less persistent than sodium. (Chapter 1 – Heating by Torch.)

Color image 1.6 Potassium produces a lavender color. (Chapter 1 – Heating by Torch.)

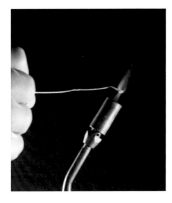

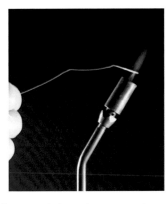

Color image 1.9 Barium produces a weak green-yellow color that is difficult to see in the presence of other flame color. (Chapter 1 – Heating by Torch.)

Color image 1.8 Lithium produces a brilliant red color. (Chapter 1 – Heating by Torch.)

Color image 1.10 Sodium produces an orange color that is persistent. (Chapter 1 – Heating by Torch.)

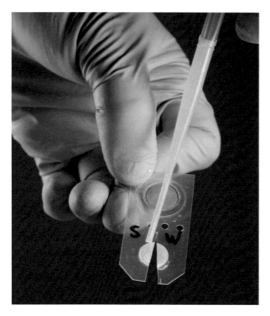

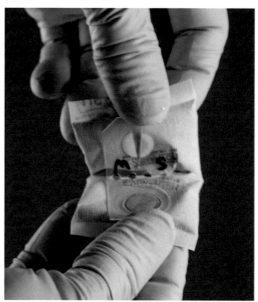

Color image 2.1 The Agri-Screen® Ticket uses acetylcholinesterase to convert indoxyl acetate in the purple pad to a blue colored reaction product that will collect on the white pad. Inhibiting compounds block this conversion and leave the test pad white. High temperature can deactivate the acetylcholinesterase and cause a false positive result. Notching the test provides a control when one side is tested with the sample (S) and the other side is tested with water (W).

Color image 2.2 The test is allowed to soak for one minute. More water or sample is added to the respective side of the white pad as needed to assure adequate moisture before folding the test for three minutes. An air-activated hand warmer can be used to assure the test reaches body temperature in cool weather; thinly gloved fingers supply adequate warmth.

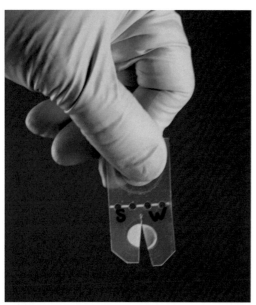

Color image 2.3 This test shows a positive result for an acetylcholinesterase inhibiting compound; the sample side was inhibited from producing a blue color and the water side was successful in converting indoxyl acetate to a blue colored reaction product. If both halves of the white test pad remain white the test is invalid. If both sides are blue, no acetylcholinesterase compound is present. (Chapter 2 – Agri-Screen® Ticket.)

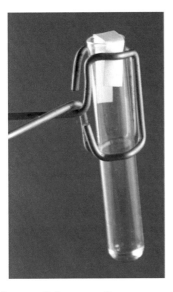

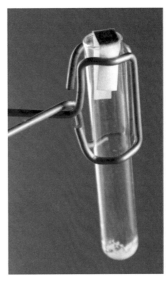

Color image 2.4 A small amount of liquid sample suspected of containing thiocyanate is mixed with solid ammonium chloride in the base of the test tube. A strip of dry lead acetate paper is placed over the opening of the test tube and held in place with a test tube clamp.

Color image 2.5 The test is heated until all liquid is boiled away and then heated a little more (200-300°C). Production of hydrogen sulfide from thiocyanate is detected by production of a black color in the lead acetate paper. (Chapter 2 – Ammonium Chloride Reagent.)

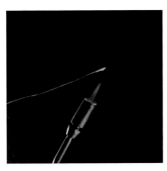

Color image 2.6 A copper wire is heated in a propane torch flame until the wire does not discolor the flame. A long coiled wire will allow longer test time before the wire becomes too hot to hold. The end of the wire can also be insulated.

Color image 2.7 With the red hot copper wire still in the flame, a cotton swab soaked with sample is brought near the air intake of the torch head. Organic molecules containing chlorine, bromine or iodine react with copper oxide in the torch flame to produce volatile copper halides that will display a characteristic green or blue flame. (Chapter 2 – Beilstein Test.)

Color image 2.8 A polypropylene swab is moistened with copper sulfate solution and used to pick up a small amount of powder. The swab is then carefully moistened with BCA so that the liquid gently mixes with the copper sulfate solution and powder. The BCA will react with the copper sulfate to form a bright green color.

Color image 2.9 Any protein that contacts the green BCA/copper sulfate complexes to form a purple color. Protein concentration as low as 1% is visible as a purple color within three minutes. (Chapter 2 – Bicinchoninic acid [BCA] Protein Test.)

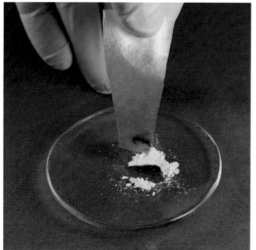

Color image 2.10 Filter paper moistened with blue Biuret solution is used to test a sample suspected of containing protein. The filter paper allows a small amount of powder to be tested. A pink-violet to purple-violet color from the original blue indicates protein. Biuret solution may also be dried on paper for storage and wetted with water before use. (Chapter 2 – Biuret Paper.)

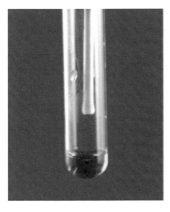

Color image 2.11 A liquid suspected of containing unsaturated hydrocarbon is about to be added to yellow-brown bromine water.

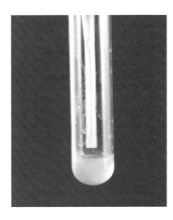

Color image 2.12 After gentle agitation the test solution clears due to reaction of bromine with unsaturated hydrocarbon compounds. (Chapter 2 – Bromine Water Solution.)

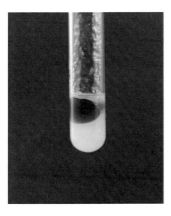

Color image 2.13 Red cobalt thiocyanate reagent solution has just been added to an oily substance suspected of containing an alkaloid.

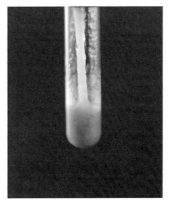

Color image 2.14 Gentle agitation produces the blue result characteristic of a positive result. (Chapter 2 – Cobalt Thiocyanate Reagent.)

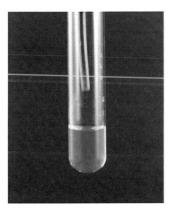

Color image 2.15 A drop of liquid suspected of containing lead is about to be added to the test solution.

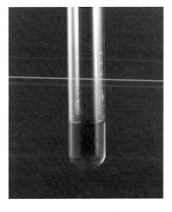

Color image 2.16 A red color develops in the test solution indicating the presence of lead. (Chapter 2 – Dithizone Reagent [Lead].)

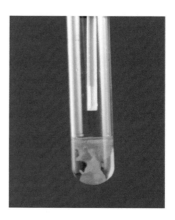

Color image 2.17 Dragendorff's Solution is shown after mixing colorless Solution A and colorless Solution B to form the clear orange test solution.

Color image 2.18 The appearance of a brilliant, opaque orange color indicates a secondary, tertiary or quaternary amine.

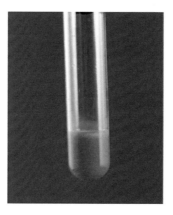

Color image 2.19 Gentle agitation uniformly distributes the brilliant, opaque orange color throughout the solution. (Chapter 2 – Dragendorff's Solution.)

Color image 2.20 Brown test solution is added to a powder suspected of containing ferrocyanide.

Color image 2.21 The appearance of a blue color confirms ferrocyanide. (Chapter 2 – Ferric Chloride Solution.)

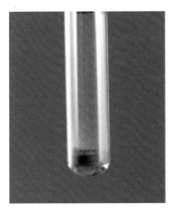

Color image 2.22 A dry sample suspected of containing bicarbonate or carbonate is shown in the test tube.

Color image 2.23 The addition of acid produces effervescence, indicating bicarbonate or carbonate. (Chapter 2 – Fizz Test.)

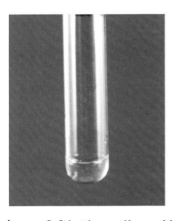

Color image 2.24 Clear yellow gold chloride test solution is shown in the test tube.

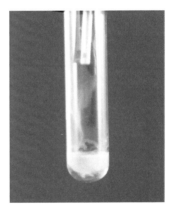

Color image 2.25 A solution suspected of containing a thiol (mercaptan) compound is added to the test solution. The appearance of an opaque yellow or orange color indicates thiol. (Chapter 2 – Gold Chloride Reagent.)

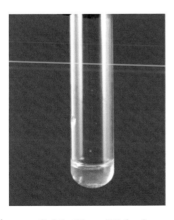

Color image 2.26 Clear 3% hydrogen peroxide (medical grade) is placed in a test tube.

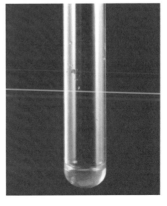

Color image 2.27 Addition of a titanium salt produces a yellow color. Molybdate can also produce a yellow color. (Chapter 2 – Hydrogen Peroxide Reagent.)

Color image 2.28 The test detects free cyanide in solution by using reagents imbedded in two simple test strips that are stored separately for field use.

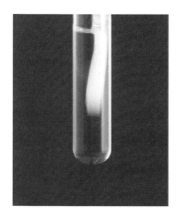

Color image 2.29 The sample is dissolved in water. The first test strip containing a buffer and chlorinating agent convert cyanide to cyanogen chloride within a few seconds of repeatedly dipping the test strip in the solution.

Color image 2.30 The second test strip contains isonicotinic acid and barbituric acid which in the presence of cyanide will produce a blue color on the test strip. (Chapter 2 – ITS Cyanide ReagentStrip™.)

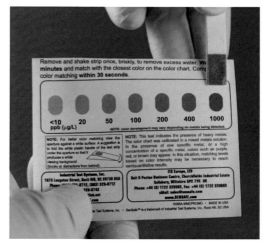

Color image 2.31 The test is a screen for heavy metals in water. Various colors are possible depending on the metal and concentration. (Chapter 2 – ITS SenSafe™ Water Metal Test.)

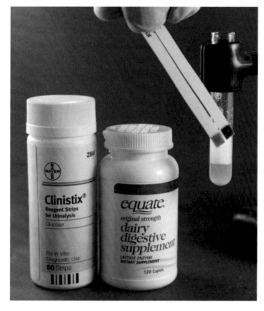

Color image 2.32 Solid and liquid samples suspected of containing dairy products may be tested for lactose in aqueous solution. Agitation of aqueous solution of protein will often cause foaming. A glucose test strip is used to determine a baseline glucose concentration for the solution.

Color image 2.33 A small amount of beta-galactosidase (lactase; Lactaid®) is mixed in the solution and a second glucose test strip is used to determine any increase in glucose. Compare the first test strip to the second. Any increase in the production of a purple color indicates a positive result. An increase in glucose indicates the presence of lactase and confirms a dairy product. (Chapter 2 – Lactase Test.)

Color image 2.34 This lead acetate test paper was exposed to sulfide to produce the black area of lead sulfide. The paper towel wicks away excess moisture.

Color image 2.35 Addition of hydrogen peroxide will clear the black color when lead sulfide is oxidized. (Chapter 2 – Lead Sulfide Paper.)

Color image 2.36 Clear blue test solution is shown in a test tube. The test solution may be performed on paper or applied directly to suspected perchlorate compounds.

Color image 2.37 A dilute solution of perchlorate is added to methylene blue test solution to produce a color change to purple. (Chapter 2 – Methylene Blue Reagent.)

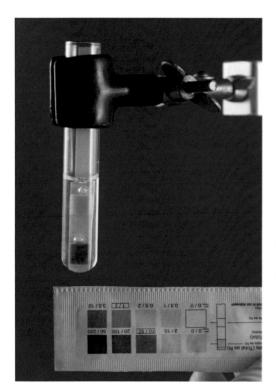

Color image 2.39 The brown test paper is suspended over a sample suspected of containing elemental mercury for 10 minutes at 25°C.

Color image 2.38 The test strip is used to confirm the presence of nitrite and/or nitrate in an aqueous solution. In this case, the solution contains greater than 200 ppm nitrate and no detectable nitrite. Note the test strip orientation diagram on the package to discern one test pad from the other. (Chapter 2 – Nitrite and Nitrate Test Strip [CZNIT600].)

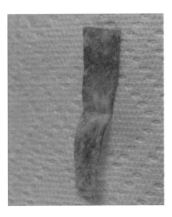

Color image 2.40 Next, the paper strip was held over the neck of a container of ammonium hydroxide to remove the original brown color. The resulting color change from brown to black/gray confirms the presence of mercury vapor; a white color would indicate a negative result. The brown portion of the strip was not exposed to the ammonia vapors. (Chapter 2 – Palladium Chloride Paper.)

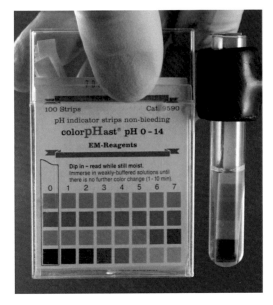

Color image 2.41 A universal pH test strip indicates this solution as pH 1. A drop of liquid sample (that must contain a little water) may be added directly to the pH test strip with a pipette. A pH test strip moistened with neutral water may be used to test air for corrosive vapor, although the accuracy is degraded. (Chapter 2 – Universal pH Test Strip.)

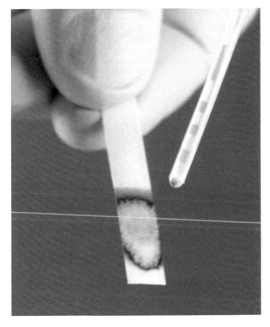

Color image 2.42 An aqueous solution suspected of containing gold(III) is applied to a potassium iodide-starch test strip to produce a blue color; gold is indicated as well as some oxidizers.

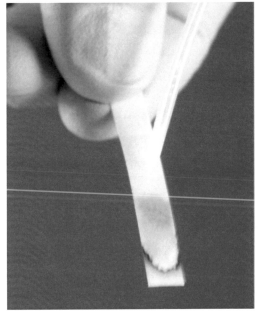

Color image 2.43 A drop of sodium sulfide solution is added to the top of the strip and has cleared the blue color at the top.

Color image 2.44 All blue color has been cleared by the sulfide solution as it reaches the lower end of the strip. (Chapter 2 – Potassium Iodide Reagent [Gold].)

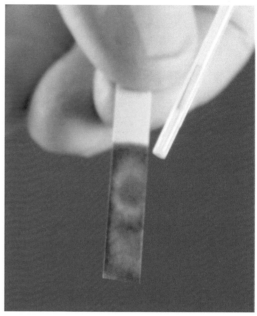

Color image 2.45. The test strip is used to detect various oxidizers (3% hydrogen peroxide in this case) with a color change from white to blue/black.

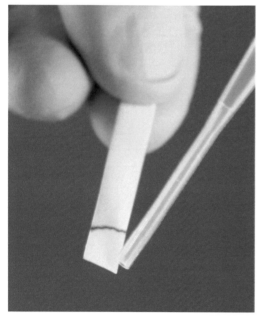

Color image 2.46 Stronger oxidizers (5% sodium hypochlorite in this case) will cause a color change from white to blue/black and then immediately bleach the blue/black color to white. This can be seen as a thin black line leading the liquid as it is absorbed through the strip. (Chapter 2 – Potassium Iodide Starch Test Strip [JD260].)

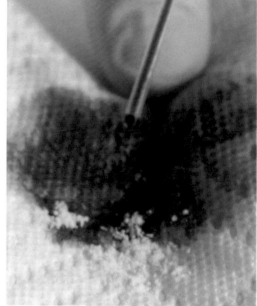

Color image 2.47 Povidone (Betadine®) is diluted 10:1 and applied to a solid sample on paper. The paper spreads the dark povidone into a thin layer that provides a better contrast between the original brown and the blue/black produced by starch. (Chapter 2 – Povidone Test for Starch.)

Color image 2.48 A solution of 2% silver nitrate in water is clear.

Color image 2.49 A liquid sample is added to the silver nitrate test solution. The immediate production of a thick, opaque white indicates chloride. A slow production of clouded white that forms evenly throughout the solution over a few minutes can indicate the presence of an organic compound containing reactive chloride. (Chapter 2 – Silver Nitrate Reagent.)

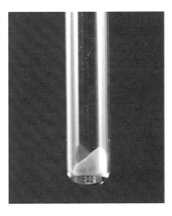

Color image 2.50 About 0.5 g of Alka-Seltzer™ is placed in a test tube as a lump or powder.

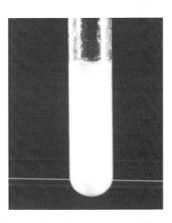

Color image 2.51 Aggressive effervescence is produced by distilled water.

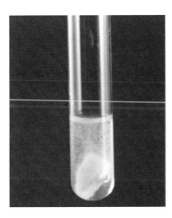

Color image 2.52 The rate of effervescence is proportional to the water content. This is a 50% solution of methanol in water.

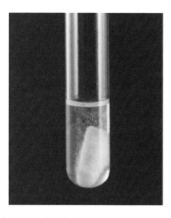

Color image 2.53 Pictured is a solution of 75% methanol in water.

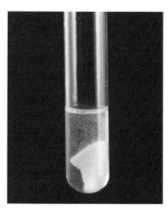

Color image 2.54 Pictured is a solution of 90% methanol in water. The test can detect lower concentrations of water. Any trail of small bubbles from the Alka-Seltzer™ indicates the presence of water. (Chapter 2 – Water Test.)

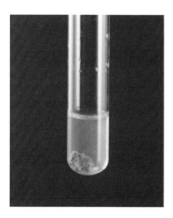

Color image 2.55 A piece of paint suspected of containing lead is exposed to 3N hydrochloric acid for a few minutes. A few drops of this solution was neutralized and then used as the sample solution.

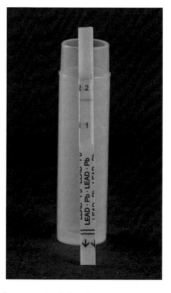

Color image 2.56 The test was performed in the plastic container shown here, which contains a buffer. The test is performed when the lateral flow immunochromatography test strip is immersed in the buffered solution. Monoclonal antibodies that are specifically attracted to lead(II) are detected at line 1 if lead(II) is present. Line 2 is a control. Although line 1 is faintly blue, lead is indicated above the transition point of 15 ppb. (Chapter 2 – Watersafe® Lead Test.)

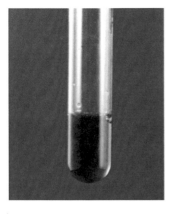

Color image 2.57 Clear, red zirconium-alizarin test solution.

Color image 2.58 Fluoride will react with the zirconium-alizarin complex and de-color the red complex to a yellow solution.

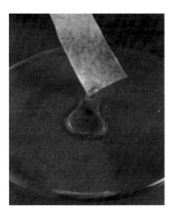

Color image 2.59 The test can be performed on paper by drying the red zirconium-alizarin test solution on filter paper and touching the paper to an aqueous solution.

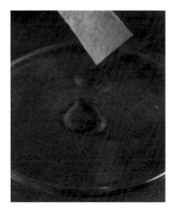

Color image 2.60 A positive result is indicated on paper when the red complex reacts to form a yellow color. (Chapter 2 – Zirconium-Alizarin Reagent.)

Instrumentation

3

Introduction

This chapter describes portable instrumentation technology that is useful for field confirmation testing. Descriptions include technology in general, but practically some instruments utilize more than one technology. Other instruments are mission-specific.

Mention of specific models is not intended to represent endorsement. Other manufacturers may have instruments available that will underperform or overperform in certain applications. Always consult the manual or manufacturer for capabilities of a specific model.

Where practical a list of material detected is included with the description of each technology. Some technologies are capable of detecting many more items than can be listed in this brief chapter, for example gas chromatography–mass spectrometry (GC-MS), Fourier transform infrared (FTIR) spectroscopy, and Raman spectroscopy.

The Department of Homeland Security has funded a guide for the selection of chemical detection equipment by first responders in two volumes. *Guide for the Selection of Chemical Agent and Toxic Industrial Material Detection Equipment for First Responders, Second Edition, Volumes I and II,* are resources provided to help determine which detection products meet your application and need. At the time of this printing the second edition is the latest version for both documents (March 2005). The guides assess commercially available equipment as tested against standards of the National Institute of Standards and Technology (NIST).

Access to these documents is limited, and they are posted at the SAVER web site, http://saver.tamu.edu/. Consult the Web site to determine accessibility or contact the SAVER office at 866-674-3251.

Chemical Detection

Electrochemical Sensors and Multisensor Monitors

Multisensor monitors use combinations of sensors to detect hazards. The most common configuration is a combustible gas sensor, oxygen sensor, carbon monoxide sensor, and hydrogen sulfide sensor (Figure 3.1). Other configurations include fewer or greater than four sensors and combinations of specialty sensors demanded by the application. Compounds that can be detected by properly configured electrochemical sensors in a multisensor monitor include:

- Amines
- Ammonia

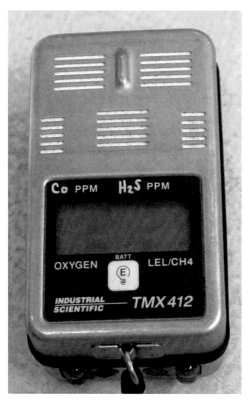

Figure 3.1 The Industrial Scientific Model TMX412 (www.indsci.com) is an example of a multisensor instrument configured to detect oxygen concentration, combustible gases and vapors, carbon monoxide, and hydrogen sulfide.

- Carbon dioxide
- Carbon monoxide
- Chlorine
- Combustible gases
- Hydrazine
- Hydrides
- Hydrogen
- Hydrogen chloride
- Hydrogen cyanide
- Hydrogen fluoride
- Hydrogen peroxide
- Hydrogen sulfide
- Nitric oxide
- Nitrogen dioxide
- Organic vapor
- Oxygen
- Phosphine
- Sulfur dioxide
- Specialized gases

All of these sensors are designed to detect gases and vapors only. Sensor contact with liquids or solids will often render the monitor inoperable. The detection principles include:

- **Combustible gases:** Flammable vapors are detected by two catalytic beads, one of which is exposed to the sample atmosphere; the other bead is a control. The temperature difference between the control bead and the sample bead is measured. Flammable vapor or gas will increase the heating on the sample bead. Quantitative readings may be possible depending on the calibration gas and related response factors. Sensitivity is 1% LEL of the calibration gas.
- **Oxygen:** Oxygen is detected by an electrochemical sensor that uses a reducible polymer to measure the concentration of oxygen in air. Sensitivity is 0.1% oxygen in air.
- **Individual gases:** Specific gases are detected by using an electrochemical sensor that contains a chemical substrate sensitive to oxidation or reduction by the target gas. A signal is generated and each sensor is calibrated to a standard gas concentration. Some sensors use filtering media to eliminate interfering gases. Sensors are often cross-sensitive to several nontargeted compounds. Sensitivity varies but is generally 1 ppm.

Combustible gas and oxygen sensors detect gross changes in target gas or vapor levels. Individual gas sensors are often used presumptively since they are cross-sensitive to several nontargeted substances. For example, most carbon monoxide sensors are cross-sensitive to alcohol. This cross-sensitivity makes verification testing with the instrument difficult, and results should be viewed as suggestive rather than definitive. Each manufacturer can produce its own electrochemical sensors, or use a sensor supplied by a vendor. Several combinations of sensor materials can be indicated, which will determine cross-sensitivities, false positives, and so on. Quantitative results are often temperature dependent. Consult the instrument manual to determine interfering substances.

Metal oxide semiconductor (MOS) sensors are sometimes encountered in detection equipment. These detectors measure small changes in temperature and electrical conductivity as the result of gas adsorption on the metal oxide surface. MOS is a nonselective device; contaminants have different thermal conductivities and responses. These are mostly useful for measuring changes from normal background reading when the contaminant is known.

Refrigerant Detector

Refrigerant detectors are a low cost and effective tool for screening and presumptive monitoring; however, they are only marginally useful in confirmation testing. Refrigerant detectors qualitatively detect part-per-million levels or less of halo-hydrocarbon refrigerant gas. The device alerts the user to an increasing concentration of gas relative to its last reset point. Relatively large concentrations of other gases or vapors may induce a false positive result. Detectors may use conductive polymer ionization, corona discharge, heated diode, or infrared methods of detection. Consult the manual to determine detection limitations.

Refrigerant detectors are designed to detect halogenated hydrocarbon gases and vapors, such as:

- 1,1,1,2,3,3,3-Heptafluoropropane
- 1,1,1,2-Tetrafluoroethane
- 1,1,1-Trichloro-2,2,2-trifluoroethane
- 1,1,1-Trichloroethane
- 1,1,2,2,2-Pentafluoroethane
- 1,1,2,2-Tetrafluoroethane
- 1,1,2-Trichloro-1,2,2-trifluoroethane
- 1,1-Dichloro-1-fluoroethane
- 1,1-Dichloroethane
- 1,1-Difluoroethane
- 1,2-Dichloro-1,1,2,2-tetrafluoroethane
- 1,2-Dichloroethane
- 1-Chloro-1,1,2,2,2-pentafluoroethane
- 1-Chloro-1,1-difluoroethane
- 2-Chloro-1,1,1,2-tetrafluoroethane
- Bromochlorodifluoromethane
- Bromoform
- Bromotrifluoromethane
- Carbon tetrachloride
- Carbon tetrafluoride
- CFC-10
- CFC-11
- CFC-110
- CFC-113
- CFC-113a
- CFC-114
- CFC-115
- CFC-12
- CFC-13
- CFC-14
- Chlorodifluoromethane
- Chloroform
- Chloromethane
- Chloropentafluoroethane
- Chlorotrifluoromethane
- Decafluorobutane
- Dibromomethane
- Dichlorodifluoromethane
- Dichloromethane
- Dichlorotetrafluoroethane
- Difluoromethane
- Ethylene dichloride
- Ethylene oxide with Halon
- Ethylidene dichloride

- Fluoroform
- Fluoromethane
- Freon 13T1
- Freon 150
- Freon 150a
- Freon-11, R-11
- Freon-12, R-12
- Halon 1211
- Halon 1211
- Halon 1301
- Halon 1301
- HCFC-141b
- HCFC-142b
- HCFC-22
- Hexachloroethane
- HFC-124
- HFC-125
- HFC-134
- HFC-134
- HFC-152a
- HFC-227ea
- HFC-23
- HFC-32
- HFC-41
- Methyl chloride
- Methyl chloroform
- Methyl fluoride
- Methylene bromide
- Methylene chloride
- PBB
- PCB
- Pentafluoroethane
- Perfluorobutane
- Polybrominated biphenyl
- Polychlorinated biphenyl
- Polychloroethene
- Polytetrafluoroethene
- Polyvinyl chloride, PVC
- PTFE, Teflon
- R-134a
- R-22
- R610
- SF-6
- Sulfur hexafluoride
- Tetrachloromethane
- Tetrafluoromethane
- Tribromomethane

- Trichlorofluoromethane
- Trichloromethane
- Trichlorotrifluoroethane
- Trifluoroiodomethane
- Trifluoromethane
- Trifluoromethyl iodide

Most refrigerant detectors are sensitive and can often detect less than 1 ppm chlorohydrocarbon and less than 100 ppm fluorohydrocarbon. For example, the TIF XP-1A Automatic Halogen Leak Detector (Figure 3.2) specifies maximum sensitivity per SAE J1627 rating criteria and is certified for R-12, R-22, and R-134a @ 0.5 oz/yr (14 gr/yr). Ultimate sensitivity is less than 0.1 oz/yr as leaking refrigerant.

Refrigerant detectors produce qualitative results only. Sensors may be cross-sensitive to other compounds, but will require larger concentrations to induce a false positive result. Older detectors, with mainly corona discharge sensors, were used to detect chlorinated hydrocarbons. New, less ozone-damaging Freons, in which fluorine replaces chlorine in a hydrocarbon molecule, still responded to the same detection techniques but are detected 20 to 100 times less sensitive than chlorinated Freons.

Exposing the sensor to a steady stream of Freon will damage the sensor, and sensor life is directly proportional to the amount of Freon that passes over the sensor. Approximate sensor service life is up to 150 hours per year. Sensors are located in the tip of the wand and

Figure 3.2 An example of a refrigerant gas detector is the TIF XP-1A Automatic Halogen Leak Detector (www.tif.com). TIF is a registered trademark of SPX Corporation. (Image courtesy of TIF.)

are easily changed if overloaded. Extra sealed sensors are essential to keeping a refrigerant detector in service.

False positive results may be caused by abrupt changes in sensor temperature caused by a sudden change in air flow past the sensor, sudden cooling, or the sensor being heated by an outside source. Keep the sensor dry, shielded from wind (while allowing continuous intake), and avoid sudden changes in temperature. The sensors are cross-sensitive to some interfering gases and vapors, but higher concentration of nonhalogenated compounds are required to induce a false positive result.

Do not assume that an initial response by a refrigerant detector immediately indicates a halogenated hydrocarbon. Do not use this technology for the detection of chlorine- and fluorine-containing chemical warfare agents as the detection threshold is above IDLH.

Electron Capture Device

An electron capture device (ECD) emits a steady stream of electrons that can be absorbed by electrophilic compounds. Often the current is produced by a small radioactive nickel-63 source that emits a steady stream of electrons. Electrophilic compounds enter the current in an inert carrier gas, absorb electrons and cause a drop in the current. An inert carrier gas such as argon or helium must be used because atmospheric oxygen will absorb electrons and be detected along with the analyte. The corresponding drop in current produces a signal that is quantitative only. When configured for a particular application, such as trace explosive material detection, ideal sensitivity is 10 pg/ml as vapor. An ECD can be used to detect gas or vapor from:

- Brominated hydrocarbon
- Chlorinated hydrocarbon
- Electrophilic compounds
- Nitrile
- Nitro compounds
- Organometals

Any compound with a strong electron affinity can be detected by ECD. Nitro compounds, such as explosives, and compounds containing halogens, such as organo-fluorine, -chlorine, -bromine, and -iodine are detected as are several others. A gas chromatographic column is needed to separate materials and to provide qualitative and quantitative measurement.

ECD cannot be used as a standalone portable air monitor, but recent advances have made possible stationary, portable devices. At this time, ECD sensitivity is less than, and the false alarm rate higher than, IMS.

Gas Chromatographic Column

A gas chromatographic column, also referred to as a gas chromatograph (GC), is an instrument used to separate chemicals in a mixture. A tube (column) contains air or an inert gas through which various molecules of a sample pass from one end to another at different rates, depending on physical and chemical characteristics of the molecule (Figure 3.3).

Figure 3.3 The Photovac Voyager Portable Gas Chromatograph with ECD and PID detectors. (Image courtesy of Photovac, Inc.)

The gas chromatograph can separate mixtures, and the individual components are detected by another instrument, thus the time taken (retention time) by portions of the sample to pass through the column become a factor in identifying the components of the sample.

Variables such as carrier gas composition, flow rate, and temperature can affect the retention time.

A GC alone can produce qualitative determinations of a mixture based on retention time and order of arrival at the end of the GC. Several types of instruments can be coupled with a GC based on the application. These technologies include:

- Electron capture detector (ECD)
- Flame photometric detector (FPD)
- Photo-ionization detector (PID)
- Thermal energy analyzer (TEA)
- Mass spectrometer (MS)

A GC is used to detect gas and vapor forms of volatile organic compounds (VOC). Sensitivity is low, but variable based on the detection equipment that can be coupled with the GC.

Field use of gas chromatography is usually limited to a particular device available at the site. Most often, these instruments use air or a simple inert gas such as nitrogen. Field GC analysis is suitable for material that is in gas or vapor phase at <450°C and with molecular weight <500. Other material may be analyzed; consult the instrument manual for detail.

Mass Spectroscopy

Mass spectroscopy (MS) is a method of measuring the mass-to-charge ratio of ions created from a sample material. Portable MS instruments (Figure 3.4) are designed as standalone detectors that sample air with any gases or vapors. A mass spectrometry instrument consists of three basic components: an ionization source, a mass analyzer, and a detector system.

The sample must first be ionized. Gases and vapors are usually ionized by electrical or chemical sources. Liquids and solids may be ionized by electrospray or matrix-assisted laser desorption/ionization (MALDI). Other ionization sources are possible. A magnetic or electrical field forces the ions into the mass analyzer.

The mass analyzer separates ions by their mass-to-charge ratio. Many types of mass analyzers and techniques are used, sometimes in series, to meet the specifications of the application.

The detector senses the induced charge or the current produced by the ion. Quantitative measurement is made by measuring the intensity of the ion flux. Signal is routed through an amplifier, signal processor, and computer to produce a visual mass spectrum. The spectrum can be used to identify one or more ions when matched with a library of known spectra.

Mass spectroscopy can generally be used to detect solid, liquid, or gas forms of:

- VOC
- TIC
- CWA
- Other compounds

Sensitivity is variable, but as low as ppb or ppt ranges. Due to the complexity of applying techniques and equipment to a particular situation, most MS instruments are found in the laboratory, and standalone field portable instrumentation is not as prevalent.

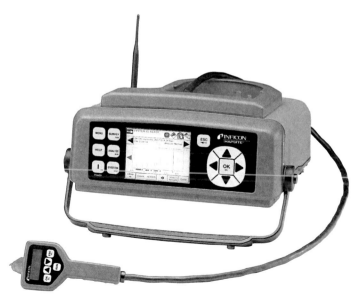

Figure 3.4 An example of a portable gas chromatograph–mass spectrometer is the HAPSITE Chemical Identification System (www.inficon.com). (Image courtesy of INFICON.)

A GC column may be coupled to the MS instrument. A GC column separates gases and vapors as they diffuse through a column of air or a specialized carrier gas. The mass analyzer will form peaks based on the time it takes (retention time) for the individual components that make up the sample to migrate to the end of the gas column. Since the retention time is known for a large number of compounds, time is used to identify the substance, and the amplitude of the peak can be used as a quantitative measurement. Additional information from the mass analyzer can be used qualitatively. A computer is used to match results to a library.

Ion Mobility Spectrometry

Ion mobility spectrometry (IMS) sorts ions by their mobility, which is affected by ion characteristics in air and electric charge. Portable IMS instruments are available (Figure 3.5).

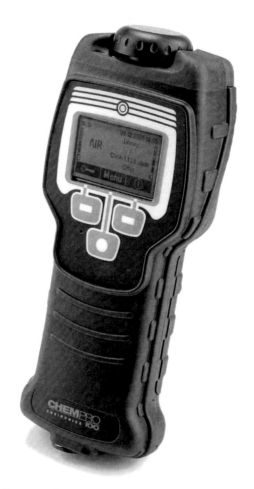

Figure 3.5 An example of a portable IMS instrument is the ChemPro100 Handheld Chemical Detector (www.environics.fi). This device also uses metal oxide semiconductor gas, temperature, and humidity sensors. (Image courtesy of Environics USA, Inc.)

Molecules are fractured into positively and negatively charged ions by one of several ionization sources, such as UV light, electron spray, radioisotope, or any other source capable of ionizing a wide range of compounds.

Mobility is determined after molecules are ionized. An "electric gate" is used to alternately admit positively and negatively charged ions into the detector tube, where they are attracted to an oppositely charged detector surface. The time of arrival of the ion is compared to the time the electric gate admitted the ion. Three characteristics affect the ion arrival time: drift caused by electric fields, diffusion to areas of lower concentration, and collision with molecules of air or gas medium.

Small, low-mass ions are likely to have fewer collisions with air molecules and are expected to arrive quickly. Larger ions will travel more slowly. Arrival time is measured in microseconds to milliseconds. Field instruments nearly always use air as a gas medium, although other gases are possible.

An electric signal is generated when ions contact the collector plates in the detector. The signal is sent through a processor and compared to representative library spectra. Ion drift, diffusion, and collision can cause ambiguity by changing the amplitude and width of peaks as well as shifting it in time. If a sample-to-library match occurs, an alarm signals or a compound name is specified.

IMS instruments are generally used to detect solid, liquid, or vapor form of:

- VOC
- CWA
- Illegal drugs
- Explosives
- Other compounds

The sensitivity is variable, but chemical warfare agents have been detected in the high ppb to low ppm range.

The instrument may draw an air sample through a thin, protective membrane that protects the detector from dust and excess water vapor; the membrane may trap water-soluble and membrane-impermeable compounds and prevent them from reaching the ionization chamber. Low-volatility material may need to be heated in order to vaporize the substance so that it can be moved into the ionization chamber.

A radioactive source is the most efficient source for field portable equipment and requires less battery performance.

IMS instruments are useful in the detection of explosives, drugs, chemical weapons, and many other compounds. However, most instruments operate within small "windows" of detection and cannot detect compounds outside the instrument's tuned parameters. Many instruments can only operate in "positive ion" or "negative ion" mode at any time, thus they cannot detect ions of charge that are opposite the current mode of operation.

Some IMS instruments are designed only to alert the operator to an alarm, not to identify the compound.

If the sample concentration is too high, the entire sample will not be ionized and whole or partially ionized fragments of molecules will also be detected. This can influence the spectral match to a library item.

Additionally, mixtures can be misrepresented in a spectrum if one component is more easily ionized than other components.

Photoionization

Photoionization detectors (PID) are nonselective monitors (Figure 3.6) used to quantify a known volatile organic compound (VOC). A PID uses a light source, usually ultraviolet (UV) lamps of various energy outputs, to produce positive ions from a sample in air. Ions are produced from a molecule if a photon from the lamp has enough energy to displace an electron from the molecule, which are associated with various energy levels. The affinity of the electron (ionization potential) for the molecule is measured in electron volts (eV).

Once a positively charged ion is formed, a negatively charged detector surface attracts the ion and a signal is generated when the two make contact. A signal processor passes the information to a display, which reads the total amount of ionized material detected. The PID is nondestructive, that is, the sample is converted from an ion back to its original state when the charge is neutralized on the detector surface.

Since the PID is a presumptive, quantitative tool, it may be calibrated with a gas standard and used to determine the concentration, but not the identity, of a known compound based on a response factor. The response factor is dependent on both the sample and the calibration gas. A PID result will be the sum of all compounds that are ionized by the UV lamp in the detection chamber as it cannot discern individual compounds in a mixture.

Sensitivity varies by model and application, but a range of a few ppb or ppm detection up to 10,000 ppm is typical. PIDs are intended to detect gas and vapor phase VOC.

The sensitivity of a PID to compounds in general can be described in the following order:

Figure 3.6 An example of a PID instrument is the MSA Sirius (www.msanorthamerica.com), which also incorporates a combustible gas sensor, oxygen sensor, and two electrochemical sensors.

- Aromatics and iodine compounds (most sensitive)
- Olefins, ketones, ethers, and sulfur compounds
- Esters, aldehydes, alcohols, and aliphatics
- Chlorinated aliphatics and ethane
- Methane (no response)

UV lamps are typically available that produce light in three wavelengths. The lamps are specified by the corresponding ionization potential the light can produce: 8.3, 10.6, and 11.7 eV. High-energy lamps produce shorter wavelength light and are typically capable of ionizing molecules with electron affinity of less than 11.7 eV. A 10.6 eV lamp is most commonly used as an economical solution to the higher cost associated with the 11.7 eV lamp. A lower energy 8.3 eV lamp is available for specialized applications. This allows the operator to screen out higher ionization potential (IP) compounds by selecting a lower IP lamp or to make detection inclusive of all compounds with an IP below 11.7 eV. A lamp with an IP greater than 11.7 eV may begin to ionize water vapor or other major components of air that would interfere with the target substance. See Table 3.1 for the IP of several representative compounds dependent on the energy produced by UV lamps.

Lamps may be changed in the field, but a period of stabilization and calibration is necessary, which would significantly slow results.

Flame Ionization

Flame ionization detectors (FID) are nonselective monitors (Figure 3.7) used to quantify a known volatile organic compound (VOC). An FID uses a flame, usually fueled by hydrogen

Table 3.1   Compounds Ionized by UV Lamps

Compound	IP (eV)	Lamp Ionization Potential		
		9.5 eV	10.6 eV	11.7 eV
Aniline	7.70	Y	Y	Y
Naphthalene	8.12	Y	Y	Y
Xylene	8.56	Y	Y	Y
Methyl bromide	10.54	N	Y	Y
Nitrogen dioxide	9.75	N	Y	Y
Phosphorous trichloride	9.91	N	Y	Y
Propane	11.07	N	N	Y
Tetraethyl lead	11.10	N	N	Y
Phosgene	11.55	N	N	Y
Hydrogen cyanide	13.60	N	N	N
Nitric acid	11.95	N	N	N
Sulfur dioxide	12.30	N	N	N
Potential Interfering Components of Air				
Nitrogen	15.58	N	N	N
Oxygen	12.08	N	N	N
Carbon dioxide	13.79	N	N	N
Water	12.59	N	N	N

Note: Y = detectable; N = not detectable

Figure 3.7 The MicroFID (www.photovac.com) is an example of a portable FID instrument. (Image courtesy of Photovac, Inc.)

in air, to produce positive ions from a sample in air. The hydrogen flame burns between 2000°C to 2500°C and produces ions from mainly organic samples. The ions are produced proportionally to the concentration of all organic compounds in the sample.

Once a positively charged ion is formed, a negatively charged detector surface attracts the ion and a signal is generated when the two make contact. A signal processor passes the information to a display that reads the total amount of ionized material detected. An oven chamber may be used between the flame and analyzer to prevent the ions from condensing before detection.

An FID instrument can detect gas and vapor phase materials:

- VOC
- Saturated hydrocarbons

- Aromatics
- Unsaturated hydrocarbons
- Chlorinated hydrocarbons
- Ketones
- Alcohols
- Some other compounds

The detection range is 1 to 50,000 ppm. Lower sensitivity is possible if the FID is coupled to a gas chromatographic column. An FID excels at ionizing larger hydrocarbon compounds through pyrolysis. Sensitivity and specificity are increased if the FID is coupled to a GC.

The sensitivity of an FID to compounds in general can be described in the following order:

- Aromatics and long-chain compounds (most sensitive)
- Short-chain compounds, including methane
- Halogen compounds (least sensitive)

FIDs are generally insensitive to water, carbon dioxide, carbon monoxide, sulfur dioxide, nitrogen oxides, and noble gases, all of which are not ionized by the hydrogen flame. The FID is destructive, that is, ions will be neutralized on the detector surface, but the original molecule will not be reconfigured.

Since the FID is a presumptive, quantitative tool, the FID may be calibrated with a gas standard and used to determine the concentration, but not the identity, of a known compound based on a response factor. The response factor is dependent on both the sample and the calibration gas.

Surface Acoustic Wave

Surface acoustic wave (SAW) instruments (Figure 3.8) use piezoelectric crystals that produce a measurable electric charge when subjected to pressure, similar to a telephone handset. A piezoelectric crystal will also slightly change shape if an electric current is applied to it.

A SAW monitor uses a flat piece of piezoelectric material subjected to alternating current that sets up waves that pulse across the surface of the piezoelectric material. This waveform becomes the baseline against which changes will be measured. A molecule on the surface of the piezoelectric material will cause the wave pattern to change slightly. A second piezoelectric substrate is used to detect the changing wave pattern and converts the wave pattern into an electric signal that can be analyzed by the computer.

The piezoelectric material is a hard and smooth crystalline surface. A thin coating of polymeric material is applied over the surface to capture vapor. The polymer contains monomer bonding sites that hold the target molecule.

A SAW instrument is designed to detect gas or vapor forms of:

- CWA
- TIC
- VOC
- Other specific compounds

Sensitivity is dependent on the design of the detector, but SAW instruments have been shown to detect chemical warfare agents below IDLH concentrations.

Figure 3.8 An example of a SAW instrument is the HAZMATCAD (www.msanorthamerica.com).

As sample passes over the coating three results can occur:

- A sample molecule may bond to the monomer and affect the SAW pattern as intended.
- A sample molecule may be absorbed into the polymer and affect the SAW pattern in a different manner. This can produce a signal different from the previous instance, which may not match the library spectrum.
- A sample molecule may not be soluble in, or able to bond to, the monomer, in which case the substance is not detected.

Once a sample molecule is bound to the polymeric surface, it must be cleared before a new test can be performed. This usually occurs with a period of heat and fresh air flushing. Some compounds other than the target may bind irreversibly to the polymer and disable the instrument. Most polymeric coatings have a service life of 1 year or less.

SAW instruments may be coupled with electrochemical cells to detect other materials, such as toxic industrial chemicals. Consult the instrument manual to identify materials that may be specifically identified or presumptively identified.

One commercially available device uses an array of three small solid-state sensors coated with specific polymers to detect chemical warfare agents. Time, amplitude, and signal patterns from all three SAW detectors are used to determine if a nerve or blister agent is present and to eliminate false alarms.

Environmental factors can affect the baseline wave pattern as well as the polymeric coating composition. Some environmental factors can be compensated with other sensors. Results are compared to a library and a result is produced based on a match. Mixtures are difficult to discern.

Flame Photometry

A flame photometry detector (FPD) detects specific frequencies of light produced by compounds being burned, usually in a hydrogen flame (Figure 3.9). An electron in a sample

Figure 3.9 The AP2C (www.proengin.com) is an FPD configured to alarm when frequencies of light are detected that correspond to sulfur and phosphorous. The alarms are presumptive for chemical warfare agents.

molecule excited to a higher energy state during combustion can drop to a lower energy state and emit light of only certain wavelength. The light emission spectrum produced is unique to individual atoms.

An FPD may use filters to remove undesirable light and to enhance detection of specific wavelengths far beyond the ability of the human eye. An FPD will not identify a specific compound, but it can identify some of the elemental "building blocks." Elementary components can presumptively be used to verify a material.

An FPD is a destructive detector and draws sample into a hydrogen flame where complex molecules degrade into simpler portions. Fragments may recombine in any number of combinations.

Light emissions are detected by a photomultiplier tube that amplifies the light signal and passes it to a signal processor. Recognition of certain wavelengths causes a positive result, usually a simple alarm.

An FPD is designed to detect gas or vapor forms of:

- CWA
- TIC
- Others

Sensitivity varies by configuration, but low ppb range is possible. Many FPDs are designed as presumptive detectors and may not detect more than one or two particular atoms. For example, the AP2C is designed to detect only sulfur and phosphorous contained in some chemical warfare agents. Response to the detector is more selective to these materials and less prone to false alarms.

Alarm thresholds are usually set at the factory. CWA detectors may have a low alarm threshold. An FPD intended for toxic industrial materials may have a higher alarm threshold.

Some FPDs use a heater device to vaporize solid and liquid samples from low-volatility materials. It is important to note that high concentration of many nontarget materials can overload the sensors and produce false positive results. Consult the instrument manual and use very low concentrations when verifying suspect material.

Too much flame energy can ionize an atom, which means the electron cannot drop to a lower energy state. The heat source must be regulated based on the atom to be detected.

Infrared Spectroscopy

Infrared spectroscopy measures the degree of absorption of infrared radiation when a wide-frequency infrared radiation (IR) beam interacts with a solid, liquid, or gas (Figure 3.10). Chemical bonds vibrate at characteristic frequencies that are affected by the type of bond and nearby atoms. Each molecule has a unique set of vibrations.

An IR spectrometer uses a source beam of IR energy with known characteristics to interrogate a sample. Certain frequencies of the source beam are absorbed, and the remainder of the beam is returned to a sensor, usually a transducer for field use, inside the instrument. A computer develops a spectrum that is the source beam minus the absorbed energy. The spectrum is compared to library spectra to determine a match. Mixtures can be determined by a simple subtraction algorithm, but complex mixtures become problematic, especially if one or more components are not in the library.

Infrared spectroscopy can detect solid, liquid, and gas forms of:

- Organic compounds
- Organometallic compounds

Sensitivity can be in the ppm range and varies by configuration. For example, an instrument configured to detect solid or liquid compounds is not likely to detect a gas. Symmetrical diatomic molecules, such as nitrogen and oxygen, are not detected since they do not absorb infrared radiation. Most other molecules will absorb infrared radiation at one or more wavelengths, including water and carbon dioxide, two common atmospheric constituents. Dilute aqueous solutions are difficult to discern.

Infrared spectrometers can be affected by noise; samples should be tested in a quiet, still area.

Figure 3.10 The MIRAN SapphIRe XL (www.thermo.com) is an example of a portable IR spectrometer capable of gas analysis. (Image courtesy of Thermo Fisher Scientific, Inc.)

Most often the mid-infrared region (wavenumbers of 4000 cm^{-1} to 400 cm^{-1}) is used for chemical analysis because it corresponds to induced changes in vibrational energy. If an instrument cannot make a "fingerprint" match, certain functional groups may be indicated by peaks in this region. Table 3.2 can be used to verify functional groups in a sample, but it does not take the place of a spectroscopist. Consult the manufacturer for reach back support or to determine if this table is compatible with the spectrum generated by the IR spectrometer.

Fourier Transform Infrared Spectrometry

Fourier Transform Infrared Spectrometry (FTIR) uses infrared radiation spectrometry as described previously with further processing of the returning signal. The returning beam is split in two to produce an interferogram and then a complex mathematical process called Fourier transform (FT) is applied. A spectrum is developed from this process.

FT reduces the sample signature by subtracting water and carbon dioxide from the spectrum. Spectral subtraction can further eliminate interference from these two

Table 3.2   Functional Groups Indicated by IR Spectroscopy

Wave Number/cm^{-1}	Indication
3000–2850	Saturated alkane
3100–3000	Unsaturated alkene or aromatic
3300	Terminal alkyne
2800 and 2700	Aldehyde (two weak peaks)
3400–3000 ~3600	Alcohol or phenol (hydrogen bonding can widen the peaks)
3450–3100	Amines (primary amines produce several peaks, secondary amines produce one peak, tertiary amines produce no peaks)
2260–2120	Alkyne (weak)
2260–2220	Nitrile
1840–1800 1780–1740	Anhydride
1815–1760	Acyl halide
1750–1715	Ester
1740–1680	Aldehyde
1725–1665	Ketone
1720–1670	Carboxylic acids
1690–1630	Amide
1675–1600	Alkene (often weak)
1690–1630	Imine (C=N, difficult to discern)
1560–1510 1370–1330	Nitro compounds
1400–1000	C-O or C-N (difficult to discern)
1480–1350	Saturated alkane or alkyl group
1000–680	Unsaturated alkene or aromatic
800–700	Chlorocarbon (difficult to discern)

substances. The ability to rapidly scan a sample with varying forms of infrared radiation and the ability to rapidly process data make FTIR more useful for confirmation testing than conventional infrared spectroscopy.

FTIR instruments (Figure 3.11) can be configured to detect solid, liquid, and gas forms of several compounds, generally hydrocarbons. Sensitivity varies by configuration. High confidence can be assumed when FTIR is used to identify pure material with a corresponding library entry. Identification becomes more difficult with mixtures and diluted aqueous solutions.

Symmetrical diatomic molecules, such as nitrogen and oxygen, are not detected since they do not absorb infrared radiation. Most other molecules will absorb infrared radiation at one or more wavelengths, including water and carbon dioxide, two common atmospheric constituents. Dilute aqueous solutions are difficult to discern. Infrared spectrometers can be affected by noise; samples should be tested in a quiet, still area.

Most often the mid-infrared region (4000 cm^{-1} to 400 cm^{-1}) is used for chemical analysis because it corresponds to induced changes in vibrational energy. Much of a field instrument's ability to confirm a substance rests with the library spectra and the ability of the software to match results. If an instrument cannot make a "fingerprint" match, certain functional groups may be indicated by peaks in this region. Table 3.2 can be used to confirm functional groups in a sample, but most instrument manufacturers provide reach back support with access to a spectroscopist.

Figure 3.11 An example of portable FTIR instrumentation is the HazMatID Ranger (www. smithsdetection.com). (Image courtesy of Smiths Detection.)

Raman Spectroscopy

Instruments using Raman spectroscopy (Figure 3.12) use an infrared laser to excite target molecules, which return a small portion of inelastically scattered light to the detector. The inelastically scattered light exhibits a quantum decrease in frequency. The scattering is influenced by several types of movement within the molecular bond. Atom-to-atom bonds are unique and are influenced in turn by neighboring atoms or functional groups. A pattern of all the inelastic scattering of the laser consisting of wave shift and intensity is processed into a spectrum that in turn is compared to a library. Some software is capable of matching several spectra from the library to determine mixtures. Mixture analysis is particularly useful for products that may vary the concentration of ingredients, for contaminated samples and for pure compounds that may have degraded over time.

Raman spectroscopy results can be affected by temperature, interfering radiation, the ability to induce strong polarization of the electron cloud around a bond, and absorption of the IR frequency rather than scattering. Large molecules tend to present more complex spectra.

The most intense Raman scattering is observed from electron clouds that can be vibrationally distorted in a symmetrical manner, making Raman spectroscopy sensitive to these compounds. In comparison, infrared absorption is most intense when asymmetrical vibration is present. When the two technologies are used in a complementary manner, analysis is enhanced.

Raman spectroscopy is preferred over IR spectroscopy when analyzing organic components in aqueous solution. IR spectroscopy detects water, and the water peak can obscure

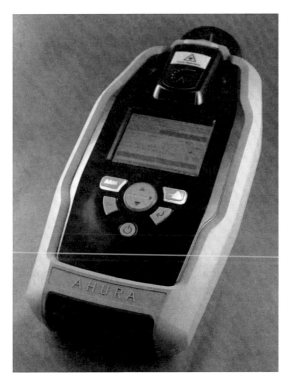

Figure 3.12 An example of a Raman spectrometer is the Ahura Scientific FirstDefender (www. ahurascientific.com).

other organic peaks. Raman spectroscopy is a poor detector of water and thus increases selectivity in aqueous solutions. Some Raman instruments can detect through most glass, translucent plastic, and thin paper.

Hydrocarbon compounds containing double bonds return a strong Raman signal. Strongly polar compounds (e.g., water, methanol), most metals and elements, and dark objects that absorb laser energy return a weak Raman signal. Proteins and other laser-fluorescent molecules are poorly detected due to noise in the signal that obscures the analyte peaks.

Field measurements made with Raman instrumentation will differ from controlled laboratory results, from which the library reference spectra are built. Unless the instrument software concludes a highly confident result, a "no match" message may be displayed after analysis. When this occurs, a second level of analysis is required by a spectroscopist. Figure 3.13 shows a report with a "no match" result and a spectrum that contains strong peaks. Spectral analysis through reach back support in the context of other field tests and observations can provide confirmation of identification or functional groups of a suspicious compound.

Table 3.3 displays peak ranges and intensity by functional groups, and Table 3.4 displays groups and intensity by peak ranges. These tables may be used to confirm the presence of functional groups in a sample; however, the tables do not consider all possibilities, especially in the context of variables that may occur in field testing. These tables do not replace the skill of a spectroscopist. Manufacturers provide reach back support and access

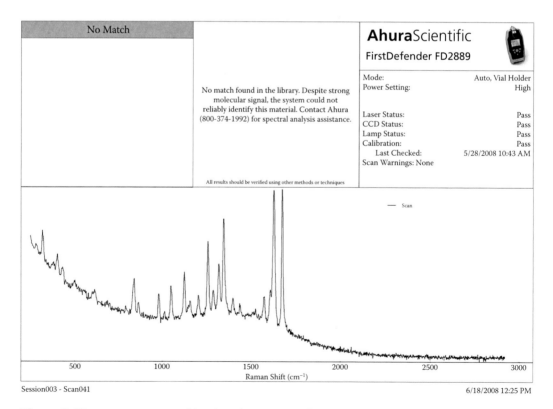

Figure 3.13 A report generated by the Ahura Scientific FirstDefender containing Raman spectrum data.

Table 3.3 Single Vibration and Group Frequencies of Peaks Commonly Identified by Raman Scattering Listed by Group

Group	Approximate Wave Number Range (cm^{-1})	Intensity
=CH$_2$	3010–3080	Strong
Acid chloride	1745–1780	Moderate
Alcohol	3210–3250	Weak
Aldehyde	2680–2740	Weak
Aldehyde	2780–2830	Weak
Aldehyde	1710–1725	Moderate
Aliphatic azo	1540–1590	Moderate
Aliphatic ester	1730–1750	Moderate
Alkyne	2070–2250	Strong
Alkyne	3250–3300	Very weak
Amide	3150–3480	Moderate
Amide	1550–1700	Strong
Amine	3150–3480	Moderate
Anhydride	1740–1830	Moderate
Aromatic azo	1365–1450	Very strong
Aromatic C-H	2870–3100	Strong
Aromatic nitrile	2200–2230	Moderate
Aromatic ring	1450–1505	Moderate
Aromatic rings	990–1100	Strong
Aromatic/hetero ring	1550–1610	Strong
Azide	2110–2160	Moderate
C=C	1625–1680	Very strong
C=N	1630–1665	Very strong
C=S	1020–1225	Strong
C=S	580–680	Strong
Carboxylate salt	1315–1435	Moderate
Carboxylic acid	1610–1740	Moderate
Carboxylic acid dimer	910–960	Weak
C-Br	505–700	Strong
C-C aliphatic chain	250–400	Strong
C-C aliphatic chains	630–1250	Moderate
C-CH$_3$	2810–2960	Strong
C-CH$_3$	1355–1385	Weak
C-Cl	550–790	Strong
C-F	720–800	Strong
CH=CH	2980–3020	Strong
CH$_2$	2770–2830	Strong
CH$_2$	1405–1455	Weak
CH$_2$	2900–2940	Strong
CH$_3$	1405–1455	Weak
C-I	490–660	Strong
C-O-C	800–950	Weak
C-S	670–780	Strong

(continued)

Table 3.3 Single Vibration and Group Frequencies of Peaks Commonly Identified by Raman Scattering Listed by Group (*Continued*)

Group	Approximate Wave Number Range (cm⁻¹)	Intensity
Diazonium salt	2200–2280	Moderate
Ester	1710–1745	Moderate
Isocyanate	2230–2270	Very weak
Isonitrile	2090–2170	Moderate
Isothiocyanate	2020–2100	Moderate
Ketone	1600–1710	Moderate
Lactone	1735–1790	Moderate
Lattice vibrations	100–210	Strong
N-CH₃	2750–2800	Weak
Nitrile	2220–2260	Moderate
Nitro	1535–1600	Moderate
Nitro	1320–1350	Very strong
O-CH₃	2790–2850	Weak
OH	2880–3530	Weak
P-H	2290–2420	Very weak
Phenol	3200–3400	Weak
Se-Se	295–340	Strong
Si-H	2080–2150	Moderate
Si-O-C	1010–1095	Weak
Si-O-C	1120–1190	Weak
Si-O-Si	460–550	Strong
Si-O-Si	1010–1095	Weak
S-S	425–550	Strong
Sulfonamide	1050–1210	Moderate
Sulfone	1050–1210	Moderate
Sulfonic acid	1145–1240	Very weak
Sulfonic acid	1025–1060	Very weak
Thiocyanate	2100–2170	Very weak
Thiol	2530–2610	Strong
Urethane	1690–1720	Moderate
Xmetal-O	150–430	Strong

Source: Adapted from Smith, E., and Dent, G. *Modern Raman Spectroscopy: A Practical Approach,* John Wiley & Sons, Ltd., West Sussex, England, May 2006.

to a spectroscopist. Consult the manufacturer to determine if these tables can be applied to an individual model of Raman instrumentation.

Modern portable instruments integrate reach back support for first responders and other users. A review by a spectroscopist provides a strong, second tier of analysis beyond that offered by an extensive library and computing power resident in the instrument. The following actual reach back support account provided by Ahura Scientific illustrates the value added by a spectroscopist. This case involved the discovery and subsequent attempt to confirm a substance as containing nitrate as well as other possible explosive compounds.

Table 3.4 Single Vibration and Group Frequencies of Peaks Commonly Identified by Raman Scattering Listed by Wavenumber

Approximate Wavenumber Range (cm^{-1})	Group	Intensity
100–210	Lattice vibrations	Strong
150–430	Xmetal-O	Strong
250–400	C-C aliphatic chain	Strong
295–340	Se-Se	Strong
425–550	S-S	Strong
460–550	Si-O-Si	Strong
490–660	C-I	Strong
505–700	C-Br	Strong
550–790	C-Cl	Strong
580–680	C=S	Strong
630–1250	C-C aliphatic chains	Moderate
670–780	C-S	Strong
720–800	C-F	Strong
800–950	C-O-C	Weak
910–960	Carboxylic acid dimer	Weak
990–1100	Aromatic rings	Strong
1010–1095	Si-O-C	Weak
1010–1095	Si-O-Si	Weak
1020–1225	C=S	Strong
1025–1060	Sulfonic acid	Very weak
1050–1210	Sulfonamide	Moderate
1050–1210	Sulfone	Moderate
1120–1190	Si-O-C	Weak
1145–1240	Sulfonic acid	Very weak
1315–1435	Carboxylate salt	Moderate
1320–1350	Nitro	Very strong
1355–1385	C-CH$_3$	Weak
1365–1450	Aromatic azo	Very strong
1405–1455	CH$_2$	Weak
1405–1455	CH$_3$	Weak
1450–1505	Aromatic ring	Moderate
1535–1600	Nitro	Moderate
1540–1590	Aliphatic azo	Moderate
1550–1610	Aromatic/hetero ring	Strong
1550–1700	Amide	Strong
1600–1710	Ketone	Moderate
1610–1740	Carboxylic acid	Moderate
1625–1680	C=C	Very strong
1630–1665	C=N	Very strong
1690–1720	Urethane	Moderate
1710–1725	Aldehyde	Moderate
1710–1745	Ester	Moderate
1730–1750	Aliphatic ester	Moderate
1735–1790	Lactone	Moderate

(continued)

Table 3.4　Single Vibration and Group Frequencies of Peaks Commonly Identified by Raman Scattering Listed by Wavenumber (Continued)

Approximate Wavenumber Range (cm^{-1})	Group	Intensity
1740–1830	Anhydride	Moderate
1745–1780	Acid chloride	Moderate
2020–2100	Isothiocyanate	Moderate
2070–2250	Alkyne	Strong
2080–2150	Si-H	Moderate
2090–2170	Isonitrile	Moderate
2100–2170	Thiocyanate	Very weak
2110–2160	Azide	Moderate
2200–2230	Aromatic nitrile	Moderate
2200–2280	Diazonium salt	Moderate
2220–2260	Nitrile	Moderate
2230–2270	Isocyanate	Very weak
2290–2420	P-H	Very weak
2530–2610	Thiol	Strong
2680–2740	Aldehyde	Weak
2750–2800	N-CH$_3$	Weak
2770–2830	CH$_2$	Strong
2780–2830	Aldehyde	Weak
2790–2850	O-CH$_3$	Weak
2810–2960	C-CH$_3$	Strong
2870–3100	Aromatic C-H	Strong
2880–3530	OH	Weak
2900–2940	CH$_2$	Strong
2980–3020	CH=CH	Strong
3010–3080	=CH$_2$	Strong
3150–3480	Amide	Moderate
3150–3480	Amine	Moderate
3200–3400	Phenol	Weak
3210–3250	Alcohol	Weak
3250–3300	Alkyne	Very weak

Source: Adapted from Smith, E., and Dent, G. *Modern Raman Spectroscopy: A Practical Approach,* John Wiley & Sons, Ltd., West Sussex, England, May 2006.

Thank you for using Ahura Scientific's reach back service. While I can not make a conclusive identification of the material under question, I believe the information I can provide will be of use to you.

You described the material as a pea green liquid in a plastic bottle. As we discussed over the phone, the data you collected are suggestive that a nitrate containing material is present. The on-board analysis you collected in the field reported a mixture of mercury(I) nitrate, sulfuric acid, and several other nitrate containing entities. While this mixture solution does explain many of the features in the data, several of these items are not readily available so it is germane to question the results.

Based on my examination of the data, I believe the particular nitrate present in the sample is ammonium nitrate. An overlay of the unknown data and a spectrum for ammonium nitrate are shown below (Figure 3.14).

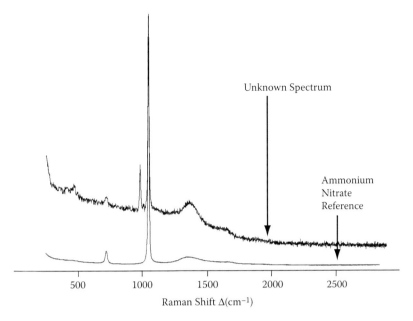

Figure 3.14 Spectra returned in spectroscopist's report.

As you can see from the plot above, ammonium nitrate corresponds well to the spectral features in the unknown at 718 cm^{-1} and 1048 cm^{-1}. The remainder of the spectrum does not contain many Raman peaks, suggesting that whatever other component is present is a very simple molecule with only a few covalent bonds. When ammonium nitrate is used as an explosive material, it is most often mixed with some sort of fuel such as kerosene. These materials have many more bonds resulting in complex spectra, so they can be ruled out as being present.

Outside of applications in explosives, ammonium nitrate is most often used as a fertilizer. Many fertilizer mixtures, including liquid fertilizers, contain ammonium nitrate mixed with one or more other simple materials such as urea, potassium sulfate, etc. Whereas I can not provide a definitive identification of the second component, the simplicity of the data and the location of the peaks are consistent with sulfate containing molecules (also likely why the on-board mixture analysis suggested that sulfuric acid was present). As such, I would speculate that the unknown is likely a fertilizer composed of ammonium nitrate and some sort of sulfate containing molecule. Whereas I believe the data is most consistent with sulfate containing materials, it is also possible that a phosphate containing material is present.

As always, we remind users that FirstDefender is not a trace detection tool. The findings provided here should always be verified with other tools at your disposal. If I can answer any further questions, please let me know.

Always consider any test and related analysis as a resource to be used by first responders and other field workers in conjunction with other situational analysis. Those responsible for response are wise to determine a course of action that considers all information and not just the results of a single test.

X-Ray Fluorescence

X-ray fluorescence (XRF) analyzers use x-rays to irradiate a sample and excite the electrons of metals and other compounds. The electrons will emit x-rays of differing wavelength (fluoresce) when returning to a normal state. An XRF instrument (Figure 3.15) detects the pattern of fluorescence and identifies elements.

XRF can generally detect elements with an atomic number of 16 (sulfur) through 94 (plutonium). Practically, the detection ability of the instrument depends on the element used to generate x-rays. In general, XRF instruments use one or more of the following x-ray–generating sources:

- Cobalt-57 for the detection of lead
- Iron-55 for the detection of calcium, chromium, potassium, sulfur, titanium, and vanadium
- Cadmium-109 for the detection of arsenic, chromium, cobalt, copper, iron, lead, manganese, mercury, molybdenum, nickel, plutonium, rubidium, selenium, strontium, uranium, vanadium, zinc, and zirconium
- Americium-241 for the detection of antimony, barium, cadmium, silver, and tin

XRF is used to detect solid and concentrated liquid forms of (although water interferes with):

- Aluminum
- Antimony

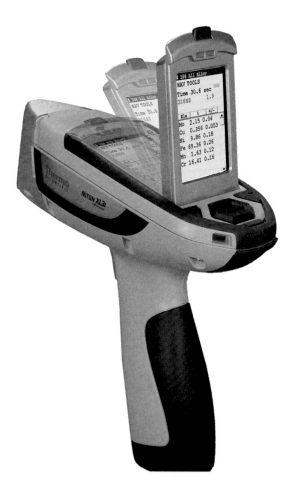

Figure 3.15 An example of a portable XRF instrument is the NITON XL3t (www.thermo. com/niton). (Image courtesy of Thermo Fisher Scientific.)

- Arsenic
- Barium
- Bromine
- Cadmium
- Chromium
- Cobalt
- Copper
- Iron
- Lead
- Magnesium
- Manganese
- Mercury
- Nickel
- Phosphorus
- Rubidium
- Selenium
- Silicon
- Silver
- Thallium
- Tin
- Titanium
- Tungsten
- Zinc
- Zirconium

Sensitivity varies by model and analyte. Typically, XRF can detect 20 to 100 ppm of most analytes. Chromium detection is less sensitive, usually in the 200 to 900 ppm range. The NITON 700 Series reliably detects 15 ppm arsenic in soil for test times of 30 to 60 seconds after sample preparation.

The XRF analyzer measures the metal content of the sample over a surface area of about 1 cm^2 to a depth of about 2 mm. Water content greater than about 20% can interfere; drying may be necessary. Pooled water definitely interferes. The most accurate field readings are obtained by drying the sample and then uniformly grinding to about 100 microns before analysis.

Some metals will interfere with the detection of other metals in mixtures. For example, iron will absorb x-ray fluorescence from copper, but will increase detection from chromium.

Radioisotope Detection

Instrumentation is necessary for field confirmation of radioactive material. A few of the basic sensors and associated instrumentation are presented here. A basic understanding of radioisotope detection is necessary to perform field confirmation tests on suspicious material.

Commercialized instruments often blend sensor types and processing equipment to meet a certain market niche, and complete coverage of all combinations is too lengthy to be presented here.

The Department of Homeland Security has funded an evaluation of field radiation detection instruments. *Results of Test and Evaluation of Commercially Available Survey Meters for the Department of Homeland Security; Results of Test and Evaluation of Commercially Available Personal Alarming Radiation Detectors and Pagers for the Department of Homeland Security;* and *Results of Test and Evaluation of Commercially Available Radionuclide Identifiers for the Department of Homeland Security* are good resources. The guides assess commercially available equipment as tested against standards of the National Institute of Standards and Technology (NIST). Access to these documents is limited to first responders and other officials. They are posted at the SAVER web site (http://saver.tamu.edu/).

Geiger-Mueller Tube

A Geiger-Mueller (GM) tube detects alpha, beta, and gamma ionizing radiation across most, but not all, energy ranges. A GM tube contains one of several possible inert gases to which a high voltage is applied when the unit is powered on. When certain ionizing radiation passes through the tube, an electrical pulse is generated. The pulse is modified by the instrument to produce a visual meter reading or an audible alarm. A GM detector is commonly called a Geiger counter (Figure 3.16). A GM tube is dependable at detecting alpha and beta radiation; high-energy gamma radiation passes largely undetected through the low-density gas in the tube, although some is detected.

A GM instrument can detect solid, liquid, and gas material producing:

- Alpha radiation
- Beta radiation
- Gamma radiation

Figure 3.16 The Inspector Alert Handheld Surface Contamination Meter (www.medcom.com) uses a halogen-quenched GM tube. (Image courtesy of International Medcom.)

Sensitivity and range are affected by several factors. Consult the manual or manufacturer to determine the sensitivity and range of energy detected. There are three basic designs for GM probes: side window or cylindrical, end window, and pancake.

The side window design is sensitive to gamma radiation. A portion of the probe may be of thin-wall construction that allows high-energy beta rays (>300 keV) to enter the tube and be detected. Some probes have a dense sleeve that can be rotated to reveal a thin wall over the GM tube. The instrument cannot discern high-energy beta radiation from gamma radiation; however, high-energy beta radiation can be determined by comparing one reading with the sidewall closed and a second reading with it open. The beta result is the window-open reading minus the window-closed reading.

The pancake probe, which resembles a spatula, contains a thin mica (or similar material) window in front of the sensor. The pancake probe detects alpha, beta, and gamma radiation. The probe must be held very close to surfaces to detect alpha radiation, and, practically, the pancake probe is most often used to detect beta radiation. Gamma and beta radiation can roughly be differentiated by flipping the probe over so that the metal back of the probe is facing the source; gamma radiation will penetrate the metal back and beta radiation will not (Figure 3.17).

The end window probe is a cylinder with a mica (or similar material) material that forms a window at the end of the cylinder. It is sensitive to gamma radiation and is most often used to detect beta-emitting surface contamination. It is capable of detecting alpha radiation, but it must be held close to the sample.

The GM detector is the simplest and most prevalent radiation monitor. Many variations exist and the GM tube is often combined with other sensors in individual models.

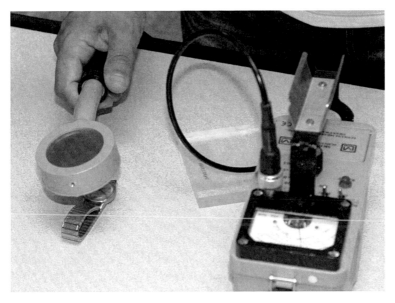

Figure 3.17 A Ludlum Model 44-9 pancake GM detector attached to a survey meter used in a training course. The operator has rotated the probe to determine if radiation from the watch will penetrate the back of the probe, which is denser than the window material on the probe that is visible. Gamma radiation will penetrate the back wall and be detected; alpha and beta radiation will only be detected through the window of the probe.

The use of distance and shielding between the sample and probe can be used to differentiate the type of ionizing radiation present, but not all radiation is detected by the instrument. Consult the instrument manual or manufacturer to determine the range of energy detected and compensation factors. Instruments should meet ANSI N42.34 performance standard.

Scintillator

Some ionizing radiation can be detected by an instrument containing a scintillator (Figure 3.18). A scintillator is a material that fluoresces briefly when struck by ionizing radiation. The scintillating material will have certain characteristics and the particular scintillator is chosen to fit the application. Scintillating materials differ in the amount, duration, and wavelength of fluorescence generated in response to ionizing radiation. These characteristics affect the selectivity and sensitivity of the instrument. Short decay time of the fluorescence means the instrument is more likely to detect an ionizing event that occurs immediately after the previous event. Low-density scintillation material may allow radiation to pass through undetected. Material that scintillates in a narrow range of energy will not detect broad ranges of ionizing radiation. Scintillators are mainly used to detect solid, liquid, and gas compounds emitting gamma radiation.

Survey meter scintillators usually consist of inorganic crystals that are dense and are able to absorb more radiation, but the decay time of the fluorescence is longer. These materials are most effective for the detection of gamma radiation:

- Sodium iodide doped with thallium (the most common) and denoted as NaI(Tl)
- Barium fluoride (BaF_2)

Figure 3.18 A Ludlum survey meter with a scintillator (in bracket with cable disconnected) and pancake probe with cover.

- Cesium iodide (CsI)
- Lanthanum bromide (LaBr$_3$)
- Lutetium iodide (LuI$_3$)
- Yttrium aluminum garnet doped with cerium (Ce:YAG)

Depending on the model, other probes may be available, such as a zinc sulfide scintillator probe, for the detection of alpha radiation.

Scintillator size and shape can make the detector more sensitive to distant gamma radiation. These instruments can be used as directional sensors and can be used to verify gamma radiation from a greater distance than would be necessary with a smaller scintillator. Figure 3.19 shows a large scintillator that can be used to "sweep" an area for far-reaching radiation.

The use of individual models is dependent on detector and sensor configuration. Not all radiation is detected by the instrument. Consult the instrument manual or manufacturer

Figure 3.19 Thermo Eberline ESM FH 40 G-L with enhanced scintillator for increased gamma sensitivity. The scintillator is most sensitive along the long axis of the sensor, which makes absorption of gamma radiation more likely. This characteristic makes the device sensitive to the direction of the source.

to determine the range of energy detected and compensation factors. Instruments should meet ANSI N42.34 performance standard.

CZT crystal is an alloy of cadmium, zinc, and telluride. It is not a scintillator, but is mentioned here as an alternative to scintillation methods. Gamma radiation striking a CZT semiconductor array will generate electrons that can be detected. Special manufacturing is required, but CZT can be applied to specialized gamma detection applications that are beyond the parameters set by scintillating materials.

Another substitute for a scintillator is an ionization (ion) chamber, a gas-filled chamber containing an anode and cathode. When gamma radiation interacts with the sensor, an electric pulse is produced. The pulse is modified through electrical circuitry and is output as a visual meter reading or an audible sound.

High-Purity Germanium

High-purity germanium (HPGe) is used for highly sensitive detection of gamma and neutron radiation (Figure 3.20). HPGe sensors detect radiation across a band of energies from which a spectrum is formed. Radionuclides can be identified by unique peak patterns. HPGe sensors are very sensitive and used for accurate detection of gamma and neutron radiation when a fast response is needed, such as screening of containers at a port.

Figure 3.20 An example of an instrument using a HPGe detector is the ORTEC Detective-Ex-100 (www.ortec-online.com), a radionuclide identifier. (Image courtesy of Advanced Measurement Technology, Inc.)

Detectors containing HPGe are designed to detect or identify:

- Gamma radiation
- Neutron radiation
- Radionuclides

Sensitivity varies, but 0 to 100 mSv/hr is typical. Consult the manual or manufacturer for details. HPGe crystals detect minute amounts of neutron and gamma ray emissions, even through heavy shielding. Some instruments use a HPGe sensor to detect fissile nuclear materials such as uranium or plutonium. Higher dose-rates are typically detected with other sensors, such as a GM tube.

Neutron Detectors

A neutron detector uses a sensor that consists of a gas-filled tube with a high voltage applied to it. The tube is filled with boron trifluoride (BF_3) or helium-3 (He-3). Neutron radiation interacts with the gas in the tube to form a detectable particle; BF_3 emits a helium-4 nucleus or He-3 emits a proton. The detectable particle creates a signal that is processed and displayed. Filters and circuitry eliminate high-energy gamma, beta, and alpha radiation so that the displayed measurement is assumed to be neutron radiation. A multichannel analyzer can be used to process a spectrum and determine the neutron source.

Neutron radiation can also be detected with a scintillator containing lithium iodide (LiI). Some instruments use a LiI scintillator and a gamma-sensitive scintillating material to detect gamma and neutron radiation. This has application in "pagers," among others, and is relative to background radiation.

A neutron-sensitive scintillating glass fiber detector contains ^6Li and Ce^{3+} in glass fibers. The lithium is used to capture neutrons and produce a tritium ion, alpha particle, and energy. The tritium ion then reacts with the cerium to cause scintillation. The flash of light is carried by the glass fiber to a photomultiplier tube. The light is converted to electronic signal and is processed. Scintillating glass fibers offer solid-state detection of neutrons.

Personal Alarming Radiation Detectors

Personal alarming radiation detectors, sometimes called radiation pagers, use technology similar to that used in survey instruments. Personal detectors are miniaturized and designed to be worn unobtrusively by a person (Figure 3.21). They can detect radiation above a background level and alert the user. They can also be used to localize a source by detecting increasing ionizing radiation. Most are designed to detect gamma and sometimes neutron radiation. These devices are not designed to routinely detect alpha or beta radiation, although some models can be adapted.

Radionuclide Identifiers

Radionuclide identifiers can identify hundreds of individual radionuclides. The device (Figure 3.22) can measure the spectrum produced by an ionizing radiation source and compare it to a library of spectra stored in the device. Spectra are defined by decay

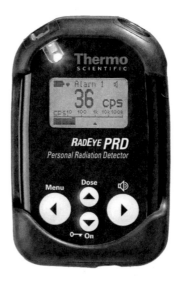

Figure 3.21 The RadEye PRD is an example of a gamma-sensitive radiation "pager" that can alert the wearer to an increase in radiation. The device does not identify the radionuclide. The RadEye PRD incorporates a high-sensitivity NaI(Tl) scintillation detector with a miniature photo-multiplier allowing the detection of very low radiation levels with particular emphasis on gamma emissions below 400 keV. For unambiguous identification of neutron sources, the RadEye N (right) incorporates an He-3 tube for neutron detection and avoids confusing spillover from gamma radiation. (Image courtesy of Thermo Scientific, www.thermo.com).

characteristics of spin and parity, gamma peak energy, alpha and beta transitions, and so on. The ability of the identifier to determine the identity of the source depends on the type and range of sensors in the device as well as the presence of a matching spectrum.

A radionuclide identifier can detect:

- Neutron radiation
- Radionuclides
- Special Nuclear Materials (SMN)

Radionuclide identifier operation requires some skill and knowledge of decay paths and other effects. Interpretation can be complex and may not be as simple as overlaying a sample's spectra on one from the library. Sometimes a skilled interpreter is necessary.

Biological Detection

Biological organisms are a complex mix of chemicals in a physical structure that is evident of life. Presumptive tests have been devised to identify characteristic compounds, but culturing the organism remains the only definitive test for viable biological organisms. Similar tests are available for biological toxins that are not alive or capable of reproducing.

Two types of tests are used for field tests for biological material, principally immunoassay and nucleic acid amplification. Antibody-based tests can be used to detect a characteristic molecule or portion of molecule that is representative of the target organism

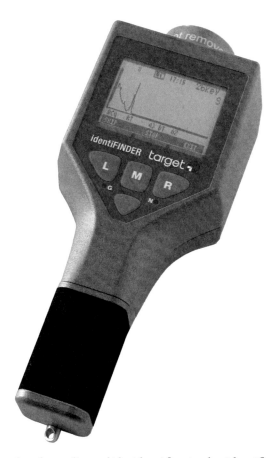

Figure 3.22 An example of a radionuclide identifier is the Identifinder (www.icxt.com), a handheld, digital gamma spectrometer. Models use a combination of NaI(Tl), GM, and 3He detectors for dose rate measurement and identification; neutron detection is optional. (Image courtesy of ICx Radiation.)

or other organisms closely related to it. Gene-based tests replicate DNA and then detect larger amounts of replicated DNA through another method. Antibody-based identifiers are quick and prone to automation. Gene-based identifiers are more exacting but require more effort and time.

Most biological field tests used by first responders have been developed to detect trace amounts of biological warfare agents. Many other field tests have been developed to detect specific material in a biological sample. For example, a food manufacturer may use one test to confirm the lack of food allergens (e.g., peanut, egg, milk, etc.), lack of food pathogens (e.g., E. coli O157:H7, salmonella, listeria, etc.) and lack of sanitation (e.g., bacteria, yeast, mold, etc.).

The very nature of biological material means that small amounts of material can have a very large effect. That small amount can be below the threshold amount the test can detect. When confirming a biological material, it is important that both sensitivity and specificity of a particular test be measured relative to a reference standard. Sensitivity and specificity are jointly determined at the threshold of detection for a positive test, because either can be made excessive at the expense of the other. In other words, considering sensitivity or specificity exclusively will result in low confidence for a test.

Table 3.5 describes performance characteristics for bacterial culture versus other detection methods. Field confirmation testing of biological material is a compromise of accuracy, time, and sensitivity.

Biological Screening Tests

Biological screening tests are often presumptive chemical tests. These tests often confirm the presence of a necessary nutrient or support media for an organism (milk, sugar, starch, blood, etc.) or a material common to many organisms (protein, lipids, certain sugars, etc.). These chemical tests are mentioned here because they are useful for testing bulk amounts of sample quickly and inexpensively, but they are not highly specific and often have sensitivity above a desired threshold. Some of these chemical tests might involve instrumentation such as an optical reader that is more sensitive than the human eye, but even these field instruments have sensitivity too high for field testing of biological warfare agents. For more information on these screening tests, see Chapter 2, "Chemical Confirmation Tests."

Immunoassay

Immunoassay tests detect and measure the highly specific binding of antibodies with their corresponding antigens by forming a three-dimensional, antigen-antibody complex. Overall, they are quick and accurate tests that can be used to detect specific molecules.

An immunoassay-based test is highly specific to an antigen in a sample by the action of its corresponding antibody. The two react to form an antigen-antibody complex, which triggers some readable result such as a color change.

Antibody-based tests can be highly specific and sensitive. Detection thresholds can be as low as 1000 to 10,000 microbial cells per milliliter. Disadvantages of antibody-based tests include false positive results due to nonspecific binding and cross-reactivity. Cross-reactivity occurs because some closely related organisms share antigens. False negative results can be caused by degradation of the antibodies over time, which also determines shelf life.

Rapid immunoassay tests are designed using combinations of monoclonal-polyclonal sandwich, antigen-down, and competitive inhibition immunoassays. Rapid immunoassays

Table 3.5   Performance Characteristics of Bacterial Culture, PCR, and Handheld Immunoassays for the Detection of *B. anthracis*

	Microbiology/ Culture	Handheld Immunoassays	Polymerase chain reaction (PCR)
Minimum limit of detection (spores)	1	100,000 to 100 million	100–1000
Assesses viability	Yes	No	No
Nonspecificity	No	Yes (near neighbor bacteria and chemicals)	Yes (near-neighbor bacteria)
Other issues	Culture may require days to complete	Susceptible to interferences[a]	Technology immature[a]

Source: Table adapted from *Update on Biodetection: Problems and Prospects,* U.S. Department of Health & Human Services, with permission.

Note: [a]This report occurred in 2002. Technology has advanced and improvements have been made to make immunoassays more specific.

were developed primarily to fill the need for a quick field test for specific material that retained an acceptable degree of selectivity and sensitivity. Most often the result is expressed through development of a color and more specifically these are called immunochromatographic tests.

Handheld immunochromatographic assays (Figure 3.23) are single-use devices that are often compared to a home pregnancy test kit. They provide a "yes or no" result based on the formation of two colored bands. The color is formed when a dye, colloidal gold, or other conjugate attaches to the antigen-antibody complex and is captured and concentrated in the band areas. One band is a control to assure the test is working; the second band represents a positive result for a specific agent. The test is designed to be qualitative; however, the degree of color change in the second band indicates a semi-quantitative measurement. These tests typically must be read after 15 minutes and before 30 minutes or results are not valid. The use of an optical reader (Figure 3.24) can estimate the concentration of biological agent present in the sample.

Some immunoassay tests can produce false negatives if too much antigen is present, a phenomenon known as Hook effect. Hook effect mainly occurs in immunoassays that mix the antigen, conjugate, and capture antibody together, as is the case in many lateral flow immunochromatographic assays. Excessive quantities of antigen compete with labeled antigen for binding sites at the test and control lines, as shown in Figure 3.25. The result is an elimination or decrease of the intensity of the colored antigen-antibody-conjugate complex at the test line and control line. Hook effect causes abrupt decrease in signal, which may be incorrectly interpreted as a negative result.

Hook effect can be eliminated by serial dilution and repeating the test. A recommended dilution procedure for the detection of biological warfare agents is a 99:1 dilution of sample with saline or phosphate buffer solution, and then repeat of the test. If still negative, repeat a 99:1 dilution of the first dilution and repeat the test. Consult the manufacturer for specific dilution procedures for specific tests. Not all handheld immunochromatographic assays are susceptible to Hook effect (Figure 3.26).

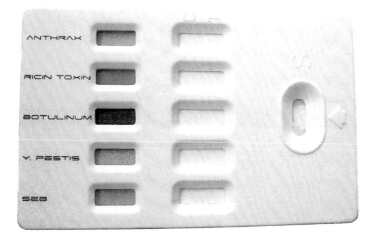

Figure 3.23 Pro Strips (www.baddbox.com) uses a multi-immunochromatographic assay system to test a single sample. Colorimetric results form as lines at the control site (C) to assure the test is functioning properly and at the test line (T) to indicate a positive result. (Image courtesy of ADVNT Biotechnologies, Inc.)

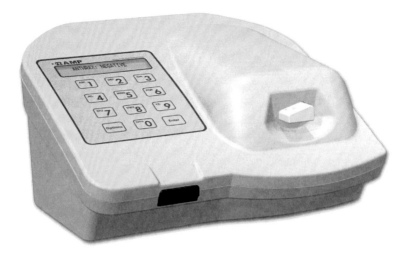

Figure 3.24 An example of a portable immunochromatographic assay with optical reader is the RAMP System (www.responsebio.com). RAMP uses a portable scanning fluorescence reader and single-use, disposable test cartridges to detect anthrax, ricin, botulinum toxin, and pox virus. The RAMP System received AOAC Official Methods Certificate 070403. (Image courtesy of Response Biomedical Corporation.)

Nucleic Acid Amplification

Nucleic acid amplification, also known as polymerase chain reaction or PCR, can replicate many copies of DNA from minute, previously undetectable amounts within a sample.

Polymerase chain reaction technology uses a polymerase, a naturally occurring enzyme, to catalyze the formation of DNA or a characteristic segment of DNA. DNA is formed from two complementary, not identical, chains of amino acid that are bound together. If the DNA is torn in half (down the center, like a zipper), polymerase can recreate the complementary chain missing from each half, which results in two strands of DNA identical to the original chain.

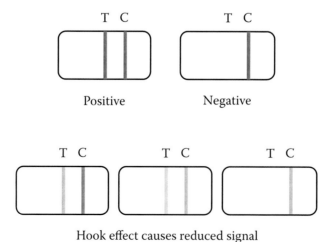

Figure 3.25 Hook effect can occur if too much antigen competes with limited detection sites.

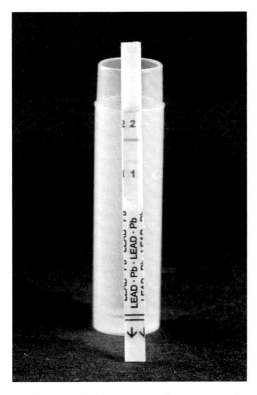

Figure 3.26 An example of a handheld immunochromatographic assay is the Watersafe Drinking Water Lead Test Kit. Lateral flow chromatography can be used to test for many analytes. When the analyte (antigen) is matched with an effective antibody and combined in a lateral flow chromatographic device, many biological and chemical substances can be identified in low concentration in the presence of material that may interfere in other tests. This test has indicated the presence of at least 15 ppm lead by forming a faint line above the number 1 (the line below the number 2 is the control to assure the test is functional). Also, see color image 2.56.

The DNA of interest is supplied by the sample and becomes the target sequence of genetic material for the test. The complementary strands of any DNA present in the sample are separated by heating. Synthetic DNA segments known as nucleic acid probes are complementary to the sequence at an end of each of the target strands. The synthetic DNA segments attach to the single-strand DNA and serve as primers for polymerase. Only genetic material that is bound to the nucleic acid probe can be replicated in the test. The polymerase enzyme begins at the primer and replicates the strand from amino acids added to the test solution.

Cooling the solution allows double-stranded DNA to remain intact. Subsequent heating and cooling cycles (thermocycling) will replicate the DNA exponentially, effectively amplifying the genetic sequence to an amount sufficient for detection (Figure 3.27). The increasing amount of target DNA can be detected in various ways. One technique uses ethidium bromide (EtBr), which will bind to double-stranded DNA and fluoresce (Figure 3.28). Over the course of thermocycling, the increase in fluorescence indicates an increasing amount of the target DNA. This method is effective even with samples that contain high amounts of nontargeted DNA since it detects an increase in fluorescence over the thermocycling periods. The fewer cycles necessary to produce a detectable increase in fluorescence, the greater the

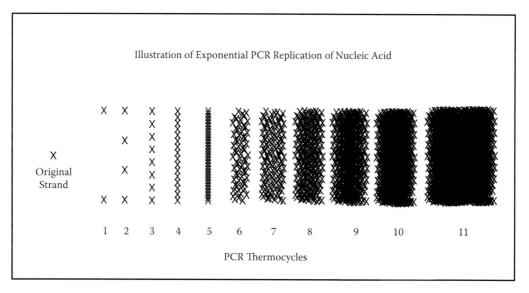

Figure 3.27 PCR provides exponential reproduction of genetic sequences to rapidly produce a detectable volume from the sample.

number of target sequences in the sample, which are duplicated exponentially by the PCR process. The cycle can occur as fast as every 17 seconds, and it is possible to produce a billion-fold (32 thermocycles) in about 10 minutes, plus set-up time.

Advantages of PCR include high sensitivity and specificity. The test is fast, requiring only a few minutes. Probes can also be designed to bind to a sequence that is common to several pathogens. Disadvantages of PCR include the need for pathogen-specific probes, which may not be available. Additionally, a specific probe must be used for each pathogen; a PCR screening test would be problematic. PCR does not work on toxins since they do not have a nucleic acid sequence.

The Advanced Nucleic Acid Analyzer (ANAA) or Mini-PCR developed by Lawrence Livermore National Laboratory was able to detect 500 colony forming units (CFUs) of E. herbicola, a vegetative bacterium used as a surrogate for Y. pestis, in 15 minutes (about 25 thermocycles). After thermocycling modifications, the ANAA was able to detect 500 CFUs of E. herbicola in 7 minutes. Ongoing software and automation improvements are intended to reduce the time required to run the tests and to protect the operator. The ANAA utilizes 10 silicon reaction chambers with thin-film resistive heaters and solid-state optics. The ANAA offers real-time monitoring, low power requirements for battery operation, and no moving parts for reliability and ruggedness.

An example of a portable PCR device is the RAZOR EX (Figure 3.29). The RAZOR EX uses freeze-dried reagents to test up to 12 samples in less than 30 minutes. Tests are available for:

- Anthrax
- Anthrax IVD
- Avian Influenza
- Botulism
- Brucella
- Campylobacter
- Cryptosporidium

- E. coli 0157
- Ebola
- Listeria
- Marburg
- Plague
- Ricin
- Salmonella
- Smallpox
- Tularemia
- Y. pestis

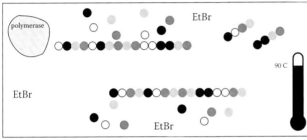

Target nucleic acid from prepared sample and added to a solution of nucleic probes (primers), high-temperature polymerase, four nucleotide bases, and ethidium bromide (EtBr).

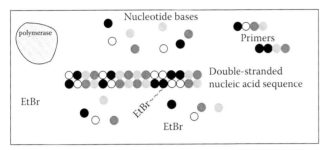

Temperature is raised to 90 C and double-stranded material separates. EtBr combined with any double-stranded DNA separates and stops fluorescing.

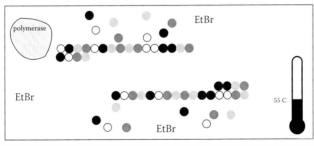

Temperature is lowered to 55 C and nucleic probes anneal to complimentary sequence of nucleic acid unique to target.

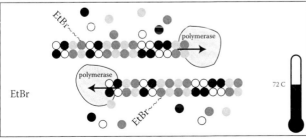

Temperature is raised to 72 C and polymerase assembles a complimentary sequence from free amino acids. Free ethidium bromide (EtBr) binds to double-stranded DNA and fluoresces. The thermocycle is repeated. Nucleic acid is exponetially copied. Increasing fluorescence for subsequent thermocycles produces a postive signal for target pathogen.

Figure 3.28 Schematic representation of PCR.

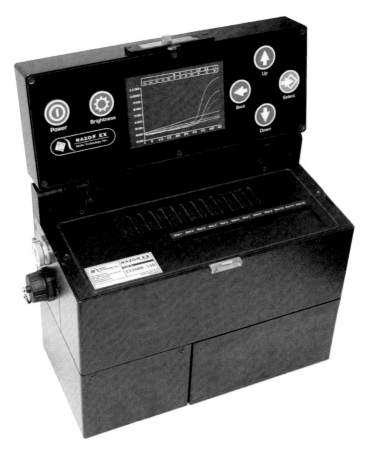

Figure 3.29 The RAZOR EX from Idaho Technology, Inc. (www.idahotech.com) is an example of portable PCR instrumentation. (Image courtesy of Idaho Technology, Inc.)

References

Aarino, P. *Expert Systems in Radiation Source Identification,* www.tkk.fi/Units/AES/projects/radphys/shaman.htm, Helsinki University of Technology, Finland, accessed December 7, 2006.

Alkema, G. *What Is Gas Chromatography?* www.gchelp.tk/, accessed June 16, 2008.

Ascher, M. *Update on Biodetection: Problems and Prospects,* U.S. Department of Health & Human Services, Washington, DC, September 2002.

Bacharach, Inc. *Instruction 19-9211Operation and Maintenance, Revision 3,* New Kensington, Pennsylvania, October 2001.

Basic Infrared Spectroscopy. www.le.ac.uk/ch/pdf/teachersworkshops.pdf, University of Leicester, Leicester, United Kingdom, May 17, 2000.

Belgrader, P, et al. PCR Detection of Bacteria in Seven Minutes, *Science* 1999, 284, 5413, 449–450.

Belgrader, P, et al. Rapid Pathogen Detection Using a Microchip PCR Array Instrument, *Clinical Chemistry* 1998, 44(10), 2191–2194.

Brockman, S. *Ametek Introduces The Ortec® Detective-Ex-100-Latest Advancement in Hand-Held Radiation Identifiers,* press release, Advanced Measurement Technology, Inc., Oak Ridge, Tennessee, May 26, 2006.

Brown, C., and Jalenak, W. From correspondence, Ahura Scientific, Inc., Chicago, Illinois, March 14, 2005.

Chasteen, T. *Photoionization Detector (PID),* www.shsu.edu/~chemistry/PID/PID.html, Sam Houston State University, Huntsville, Texas, June 16, 2008.

Constellation Technology. *The CT-1128 Portable GC-MS Chemical Detection System,* www.gcms.de/download/ct1128.pdf, Largo, Florida, December 9, 2004.

Course notes. Advanced Chemical/Biological Integrated Response Course, Dugway, Utah, August 2007.

Fatah, A., et al. *Guide for the Selection of Chemical Agent and Toxic Industrial Material Detection Equipment for First Responders, Second Edition, Guide 100-04, Volume II,* National Institute of Standards and Technology, Gaithersburg, Maryland, March 2005.

FemtoScan Corporation, *Environmental Vapor Monitor (EVM) II,* www.femtoscan.com/evm.htm, July 19, 1999.

Fruchey, I., and Emanuel, P. *Market Survey: Biological Detectors 2005 Edition,* SA-ECBC-2005-01-MSRPT, Edgewood Chemical Biological Center and Critical Reagents Program, Aberdeen, Maryland, March 2005.

Geisberg, M., Personal communication, Silver Lake Research Corporation, Monrovia, California, June 30, 2008.

Goldsby, R. A., et al. *Enzyme-Linked Immunosorbent Assay, Immunology, 5th ed.,* W. H. Freeman, New York, 2003.

Guevremont, R., and Purves, R.W. Atmospheric Pressure Ion Focusing in a High-Field Asymmetric Waveform Ion Mobility Spectrometer, *Review of Scientific Instruments,* 1999, 70, 1370.

Haupt, S., Rowshan, S., and Sauntry, W. *Applicability of Portable Explosive Detection Devices in Transit Environments,* TCRP Report 86, Volume 6, Federal Transit Administration, Washington, DC, January 2004.

Higuchi, R., Dollinger, G., Walsh P. S., and Griffith, R. *Simultaneous Amplification and Detection of Specific DNA Sequences,* Biotechnology, 1992 April 10(4).

Higuchi, R., Fockler, C., Dollinger, G., and Watson, R. *Kinetic PCR Analysis: Real-Time Monitoring of DNA Amplification Reactions,* Biotechnology, New York, 1993 Sep, 11(9), 1026–1030.

Idaho Technology, Inc. RAZOR EX System, www.idahotech.com/RAZOREX/index.html, Salt Lake City, Utah, July 1, 2008.

IEEE Standards. *American National Performance Criteria for Hand-held Instruments for the Detection and Identification of Radionuclides,* N42.34-2003, The Institute of Electrical and Electronics Engineers, Inc., New York, New York, January 30, 2004.

Inficon. *HAPSITE® SituProbe Purge and Trap GC/MS System,* Syracuse, New York, March 11, 2004.

Innov-X Systems. *Product Information,* www.innovx.com, Woburn, Massachusetts, June 18, 2008.

Janeway, C.A., Jr., et al. *Immunobiology, 5th ed.,* Garland Publishing, 2001.

Keyser, R., Twomey, T., and Upp, D. *An Improved Handheld Radioisotope Identifier (RID) for Both Locating and Identifying Radioactive Materials,* Health Physics Society, McLean, Virginia, February 13, 2005.

Lachish, U. *Semiconductor Crystal Optimization of Gamma Detection,* Guma Science, Rehovo, Israel, March 1998.

Longworth, T., Barnhouse, J., and Ong, K. *Domestic Preparedness Program: Testing of MIRAN SapphIRe Portable Ambient Air Analyzers Against Chemical Warfare Agents, Summary Report,* Research and Technology Directorate, Soldier and Biological Chemical Command, AMSSB-RRT, Aberdeen Proving Ground, Maryland, July 2000.

MSA. *Hazmatcad™ Hazardous Material Chemical Agent Detector, A Guide for First Responders,* Pittsburgh, Pennsylvania, March 5, 2008.

MSA. *MSA Sirius® MultiGas Detector Operating Manual, Revision 2,* Pittsburgh, Pennsylvania, September 18, 2007.

Mullis, K. B., and Faloona, F. A. *Specific Synthesis of DNA In Vitro via a Polymerase-Catalyzed Chain Reaction,* Methods in Enzymology, 1987, 155, 335–350.

Neogen, Inc. *Food Safety,* www.neogen.com/FoodSafety/FS_Industry.asp?Ind_Cat=102, Lansing, Michigan, June 30, 2008.

NEWMOA Technology Review Committee Advisory Opinion. *Innovative Technology: X-Ray Fluorescence Field Analysis,* September 21, 1999.

NITON Corporation. *Detecting Arsenic in Soil Using Field Portable X-Ray Fluorescence (XRF)*, NITON Corporation, Billerica, Massachusetts, May 2000.

Nucsafe, Inc. *Selecting Your Neutron Detector*, www.nucsafe.com/Technology/selecting_neutron_detector.htm, Oakridge, Tennessee, January 24, 2008.

Omega Engineering, Inc. *Electronic Halogen Leak Detector*, www.omega.com/techref/ph-4.html, May 31, 2006.

Ortec®, Detective-EX and Detective-DX HPGe-based Hand-Held Radioisotope Identifiers, www.ortec-online.com/detective-ex.htm, January 15, 2008.

Photovac, Inc. MicroFID® product brochure, Waltham, Massachusetts, June 7, 2007.

Photovac, Inc. *Photovac Voyager Portable GC User's Guide*, Syracuse, New York, May 23, 2001.

Pibida, L., Karam, L., and Unterweger, M. *Results of Test and Evaluation of Commercially Available Survey Meters for the Department of Homeland Security, Version 1.3*, National Institute of Standards and Technology, Gaithersburg, Maryland, May 2005.

Pibida, L., Karam, L., and Unterweger, M. *Results of Test and Evaluation of Commercially Available Personal Alarming Radiation Detectors and Pagers for the Department of Homeland Security, Version 1.3*, National Institute of Standards and Technology, Gaithersburg, Maryland, May 2005.

Pier, G., Lyczak, J., and Wetzler, L. *Immunology, Infection, and Immunity*, ASM Press, 2004.

Proengin, S.A. *AP2C*, www.proengin.fr/fp_ap2c.htm, June 17, 2008.

RAE Systems. *Comparison of Photoionization Detectors (PIDs) and Flame Ionization Detectors (FIDs)*, Application Note 226, Revision 2, San Jose, California, January 2004.

RAE Systems. *Photoionization Detectors*, www.gotgas.com/pdf/photoionization.pdf, San Jose, California, May 27, 2004.

RAE Systems. Technical Note TN-144, Revision 1, San Jose, California, March 1999.

RAE Systems. Technical Note TN-152, Revision 2, San Jose, California, February 2000.

Response Biomedical Corporation. *Biodefense Product Catalogue*, Vancouver, BC, January 31, 2005.

RKI Instruments, Inc. *Product Sheet P001-9906*, Union City, California, September 29, 1999.

SA Scientific, *SAS™ Legionella Test FAQ*, http://www.sascientific.com/Documents/LegionellaFAQ_10-07.pdf, San Antonio, Texas, October 25, 2007.

Saiki R. K., Bugawan T. L., Horn G. T., Mullis K. B., Erlich H. A. *Analysis of Enzymatically Amplified Beta-globin and HLA-DQ Alpha DNA with Allele-Specific Oligonucleotides Probes*, Nature, 1986 Nov 13-19, 324(6093):163-6.

Sandia National Laboratories, *Miniature Ion Mobility Spectrometer*, http://www.sandia.gov/mstc/technologies/microsensors/IMS.html, November 16, 2006.

Santillán, J. *FirstDefender Training Kit*, Ahura Corporation, Wilmington, Massachusetts, November, 2006.

Signal Group, Ltd., *Model 3030PM Portable Heated VOC Analyser Operating Manual*, Camberley, England, June 4, 2003.

Smith, E and Dent, G. *Modern Raman Spectroscopy - A Practical Approach*, John Wiley & Sons, Ltd., West Sussex, England, May, 2006.

Smiths Detection, Inc., HazMatID™ System User's Guide, Revision 5, Danbury, Connecticut, June 17, 2004.

Smiths Detection, Inc., *Sabre 4000*, http://www.smithsdetection.com/eng/1384.php, Danbury, Connecticut, November 14, 2007.

SPX Corporation, *TIF XP-1A Automatic Halogen Leak Detector Owner's Manual*, Owatonna, Minnesota, March 7, 2003.

Test Products International, Inc., *750a Instruction Manual*, Beaverton, Oregon, 2001.

U.S. Department of Justice, *An Introduction to Biological Agent Detection Equipment for Emergency First Responders*, NIJ Guide 101–00, Office of Justice Programs, National Institute of Justice, December, 2001.

U.S. EPA Region I, *Standard Operating Procedure for Elemental Analysis Using the X-MET 920 Field X-Ray Fluorescence Analyzer*, October 1996.

U.S. EPA, *Method 6200 Field Portable X-Ray Fluorescence Spectrometry for the Determination of Elemental Concentrations in Soil and Sediment,* http://www.epa.gov/SW-846/pdfs/6200.pdf, Washington, D.C., February, 2007.

Verkouteren, J. and Fatah, A. et. al. *IMS-based Trace Explosives Detectors for First Responders,* NISTIR 7240, National Institute of Standards and Technology, U.S. Department of Commerce, Washington, DC, January 2005.

Appendix

Drug Confirmation Testing

The following material as it relates to field drug tests is taken from the U.S. Department of Justice, which has established a National Institute of Justice Standard 0604.01, *Color Test Reagents/Kits for Preliminary Identification of Drugs of Abuse.* This standard describes approved reagents and test methods for the identification of some illegal drugs and commonly mistaken compounds. The test result matrix for illegal drugs and commonly mistaken substances are shown in Table A.1.

The test reagents can be used individually to confirm or deny the identity of a substance based on the results of a test in the drug screen; however, more precise identification is made through use of several tests as shown in the result matrix. All tests are presumptive, and laboratory testing is required to prove identity.

These tests are sensitive. A few hundred micrograms is often sufficient to provide a colorimetric result. For more information refer to Color Test Reagents/Kits for Preliminary Identification of Drugs of Abuse, NIJ Standard–0604.01. Individual reagents from this drug screen are often packaged into presumptive field test kits (Figure A.1). See Chapter 2, "Chemical Confirmation Tests," to conduct noncommercial field tests that include more than the 12 drug screen tests. For example, aspirin can be identified with a clear aqueous solution of ferric nitrate that will turn deep purple in contact with aspirin.

Since some reagents have uses other than the drug screen, this book refers to them by their traditional name, for example, Cobalt Thiocyanate Reagent. The U.S. Department of Justice refers to the test reagents by the name used in the standard, for example, Cobalt Thiocyanate. The drug test reagents are also designated alphanumerically, for example, A.1, for ease of use in the matrix. The terms are interchangeable and are shown in Table A.2, "Drug Test Reagents." To find a drug screen reagent in this book, use Table A.2 to cross reference the U.S. Department of Justice reagent name to the name of the chemical confirmation test found in Chapter 2. Lastly, colorimetric test results are listed in Table A.3, "Final Colors Produced by Reagents A.l to A.12 by Various Drugs and Other Substances."

Table A.1   Screen for Drugs and Commonly Mistaken Material

(+) Indicates that a color reaction occurs[a]	Reagent											
	A.1	A.2	A.3	A.4	A.5	A.6	A.7	A.8	A.9	A.10	A.11	A.12
Acetaminophen	-	-	-	+	-	+	-	+	-	-	-	-
Alprazolam	-	-	-	-	-	-	-	-	-	-	-	-
Aspirin	-	-	-	+	+	-	-	-	+	-	-	-
Baking Soda	-	-	-	-	-	-	-	+	-	-	+	-
Brompheniramine Maleate	+	-	-	+	-	-	-	-	-	-	-	-
Chlordiazepoxide HCl	+	-	-	-	-	-	-	-	-	-	-	-
Chlorpromazine HCl	+	-	-	+	+	+	-	+	+	+	-	-
Contac®	-	-	-	+	-	-	-	-	+	+	-	-
Diazepam	-	-	-	-	-	-	-	-	-	-	-	-
Doxepin HCl	+	-	-	+	+	+	-	-	+	+	-	-
Dristan®	-	-	-	+	+	+	-	+	+	+	-	-
Ephedrine HCl	+	-	-	-	-	-	-	-	-	-	-	-
Excedrin®	-	-	-	+	+	+	-	+	+	+	+	-
Hydrocodone tartrate	+	-	-	-	-	-	-	-	-	+	-	-
Mace[b]	-	-	+	+	+	+	-	-	+	+	-	-
Meperidine HCl	+	-	-	-	+	-	-	-	-	-	-	-
Methaqualone	-	-	-	+	-	-	-	-	-	-	-	-
Methylphenidate HCl	+	-	-	+	+	-	-	-	-	-	-	+
Nutmeg[b]	-	-	+	-	-	-	-	-	-	+	-	-
Phencyclidine HCl	+	-	-	-	-	-	-	-	-	-	-	-
Propoxyphene HCl	+	-	-	+	+	-	-	-	+	+	-	-
Pseudoephedrine HCl	+	-	-	-	-	-	-	-	-	-	-	-
Quinine HCl	+	-	-	+	-	-	-	-	-	-	-	-
Salt	-	-	-	+	-	-	-	-	-	-	-	-
Sugar	-	-	-	-	+	-	-	-	-	-	-	-
Tea[b]	-	-	+	-	-	-	-	-	-	-	+	-
Tobacco	-	-	-	-	-	-	-	-	-	-	+	-

Source: Table reproduced from *Color Test Reagents/Kits for Preliminary Identification of Drugs of Abuse,* NIJ Standard–0604.01, U.S. Department of Justice.

Notes: [a] Substances that gave no colors with these reagents are: D-galactose, glucose, mannitol, oregano, rosemary, and thyme.

[b] Tea, mace, and nutmeg may interfere with the Duquenois test but not the Duquenois-Levine modified test (A.3).

Table A.2   Drug Test Reagents

U.S. Dept. of Justice Alphanumeric Test Name	U.S. Dept. of Justice Reagent Name	Chemical Confirmation Test Name (Chapter 2)
A.1	Cobalt Thiocyanate	Cobalt Thiocyanate Reagent
A.2	Dille-Koppanyi Reagent, Modified	Dille-Koppanyi Reagent, Modified
A.3	Duquenois-Levine Reagent, Modified	Duquenois-Levine Reagent, Modified
A.4	Mandelin Reagent	Mandelin Reagent

Table A.2   Drug Test Reagents (*Continued*)

U.S. Dept. of Justice Alphanumeric Test Name	U.S. Dept. of Justice Reagent Name	Chemical Confirmation Test Name (Chapter 2)
A.5	Marquis Reagent	Marquis Reagent
A.6	Nitric Acid	Nitric Acid Reagent
A.7	Para-Dimethylaminobenzaldehyde (p-DMAB)	Dimethylaminobenzaldehyde (p-DMAB)
A.8	Ferric Chloride	Ferric Chloride Reagent
A.9	Froede Reagent	Froede Reagent
A.10	Mecke Reagent	Mecke Reagent
A.11	Zwikker Reagent	Zwikker Reagent
A.12	Simon's Reagent	Simon's Reagent

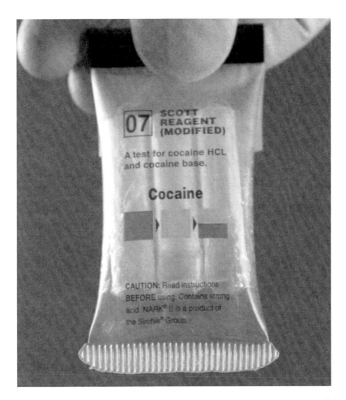

Figure A.1 This presumptive field cocaine test kit displays a positive (blue) result for the first of three steps in the test. This test uses cobalt thiocyanate to detect cocaine by colors produced first in an acid solution and then in a basic solution. This test was conducted on dibucaine extracted from a 1% topical medicine. Alkaloids are derivatives of amino acids and most alkaloids are detected by cobalt thiocyanate, Reagent A.1 in Table A.2. Alkaloids include the drugs ending in "-caine" such as cocaine, lidocaine, xylocaine, as well as strychnine, morphine, atropine, ephedrine, caffeine, ricinine, and many others. Cobalt thiocyanate will react with other material so a positive result must be confirmed with other test methods or use of a matrix similar to the drug screen.

Table A.3 Final Colors Produced by Reagents A.l to A.12 by Various Drugs and Other Substances

Reagent	Analyte	Solvent	Color
A.1	Benzphetamine HCl	Chloroform	Brilliant greenish blue
A.1	Brompheniramine maleate	Chloroform	Brilliant greenish blue
A.1	Chlordiazepoxide HCl	Chloroform	Brilliant greenish blue
A.1	Chlorpromazine HCl	Chloroform	Brilliant greenish blue
A.1	Cocaine HCl	Chloroform	Strong greenish blue
A.1	Diacetylmorphine HCl	Chloroform	Strong greenish blue
A.1	Doxepin HCl	Chloroform	Brilliant greenish blue
A.1	Ephedrine HCl	Chloroform	Strong greenish blue
A.1	Hydrocodone tartrate	Chloroform	Brilliant greenish blue
A.1	Meperidine HCl	Chloroform	Strong greenish blue
A.1	Methadone HCl*	Chloroform	Brilliant greenish blue
A.1	Methylphenidate HCl	Chloroform	Brilliant greenish blue
A.1	Phencyclidine HCl	Chloroform	Strong greenish blue
A.1	Procaine HCl*	Chloroform	Strong greenish blue
A.1	Propoxyphene HCl*	Chloroform	Strong greenish blue
A.1	Pseudoephedrine HCl	Chloroform	Strong greenish blue
A.1	Quinine HCl	Chloroform	Strong blue
A.2	Amobarbital	Chloroform	Light purple
A.2	Pentobarbital*	Chloroform	Light purple
A.2	Secobarbital*	Chloroform	Light purple
A.3	Mace crystals		Strong reddish purple to very light purple
A.3	Nutmeg extract		Pale reddish purple to light gray purplish red
A.3	Tea extract		Light yellow green
A.3	THC*	Ethanol	Gray purplish blue to light purplish blue to deep purple
A.4	Acetaminophen	Chloroform	Moderate olive
A.4	Aspirin powder		Grayish olive green
A.4	Benzphetamine HCl*	Chloroform	Brilliant yellow green
A.4	Brompheniramine maleate	Chloroform	Strong orange
A.4	Chlorpromazine HCl	Chloroform	Dark olive
A.4	Cocaine HCl*	Chloroform	Deep orange yellow
A.4	Codeine*	Chloroform	Dark olive
A.4	Contac® powder		Strong yellow
A.4	d-Amphetamine HCl*	Chloroform	Moderate bluish green

Table A.3   Final Colors Produced by Reagents A.1 to A.12 by Various Drugs and Other Substances (*Continued*)

Reagent	Analyte	Solvent	Color
A.4	d-Methamphetamine HCl*	Chloroform	Dark yellowish green
A.4	Diacetylmorphine HCl*	Chloroform	Moderate reddish brown
A.4	Dimethoxy-meth HCl	Chloroform	Dark olive brown
A.4	Doxepin HCl	Chloroform	Dark reddish brown
A.4	Dristan powder		Grayish olive
A.4	Excedrin® powder		Dark olive
A.4	Mace crystals		Moderate olive green
A.4	MDA HCl	Chloroform	Bluish black
A.4	Mescaline HCl*	Chloroform	Dark yellowish brown
A.4	Methadone HCl	Chloroform	Dark grayish blue
A.4	Methaqualone	Chloroform	Very orange yellow
A.4	Methylphenidate HCl	Chloroform	Brilliant orange yellow
A.4	Morphine monohydrate*	Chloroform	Dark grayish reddish brown
A.4	Opium*	Chloroform	Dark brown
A.4	Oxycodone HCl	Chloroform	Dark greenish yellow
A.4	Procaine HCl	Chloroform	Deep orange
A.4	Propoxyphene HCl	Chloroform	Dark reddish brown
A.4	Quinine HCl	Chloroform	Deep greenish yellow
A.4	Salt crystals		Strong orange
A.5	Aspirin powder		Deep red
A.5	Benzphetamine HCl*	Chloroform	Deep reddish brown
A.5	Chlorpromazine HCl	Chloroform	Deep purplish red
A.5	Codeine*	Chloroform	Very dark purple
A.5	d-Amphetamine HCl*	Chloroform	Strong reddish orange to dark reddish brown
A.5	d-Methamphetamine HCl*	Chloroform	Deep reddish orange to dark reddish brown
A.5	Diacetylmorphine HCl*	Chloroform	Deep purplish red
A.5	Dimethoxy-meth HCl	Chloroform	Moderate olive
A.5	Doxepin HCl	Chloroform	Blackish red
A.5	Dristan® powder		Dark grayish red
A.5	Excedrin® powder		Dark red
A.5	LSD chloroform		Olive black
A.5	Mace crystals		Moderate yellow
A.5	MDA HCl*	Chloroform	Black

(continued)

Table A.3   Final Colors Produced by Reagents A.l to A.12 by Various Drugs and Other Substances (*Continued*)

Reagent	Analyte	Solvent	Color
A.5	Meperidine HCl	Chloroform	Deep brown
A.5	Mescaline HCl*	Chloroform	Strong orange
A.5	Methadone HCl	Chloroform	Light yellowish pink
A.5	Methylphenidate HCl	Chloroform	Moderate orange yellow
A.5	Morphine monohydrate*	Chloroform	Very deep reddish purple
A.5	Opium* powder		Dark grayish reddish
A.5	Oxycodone HCl*	Chloroform	Pale violet
A.5	Propoxyphene HCl	Chloroform	Blackish purple
A.5	Sugar crystals		Dark brown
A.6	Acetaminophen	Chloroform	Brilliant orange yellow
A.6	Codeine*	Chloroform	Light greenish yellow
A.6	Diacetylmorphine HCl*	Chloroform	Pale yellow
A.6	Dimethoxy-meth HCl	Chloroform	Very yellow
A.6	Doxepin HCl	Chloroform	Brilliant yellow
A.6	Dristan® powder		Deep orange
A.6	Excedrin® powder		Brilliant orange yellow
A.6	LSD	Chloroform	Strong brown
A.6	Mace crystals		Moderate greenish yellow
A.6	MDA HCl	Chloroform	Light greenish yellow
A.6	Mescaline HCl*	Chloroform	Dark red
A.6	Morphine monohydrate*	Chloroform	Brilliant orange yellow
A.6	Opium* powder		Dark orange yellow
A.6	Oxycodone HCl	Chloroform	Brilliant yellow
A.7	LSD*	Chloroform	Deep purple
A.8	Acetaminophen	Methanol	Dark greenish yellow
A.8	Baking soda powder		Deep orange
A.8	Chlorpromazine HCl	Methanol	Very orange
A.8	Dristan® powder		Moderate purplish blue
A.8	Excedrin® powder		Moderate purplish blue
A.8	Morphine monohydrate*	Methanol	Dark green
A.9	Aspirin powder		Grayish purple
A.9	Chlorpromazine HCl	Chloroform	Very deep red
A.9	Codeine*	Chloroform	Very dark green
A.9	Contac® powder		Moderate olive brown

Table A.3   Final Colors Produced by Reagents A.1 to A.12 by Various Drugs and Other Substances (*Continued*)

Reagent	Analyte	Solvent	Color
A.9	Diacetylmorphine HCl*	Chloroform	Deep purplish red
A.9	Dimethoxy-meth HCl	Chloroform	Very yellow green
A.9	Doxepin HCl	Chloroform	Deep reddish brown
A.9	Dristan® powder		Light bluish green
A.9	Excedrin® powder		Brilliant blue
A.9	LSD chloroform		Moderate yellow green
A.9	Mace crystals		Light olive yellow
A.9	MDA HCl*	Chloroform	Greenish black
A.9	Morphine monohydrate*	Chloroform	Deep purplish red
A.9	Opium* powder		Brownish black
A.9	Oxycodone HCl	Chloroform	Strong yellow
A.9	Propoxyphene HCl	Chloroform	Dark grayish red
A.9	Sugar crystals		Brilliant yellow
A.10	Chlorpromazine HCl	Chloroform	Blackish red
A.10	Codeine*	Chloroform	Very dark bluish green
A.10	Contac® powder		Moderate olive brown
A.10	Diacetylmorphine HCl*	Chloroform	Deep bluish green
A.10	Dimethoxy-meth HCl	Chloroform	Dark brown
A.10	Doxepin HCl	Chloroform	Very dark red
A.10	Dristan® powder		Light olive brown
A.10	Excedrin® powder		Dark grayish yellow
A.10	Hydrocodone tartrate	Chloroform	Dark bluish green
A.10	LSD chloroform		Greenish black
A.10	Mace crystals		Dark grayish olive
A.10	MDA HCl*	Chloroform	Very dark bluish green
A.10	Mescaline HCl*	Chloroform	Moderate olive
A.10	Morphine monohydrate*	Chloroform	Very dark bluish green
A.10	Nutmeg leaves		Brownish black
A.10	Opium* powder		Olive black
A.10	Oxycodone HCl	Chloroform	Moderate olive
A.10	Propoxyphene HCl	Chloroform	Deep reddish brown
A.10	Sugar crystals		Brilliant greenish yellow
A.11	Baking soda powder		Light blue
A.11	Excedrin® powder		Light green

(continued)

Table A.3   Final Colors Produced by Reagents A.1 to A.12 by Various Drugs and Other Substances (*Continued*)

Reagent	Analyte	Solvent	Color
A.11	Pentobarbital*	Chloroform	Light purple
A.11	Phenobarbital*	Chloroform	Light purple
A.11	Secobarbital*	Chloroform	Light purple
A.11	Tea leaves		Moderate yellow green
A.11	Tobacco leaves		Moderate yellowish green
A.12	d-Methamphetamine HCl*	Chloroform	Dark blue
A.12	Dimethoxy-meth HCl*	Chloroform	Deep blue
A.12	MDMA HCl	Chloroform	Dark blue
A.12	Methylphenidate HCl	Chloroform	Pale violet

Source: Adapted from Color Test Reagents/Kits for Preliminary Identification of Drugs of Abuse, NIJ Standard–0604.01.

Notes: *Usual kit reagent for that particular drug.

Explosive Material Confirmation Testing

Explosive material can vary in appearance and be solid or liquid, brightly colored or dull and drab, organic or inorganic. Explosives may be confirmed by verifying energetic characteristics. Most often, a small amount of material placed in a flame should explode or burn quickly and intensely like a flare.

Explosive material can be confirmed by testing that indicates it is:

- Shock sensitive
- Thermally sensitive
- A combination of fuel and oxidizer
- Nitrate containing
- Perchlorate containing
- Peroxide containing
- Any substance containing a combination of groups listed in Table A.4
- A substance producing a positive result from a commercially available colorimetric explosive screening test

To determine if a substance is shock sensitive, place a milligram of sample (equivalent to 5 to 10 grains of table salt) on a hard, inert surface and give it a good whack with a hammering device. Since you are suspecting a crisp and powerful explosion, use material for the hammer and anvil that will not fragment when an explosive wave passes through it. Perform this test in a manner that controls any material that may be ejected when struck. Preferably you should use a device that you can operate remotely so you are removed from the area of any possible fragmentation. Wear appropriate protective equipment such as a face shield and ear protection.

If the first test is negative for shock-sensitive materials, work your way up to a larger volume, never exceeding a gram (the size of an M&M). The larger the sample the farther away you should be from it. A remote hammer-and-anvil device will greatly improve safety. Not all explosives are shock sensitive, so a negative result does not rule out explosives. Some explosives such as plastic C-4 can be safely pounded into shape and when ignited will burn gently; but if shocked while burning, a violent explosion will occur. A gram of exploding C-4 will most likely cause injury. If you suspect a material such as C-4, you should use a technology that does not require simultaneous shocking and burning since a relatively large sample will be required and the test itself would present a hazard.

To determine thermal sensitivity, place a small amount on a wire and place it in a torch flame. Explosion or a flaring flame indicates explosive tendency. Some explosive materials require increased pressure to explode and when burned in an open flame may appear as ordinary organic material. Do not rely on a flame test alone to confirm or deny the material as explosive. Combinations of fuel and oxidizer may appear to burn intensely, like a road flare, or with a brighter-than-expected flame.

The remainder of the explosive traits, such as the presence of nitrate, peroxide, perchlorate, and other functional groups may be determined by selecting specific tests from Chapter 2, "Chemical Confirmation Tests." Use Table A.4 to determine which tests from Chapter 2 are applicable.

Unfortunately, there is no universal colorimetric test for the universal detection of nitro or organic nitrate groups. Some colorimetric indicators have been used in the past, but have been found to be unstable for field use, toxic or carcinogenic. Some commercially available colorimetric test kits are available. These kits will detect many of the more common commercial explosive materials based on organic structures, functional groups, and degradation products. These are very useful in presumptive testing such as post-blast forensics, but they may also produce false positives in the case of unknown material. These kits are useful in the detection of peroxide-based and chlorate-based improvised explosive material.

Some explosive materials can be identified in very small amounts. IMS technology is particularly useful in the identification of explosives as evidenced by the deployment of desktop IMS units in airports. However, some explosives cannot be detected by IMS and those that can be detected often require more than one test since most IMS instruments cannot operate across the entire range of parameters necessary to detect all explosives. When operated as designed, IMS instruments are capable of detecting a few nanograms of invisible explosive particles. A 50 micron particle of RDX, which is barely visible with unaided vision, weighs about 100 ng, more than enough to be reliably detected.

Portable IMS units designed for detection of trace explosives have several advantages over other technologies. They are fast, relatively inexpensive, are adaptable to changing environments and can be used by operators without extensive training. IMS units are limited in the detection of explosives in that they often cannot detect the slight amount of vapor emitting from explosives that typically have very low vapor pressure, but when characterizing a visible sample of suspected explosive the instrument can be switched to particle mode.

IMS is not as definitive for identification of explosive compounds as some other technologies, but the other technologies are often not available or not practical for field application. False positives are possible from common, nonhazardous material. False negatives are possible from interfering material that may be preferentially ionized; these masking agents, whether present intentionally or coincidentally, prevent the explosive compound from being ionized. Other factors may be present in individual cases that prevent

Table A.4   Characteristic Composition of Explosives

Explosive	Hydrocarbon	Nitro	Organic Nitrate	Nitrate Anion	Perchlorate	Chlorate	Peroxide	Liquid Oxygen	Oxygen	Metal	Picrate	Unstable Structure
Acetylides of heavy metals										•		•
Aluminum containing polymeric propellant	•									•		
Aluminum ophorite explosive					•					•		
Amatex	•	•		•								
Amatol	•	•		•								
Amm nitrate exp. mixtures (cap sensitive)				•								
Amm nitrate exp. mixtures (non-cap sensitive)				•								
Amm perchlorate composite propellant	•				•							
Amm perchlorate explosive mixtures	•				•							
Amm salt lattice w/substituted inorganic salts												•
Ammonal	•	•		•						•		
Ammonium picrate	•	•									•	
ANFO	•			•								
Aromatic nitro-compound explosive mixtures	•	•										
Azide explosives												•
Baranol	•	•		•						•		
Baratol	•	•		•								
BEAF	•	•							•			
Black powder	•			•								
Black powder based explosive mixtures	•			•								
Blasting agents, nitro-carbo-nitrates	•	•		•								
BTNEC	•	•										
BTNEN	•	•										
BTTN	•		•	•								
Butyl tetryl	•	•										
Calcium nitrate explosive mixture	•			•								
Cellulose hexanitrate explosive mixture	•											
Chlorate explosive mixtures	•					•						
Composition A	•	•										
Composition B	•	•										
Composition C A61	•	•										
Copper acetylide										•		•
Cyanuric triazide												•

Table A.4   Characteristic Composition of Explosives (*Continued*)

Explosive	Hydrocarbon	Nitro	Organic Nitrate	Nitrate Anion	Perchlorate	Chlorate	Peroxide	Liquid Oxygen	Oxygen	Metal	Picrate	Unstable Structure
Cyclonite	•	•										
Cyclotetramethylenetetranitramine	•	•										
Cyclotol	•	•										
Cyclotrimethylenetrinitramine	•	•										
DATB	•	•										
DDNP	•	•										
DEGDN	•		•									
Dimethylol dimethyl methane dinitrate composition	•		•									
Dinitroethyleneurea	•	•										
Dinitroglycerine	•	•										
Dinitrophenol	•	•										
Dinitrophenolates	•	•										
Dinitrophenyl hydrazine	•	•										
Dinitroresorcinol	•	•										
Dinitrotoluene-sodium nitrate explosive mixtures	•	•		•								
DIPAM	•	•										
Dipicryl sulfone	•	•										
Dipicrylamine	•	•										
DNPA	•	•										
DNPD	•	•										
Dynamite	•	•										
EDDN	•											
EDNA	•	•										
Ednatol	•	•										
EDNP	•	•										
EGDN	•		•									
Erythritol tetranitrate explosives	•											
Esters of nitro-substituted alcohols	•	•										
Ethyl tetryl	•											
Exp. mix containing sensitized nitromethane	•	•										
Exp. mix containing tetranitromethane (nitroform)	•	•										
Exp. mix containing tetranitromethane (nitroform)	•	•										
Exp. mix of oxy-salts and hydrocarbons	•											
Exp. mix of oxy-salts and nitro bodies		•										

(*continued*)

Table A.4   Characteristic Composition of Explosives (*Continued*)

Explosive	Hydrocarbon	Nitro	Organic Nitrate	Nitrate Anion	Perchlorate	Chlorate	Peroxide	Liquid Oxygen	Oxygen	Metal	Picrate	Unstable Structure
Exp. mix of oxy-salts and water-insoluble fuels	•											
Exp. mix of oxy-salts and water-soluble fuels	•											
Explosive mix containing sensitized nitromethane	•	•										
Explosive nitro compounds of aromatics	•	•										
Explosive organic nitrate mixtures	•			•								
Flash powder										•		
Fulminate of mercury	•									•		
Fulminate of silver										•		
Fulminating gold										•		
Fulminating mercury										•		
Fulminating platinum										•		
Fulminating silver										•		
Gelatinized nitrocellulose	•	•										
Gem-dinitro aliphatic Explosive mixtures	•	•										
Guanyl nitrosamino guanyl tetrazene	•	•										
Guanyl nitrosamino guanylidene hydrazine	•	•										
Guncotton	•	•										
Heavy metal azides										•		•
Hexanite	•	•										
Hexanitrodiphenylamine	•	•										
Hexanitrostilbene	•	•										
Hexogen	•	•										
Hexogene or octogene and nitrated N-methylaniline	•	•										
Hexolites	•	•										
HMTD	•						•					
HMX	•	•										
HNIW	•		•									
Hydrazinium nitrate/hydrazine/aluminum explosive system				•						•		
Hydrazoic acid												•
KDNBF	•	•										
Lead azide										•		•
Lead mannite	•									•		
Lead mononitroresorcinate	•	•								•		

Table A.4 Characteristic Composition of Explosives (*Continued*)

Explosive	Hydrocarbon	Nitro	Organic Nitrate	Nitrate Anion	Perchlorate	Chlorate	Peroxide	Liquid Oxygen	Oxygen	Metal	Picrate	Unstable Structure
Lead picrate		•								•	•	
Lead salts, explosive										•		
Lead styphnate	•	•								•		
Liquid nitrated polyol and trimethylolethane	•											
Liquid oxygen explosives	•							•				
Magnesium ophorite explosives					•					•		
Mannitol hexanitrate	•	•										
MDNP	•	•										
MEAN	•											
Mercuric fulminate										•		
Mercury oxalate										•		•
Mercury tartrate	•									•		
Methyl nitrate	•		•									
Metriol trinitrate	•		•									
Minol-2	•	•		•						•		
MMAN	•	•		•								
Mononitrotoluene-nitroglycerin mixture	•	•										
NIBTN	•	•										
Nitrate explosive mixtures	•			•								
Nitrate sensitized with gelled nitroparaffin	•	•		•								
Nitrated carbohydrate explosive	•	•										
Nitrated glucoside explosive	•	•										
Nitrated polyhydric alcohol explosives	•	•										
Nitric acid and a nitro aromatic explosive	•	•		•								
Nitric acid and carboxylic fuel explosive	•			•								
Nitric acid explosive mixtures	•			•								•
Nitro aromatic explosive mixtures	•	∘										
Nitro compounds of furane explosive mixtures	•	•										
Nitrocellulose explosive	•	•										
Nitroderivative of urea explosive mixture	•		•									
Nitrogelatin explosive	•	•										
Nitrogen trichloride												•
Nitrogen tri-iodide												•
Nitroglycerine	•	•										

(*continued*)

Table A.4 Characteristic Composition of Explosives (*Continued*)

Explosive	Hydrocarbon	Nitro	Organic Nitrate	Nitrate Anion	Perchlorate	Chlorate	Peroxide	Liquid Oxygen	Oxygen	Metal	Picrate	Unstable Structure
Nitroglycide	•	•										
Nitroglycol	•		•									
Nitroguanidine explosives	•	•										
Nitronium perchlorate propellant mixtures					•							•
Nitroparaffins (explosive)	•	•		•								
Nitrostarch	•	•										
Nitro-substituted carboxylic acids	•	•										
Nitrourea	•	•										
Octogen	•	•										
Octol	•	•										
Organic amine nitrates	•		•									
PBX	•		•									
Penthrinite composition	•	•										
Pentolite	•	•										
Perchlorate explosive mixtures					•							
Peroxide based explosive mixtures							•					
PETN	•		•									
Picramic acid and its salts	•	•										
Picramide	•	•										
Picrate explosives	•	•										
Picrate of potassium explosive mixtures	•	•								•		
Picratol	•	•										
Picric acid (manufactured as an explosive)	•	•										
Picryl chloride	•	•										
Picryl fluoride	•	•										
PLX	•	•										
Polynitro aliphatic compounds	•	•										
Polyolpolynitrate-nitrocellulose explosive gels	•	•										
Potassium chlorate and lead sulfocyanate explosive						•				•		
Potassium nitrate explosive mixtures	•			•								
Potassium nitroaminotetrazole			•									•
PYX	•	•										
RDX	•	•										
Salts of organic amino sulfonic acid explosive mixture	•											
Silver acetylide										•		•

Table A.4   Characteristic Composition of Explosives (*Continued*)

Explosive	Hydrocarbon	Nitro	Organic Nitrate	Nitrate Anion	Perchlorate	Chlorate	Peroxide	Liquid Oxygen	Oxygen	Metal	Picrate	Unstable Structure
Silver azide										•		•
Silver fulminate										•		
Silver oxalate explosive mixtures										•		•
Silver styphnate	•	•								•		
Silver tartrate explosive mixtures	•									•		
Silver tetrazene										•		•
Smokeless powder	•	•										
Sodatol	•	•				•						
Sodium amatol	•	•		•								
Sodium azide explosive mixture												•
Sodium dinitro-ortho-cresolate	•	•										
Sodium nitrate explosive mixtures				•								
Sodium nitrate-potassium nitrate explosive mixture				•								
Sodium picramate	•	•										
Styphnic acid explosives	•	•										
T4	•	•										
Tacot	•	•										•
TATB	•	•										
TATP	•						•					
TEGDN	•		•									
Tetranitrocarbazole	•	•										
Tetrazene	•		•									•
Tetryl	•	•										
Tetrytol	•	•										
TMETN	•	•										
TNEF	•	•										
TNEF	•	•										
TNEOC	•	•										
TNEOF	•											
TNT	•	•										
Torpex	•	•								•		
Tridite	•	•										
Trimethylol ethyl methane trinitrate composition	•	•										
Trimethylolethane trinitrate-nitrocellulose	•	•										
Trimonite	•	•										
Trinitroanisole	•		•									
Trinitrobenzene	•	•										
Trinitrobenzoic acid	•	•										

(continued)

Table A.4   Characteristic Composition of Explosives (*Continued*)

Explosive	Hydrocarbon	Nitro	Organic Nitrate	Nitrate Anion	Perchlorate	Chlorate	Peroxide	Liquid Oxygen	Oxygen	Metal	Picrate	Unstable Structure
Trinitrocresol	•	•										
Trinitro-meta-cresol		•										
Trinitronaphthalene	•	•										
Trinitrophenetol	•	•										
Trinitrophloroglucinol	•	•										
Trinitroresorcinol	•	•										
Tritonal	•	•								•		
Urea nitrate				•								
Water-in-oil emulsion explosive compositions	•											

Source: Reproduced from *Emergency Characterization of Unknown Materials,* CRC Press, 2007, with permission.

accurate analysis of explosive material for which the IMS instrument was designed to identify. Identification will depend on matching library spectra resident in the instrument. Additionally, IMS instruments require strict maintenance, calibration, and operating procedures in order to analyze a result with a high level of confidence.

Many IMS instruments are optimized for the detection of explosives RDX, TNT, PETN, and many other commercial explosives with related structures. In order to realize the full analyzing potential of an IMS instrument, the ability to reconfigure the instrument and also understand its limitations is necessary. IMS is generally not optimized for detection of gun powders, most notably black powder. Since sampling in particle mode requires heating of the sample to 180°C (356°F) or higher, care must be taken to test very small amounts to avoid explosive decomposition. A milligram (approximately five grains of table salt) is generally well above the detection limit for many explosives.

Those considering procurement of an IMS instrument for explosive material identification may want to refer to IMS-Based Trace Explosives Detectors for First Responders, Document NISTIR 7240 from the U.S. Department of Commerce. While the publication does not recommend or evaluate individual IMS instruments, the document provides acceptance criteria for instrument performance for the purpose of trace explosive detection.

Raman spectroscopy can also be used to identify many explosive materials. Raman spectroscopy would be expected to excel at detection of explosive materials that are organic or polyatomic inorganics such as perchlorates and nitrates. It provides a less robust response to organic molecules that are small and highly polar and those containing only single bonds, such as aliphatics, sugars, starches and cellulose. This is not to say these compounds cannot be detected with Raman spectroscopy. Identification will depend on matching library spectra resident in the instrument.

Directing a laser from a Raman instrument can ignite or explode some explosive materials. Dark materials are more susceptible to heating by the laser. Black powder and silver azide are known to be sensitive to the laser; however, black plastic, latex paint, and cardboard can also ignite. Since many explosives are heat sensitive and since the instrument must contact or be very close to the material, it could be an expensive and dangerous proposition to cause the sample to explode. If the instrument can detect through glass, it might be possible to collect a sample and dissolve it in a liquid to absorb the heat generated during analysis. Water, methanol, and hexane would be good choices since they have low interference with Raman. Keep in mind that the solvent may be flammable and the lower the concentration of the sample in solvent, the less reliable the result.

Fluorescent material can interfere with Raman instruments to the point of making the test useless. Fluorescence can be caused by brightly colored material (especially blues, greens, and black), some low-quality glass, and some biological material. Improvements in lasers and data analysis software are expected to reduce these problems.

Infrared spectroscopy instruments may be helpful in identifying explosive compounds, but the problem of heating from the infrared beam is similar, but not as intense, as Raman spectroscopy. Consult the instrument manual to determine the power of the light source, which might range from 1 to 300 mW. Typical IR limitations apply, and identification will depend on matching library spectra resident in the instrument. Since the explosive material may be produced in a clandestine manner to include mixtures of undetermined specification, robust data analysis software is required to determine impure compounds.

References

Haupt, S., Rowshan, S., and Sauntry, W. *Applicability of Portable Explosive Detection Devices in Transit Environments,* TCRP Report 86, Volume 6, Federal Transit Administration, Washington, DC, January 2004.

Houghton, R. *Emergency Characterization of Unknown Materials,* CRC Press, Boca Raton, Florida, 2007.

United States Department of Justice. *Color Test Reagents/Kits for Preliminary Identification of Drugs of Abuse, NIJ Standard–0604.01,* National Institute of Justice, National Law Enforcement and Corrections Technology Center, Rockville, MD, July 2000.

Index

Page numbers followed by f indicate figure; those followed by t indicate table.